Cultivated Therapeutic Landscapes

Cultivated Therapeutic Landscapes provides an in-depth and critical exploration of the impact of gardens and gardening on health and wellbeing. In this book we explore the ways in which gardens and gardening prevent illness and restore wellbeing, and how they improve social and health equity via traditional and innovative mechanisms and across a range of sites.

Therapeutic landscapes are relational, reciprocal, and evolving. In this book, leading scholars from across the globe demonstrate how therapeutic landscapes research and practice is expanded through and around the processes of cultivation. Deliberately interdisciplinary, the book explores how tending and caring for green spaces, collectively and individually, works to prevent and restore health and wellbeing, as well as impact upstream factors determining social justice and equity. A unique combination of academics, clinicians, and practitioners deliver theoretical and practical insights into wide-ranging health-enabling factors, based on new evidence and autoethnographic experiences in home gardens, school, and community gardens, clinical settings, public green spaces, and sites of conservation and wildness. This book pushes concepts of cultivation and horticulture into underexplored spatial, ontological, and wellbeing territories. Despite long-term practical interest, therapeutic horticulture is only now establishing a strong theoretical and research foundation.

This book provides much-needed critical insights into the impact on the key drivers of health, wellbeing, and social equity, with a focus on practical skills for utilising horticulture or designing for particular health needs. It will be of interest to students, scholars, and practitioners in the areas of health geography; cultural geography; cultural studies; therapeutic horticulture; environmental studies; community development and planning; landscape architecture; social work; health studies; and health policy.

Pauline Marsh is a health geographer and senior lecturer in the Wicking Dementia Research and Education Centre at the University of Tasmania. Pauline was awarded a PhD in indigenous film theory and over the past decade her research has primarily been in therapeutic landscapes, specifically exploring how being in nature improves quality of life and health equities. She is particularly interested in the benefits of gardens and the outdoors for people with cognitive, emotional, and physical health challenges, as well as the impacts that gardeners and gardens have on the health and wellbeing of whole communities and the planet. She utilises methods of participatory action research, story-gathering, and filmmaking, and publishes in a range of academic journals. As a practitioner-academic, one of her greatest achievements is the co-founding of DIGnity Supported Community Gardening, a unique therapeutic horticulture model that operates in community gardens. Pauline is also a beekeeper and mother of two spectacular adults. She lives in a small beachside community at the bottom of the world.

Allison Williams is a professor at McMaster University, in Hamilton, Ontario, in Canada. She is a social and health geographer with research interests in carer-employees, quality of life, critical policy/programme evaluation, and therapeutic landscapes. Her research focuses on improving workplace practices for supporting employees with adult care responsibilities. Allison has received various awards for her work, having had three five-year Research Chairs funded by the Canadian Institutes of Health Research. She currently holds a McMaster University Faculty of Science Research Chair and leads a Partnership Grant made up of more than 50 collaborators which mobilises carer-friendly workplaces. Allison has supervised more than 30 graduate trainees. Allison works with UN Women across the world, specifically engaging in research on United Nations Sustainable Development Goal 5: Gender Equity, and Target 5-4: Value unpaid care and promote shared domestic responsibilities. In addition to co-editing the Routledge Geographies of Health Series, Allison has authored five books, 30 book chapters, and more than 180 peer-reviewed journal articles. Her most recent book is entitled *Geography, health and sustainability: Gender matters globally* (Routledge, 2021). Allison has been Principal Investigator of over 25 research projects and serves on a range of national and international adjudication committees. She is a mother of two beautiful children and cares for her 90-year-old uncle, together with her ageing parents, both of whom are in their 80s.

Geographies of Health
Edited by

Allison Williams,
Professor, School of Geography and Earth Sciences, McMaster University, Canada,

Susan Elliott,
Professor, Department of Geography and Environmental Management and School of Public Health and Health Systems, University of Waterloo, Canada

There is growing interest in the geographies of health and a continued interest in what has more traditionally been labelled medical geography. The traditional focus of 'medical geography' on areas such as disease ecology, health service provision, and disease mapping (all of which continue to reflect a mainly quantitative approach to inquiry) has evolved to a focus on a broader, theoretically informed epistemology of health geographies in an expanded international reach. As a result, we now find this subdiscipline characterised by a strongly theoretically informed research agenda, embracing a range of methods (quantitative; qualitative, and the integration of the two) of inquiry concerned with questions of: risk; representation and meaning; inequality and power; and culture and difference, among others. Health mapping and modelling has simultaneously been strengthened by the technical advances made in multilevel modelling, advanced spatial analytic methods, and GIS, while further engaging in questions related to health inequalities, population health, and environmental degradation.

This series publishes superior quality research monographs and edited collections representing contemporary applications in the field; this encompasses original research as well as advances in methods, techniques, and theories. The *Geographies of Health* series will capture the interest of a broad body of scholars, within the social sciences, the health sciences and beyond.

Public Health, Disease and Development in Africa
Edited by Ezekiel Kalipeni, Juliet Iwelunmor, Diana Grigsby-Toussaint, and Imelda K. Moise

Blue Space, Health and Wellbeing
Hydrophilia Unbounded
Edited by Ronan Foley, Robin Kearns, Thomas Kistemann, and Ben Wheeler

Cultivated Therapeutic Landscapes
Gardening for Prevention, Restoration, and Equity
Edited by Pauline Marsh and Allison Williams

For a full list of titles in this series, please visit https://www.routledge.com/Geographies-of-Health-Series/book-series/GHS

Cultivated Therapeutic Landscapes
Gardening for Prevention, Restoration, and Equity

Edited by Pauline Marsh and Allison Williams

Routledge
Taylor & Francis Group
LONDON AND NEW YORK

First published 2023
by Routledge
4 Park Square, Milton Park, Abingdon, Oxon OX14 4RN

and by Routledge
605 Third Avenue, New York, NY 10158

Routledge is an imprint of the Taylor & Francis Group, an informa business

British Library Cataloguing-in-Publication Data
A catalogue record for this book is available from the British Library

Library of Congress Cataloging-in-Publication Data
Names: Marsh, Pauline (Health geographer), editor. | Williams, Allison, 1965- editor.
Title: Cultivated therapeutic landscapes / edited by Pauline Marsh and Allison Williams.
Identifiers: LCCN 2023005093 (print) | LCCN 2023005094 (ebook) | ISBN 9781032409924 (hardback) | ISBN 9781032409955 (paperback) | ISBN 9781003355731 (ebook)
Subjects: LCSH: Gardening--Therapeutic use--Case studies. | Gardening--Social aspects--Case studies. | Gardening--Psychological aspects--Case studies. | Community gardens--Case studies. | Public spaces--Case studies.
Classification: LCC RM735.7.G37 C85 2022 (print) | LCC RM735.7.G37 (ebook) | DDC 615.8/515--dc23/eng/20230405
LC record available at https://lccn.loc.gov/2023005093
LC ebook record available at https://lccn.loc.gov/2023005094

ISBN: 978-1-032-40992-4 (hbk)
ISBN: 978-1-032-40995-5 (pbk)
ISBN: 978-1-003-35573-1 (ebk)

DOI: 10.4324/9781003355731

Typeset in Times New Roman
by SPi Technologies India Pvt Ltd (Straive)

Contents

Figures

Tables

Boxes

Acknowledgements

The editors would firstly like to thank the individual chapter authors for their generous contributions of time, expertise, passion, and knowledge. We strove to create a book that was interdisciplinary, and that fused theory, research, and practice, and this has been achieved only by the willingness of the authors to participate. For some, this work was done outside of paid employment, for others amid competing demands of academic research and teaching loads. We are incredibly grateful to all, and we hope that this book supports your efforts to raise the profile of the positive impacts of cultivated therapeutic landscapes into the future.

The editors would also like to acknowledge the excellent editorial contribution of Dr Jacqueline Fox in compiling this book. Dr Fox's skills and diligent attention to detail enabled our multifarious points of view, methodologies, and approaches to coalesce into a harmonious collection.

Pauline Marsh would like to thank the Australian Government, which provided partial financial support via funding from under the National Environmental Science Program as part of the Sustainable Communities and Waste Hub. She would also like to express her appreciation to the University of Tasmania, Departments of Rural Health, and the Wicking Dementia Research and Education Centre for allowing her the privilege of time to work on this book.

Allison Williams would like to thank McMaster University for the time provided for this project, which took place at the tail end of a sabbatical leave. Gratitude to the School of Earth, Environment & Society, and to the Faculty of Science, who granted a course release via a Faculty of Science Chair Award. Allison is most appreciative of the patience and support provided by her family – her son Cameron, her daughter Sarah and her partner, Saeid; thank you all for allowing me the time for and attention to this work!

Contributors

Alejandra Aguilar is a lecturer in Occupational Therapy at the University of South Australia. She has an interest in researching how meaningful activities such as therapeutic gardening can have an impact on individuals and their lives. Ali has also engaged in therapeutic gardening with her patients as an occupational therapist. She has published qualitative research within the community gardening research space.

Amy Baker is a researcher and lecturer in the Occupational Therapy Program at the University of South Australia. A key area of Amy's research focuses on the connection between nature and wellbeing. Amy is a member of University of South Australia's Outdoors, Nature and Health Network and Working Group, as well as the University's National Enterprise for Rural Community Wellbeing. Amy is a member of Therapeutic Horticulture Australia's Research Sub-Committee. Amy has contributed to a range of studies and initiatives focused on the health and wellbeing of people who live rurally. In her spare time, Amy is also a certified nature and forest bathing guide and a member of several conservation groups in South Australia.

Sara Barron is a lecturer in Urban Horticulture at the University of Melbourne. Sara has research expertise in urban forests, urban design, and landscape architecture. Her doctoral research examined the potential benefits of future urban forests in dense communities facing climate change. Since 2005, Sara has collaborated on policy-changing research and design projects for sustainable urban communities and climate change with both the Design Centre for Sustainability and the Collaborative for Advanced Landscape Planning at UBC. She currently is a member of the Green Infrastructure Research Group (GiRG) at the University of Melbourne.

Penny Cook is Professor of Public Health at the University of Salford, and leads a research group, 'Protecting the Public's Health'. She investigates the health and wellbeing benefits of greenspace, and public health topics such as physical activity and impacts of alcohol and tobacco. She also has expertise in asset-based community development. Her previous research includes investigating the overlap of health and environmental indicators, identifying health benefits of urban parks, quantifying access to country parks, and evaluating the socioeconomic, psychosocial, and environmental factors that promote sedentary behaviour.

Karolina Doughty is an assistant professor in Cultural Geography at Wageningen University & Research in the Netherlands. Her research interests fall broadly within health and cultural geographies, with a particular interest in the nexus of relations between wellbeing, sensory experience, and place, with a particular attention to sound. Her work has contributed to literatures on therapeutic landscapes, geographies of sound, and everyday mobilities.

Ronan Foley is an Associate Professor in Health Geography at Maynooth University, Ireland, with specialist expertise in therapeutic landscapes and the health-enabling potential of green and blue space. His current research focuses on relationships between green/blue, health and place, including two authored/co-edited books and journal articles on therapeutic landscapes and blue space, health and wellbeing. He co-edited a special issue on healthy blue space for Health & Place (2015) and is a member of the Editorial Board of *Health & Place*. He collaborates on a number of ongoing green/blue research projects in Ireland, the UK, Spain, Germany, New Zealand and Australia, with a particular interest in in-situ methodologies and the accretive potential of everyday outdoor practices for health and wellbeing.

Bruce French AO is an Australian agricultural scientist. He founded Food Plants International in 1999 with the goal of documenting food plants around the world. Bruce was made Officer of the Order of Australia in the 2016 Australia Day Honours list, for 'distinguished service to agricultural science through the provision of edible plant information for improved food security, nutrition, and improved health outcomes for people in developing countries'. He was the 2022 Tasmanian Senior Australian of the Year.

Selma Lunde Fjaestad has an interest in how the environment, in particular nature, can be used to improve wellbeing and encourage pro-environmental behaviours. She sees nature to be an essential gateway to positive mental health and an effective tool in connecting communities and humanity as a whole. Although she was born in Norway, she has spent the past seven years in Melbourne. Selma completed her Bachelor of Psychology and Psychophysiology at Swinburne University of Technology, including one semester of Neuroscience at Vrije University of Amsterdam.

Sumita Ghosh is an architect and urban planner and is currently working as an Associate Professor in Planning at the Faculty of Design, Architecture, and Building (DAB), University of Technology Sydney (UTS), Australia. Sumita has led collaborative research projects on green infrastructure and sustainable community development with universities and industries. Her research interests focus on green infrastructure planning and policy (urban agriculture, urban forests, and green roofs), urban sustainability performance assessment, and spatial analysis. Sumita had previously worked in local councils, research organisations, and universities in New Zealand and Australia. Sumita teaches doctoral and postgraduate planning students in UTS.

Michael Hardman is a senior lecturer in Urban Geography at the University of Salford and is co-lead of the Salford Care and Urban Farm Research Hub. Mike's research focuses on urban greening, particularly radical ways of juxtaposing nature with the city. He has led an array of projects funded by charities, businesses, local and national government, international funders, and other bodies around concepts such as urban agriculture and green infrastructure. His recent work has involved critically exploring the upscaling of urban farming, to conducting ethnographic work with guerrilla gardeners who colonise land without permission.

Michelle Howarth is a senior lecturer in Nursing and Deputy Director for Post Graduate Research at the University of Salford, School of Health & Society. Michelle has significant research expertise in evaluating the impact of nature-based interventions and has led over 8 funded projects. Michelle has worked with the Royal Horticultural Society (RHS) to evaluate their first Wellbeing programme and is on the Board of Trustees for Social Farms & Gardens. Michelle also leads the UK National Social Prescribing Network Special Interest Group and is an active member of the UK Green Care Coalition promoting the use of nature-based interventions for health wellbeing.

Jonathan Kingsley is a senior lecturer in Health Promotion (Swinburne University of Technology). Jonathan sees the natural environment as central to health and having the capacity to bridge health inequalities (the basis of his Honours, Masters, PhD, and previous Visiting Academic position at Cambridge University). Jonathan views himself as not only an academic but also an activist, having experience in programme management in the community and NGO sector, winning community engagement environment awards and sitting on multiple steering committees related to Indigenous and environmental health.

Robin Kearns is Professor of Geography in Te Kura Mātai Taiao/ School of Environment at the Waipapa Taumata Rau/ University of Auckland. He has long-standing interests in the connections between place and health/ wellbeing. His most recent book is the co-edited *Blue Space, Health and Wellbeing: Hydrophilia Unbounded* (Routledge, 2019).

Dave Kendal researches and teaches about the reciprocal effects of people on nature (mostly plants, but increasingly wildlife) in cities and beyond, the drivers and effects of environmental management in urban green space and conservation contexts, and outcomes including health and wellbeing, biodiversity, and ecosystem services. Dave worked as a postdoctoral ecologist at the Australian Research Centre for Urban Ecology, a division of the Royal Botanic Gardens Melbourne and was Research Fellow in Urban Greening at the School of Ecosystem and Forest Sciences at the University of Melbourne, and knowledge broker for two iterations of the National Environmental Science Program, funded by the Australian Government.

Kate Lee is an environmental psychology research fellow and her applied interdisciplinary work explores qualities of city and business environments associated with wellbeing, including urban green spaces and greening at work. Kate has explored a range of topics spanning individuals, communities, and organisations, including stress and performance at work, creativity, green roof visits and views, and work breaks. Kate has also explored collaboration for flourishing green cities, working alongside government, industry, and the community. She is currently a member of the green infrastructure research group and environmental social science group at the University of Melbourne.

Jessica L Mackelprang is a senior lecturer in Psychology (Swinburne University of Technology) and an Affiliate Member of the Harborview Injury Prevention and Research Center (USA). A Clinical Psychologist by training, she has delivered psychotherapy in community health and hospital settings to individuals across the lifespan who are affected by life-altering injury or chronic illness in addition to mental health difficulties. Her research aims to identify antecedents and outcomes associated with psychological and physical trauma among populations that have been marginalised, with a focus on supporting community members affected by homelessness.

Suzanne Mallick is a researcher and teacher with the Centre for Rural Health. She is social scientist with extensive experience in state and federal government policy and programme work and service sector evaluation and design. She has a particular research interest in the health and wellbeing of migrants and refugees; qualitative research methods and community development practice; and health workforce planning. She has published her research in international and national journals and written policy and programme publications for government. Her key interests are rural health, health benefits of economic and social participation and the wellbeing of migrants and refugees.

Pauline Marsh is a Health Geographer with the Wicking Dementia Research and Education Centre, at the University of Tasmania. Her research explores how being in nature improves our quality of life and she is particularly interested in the therapeutic benefits of gardens and the outdoors for people with cognitive, emotional, and physical health challenges. She utilises methods of participatory action research, story-gathering and filmmaking and publishes in a range of academic journals. One of her greatest achievements is the co-founding of DIGnity Supported Community Gardening. She is a lead investigator on the Nature Connection Project (NESP), and Co-lead of the Healthy Landscapes Research Group and the Venture Out Dementia Research Group.

Anthea Maynard is a microbiologist, with a master of public health and tropical medicine. She has worked in community nutrition in Zimbabwe, intergenerational faith formation in Tasmania and Aboriginal health promotion in Townsville. Anthea is currently working in three roles: a Primary Health Consultant for Primary Health Tasmania, a team leader for Launceston City Baptist church, and a coordinator of Food Plants International. These roles allow Anthea to pursue her passion for preventative health, intergenerational community, spiritual wellbeing, community nutrition and ecological restoration

Alice McSherry is a doctoral candidate in human geography at the University of Auckland/Waipapa Taumata Rau. Her current research focuses on human plant relationships in the context of folk herbalism, ancestral healing, and decolonisation. If not reading or writing, Alice is found in the garden, tending to the plants around her and muttering spells of gratitude to the Earth.

Maddison Miller is a Darug woman and research fellow at the University of Melbourne and the Arthur Rylah Institute for Environmental Research, Department of Environment, Land, Water and Planning. Her research explores ways in which to bring together Indigenous and western ecological knowledges to foster healthy Country outcomes in land management and research. Maddi uses storytelling as a mechanism for the weaving together of these knowledges, to create new ways of knowing grounded in Indigenous ontologies. Maddi has a strong background in archaeology and cultural heritage management, as well as embedding Indigenous knowledge in urban research through her position as co-chair of the Indigenous Advisory Group to the Clean Air and Urban Landscapes Hub.

Renae Riviere: With a background in outdoor and environmental education, Renae joined Conservation Volunteers Australia (CVA) to further develop her interest in inspiring positive outcomes for our environment, by connecting people to nature through conservation volunteering. As the Director of CVA's Revive Campaign, her focus is on engaging the community in activities that benefit the health and function of coastal, riparian, and wetland habitats, whilst simultaneously improving the health and social connection of the people who volunteer, live, work, and play in these areas. She is particularly interested in the role that volunteering and spending time in nature can play in the well-being and social connectedness of marginalised and/or isolated segments of the community, and is currently studying a Bachelor of Arts, majoring in Anthropology and Archaeology and hopes to use this to further explore her interest in this area.

Emily Rugel is a postdoctoral fellow at Simon Fraser University and a visiting researcher at the University of Sydney, where her work explores health-promoting community design across the lifespan with the aim of developing evidence that can be embedded in sustainability plans and integrated in policies that advance equity. She received her doctorate from the University of

British Columbia, where she developed a regional model of access to natural spaces and applied it to prescription and health-survey data to clarify pathways linking urban nature to social ties and mental health. In addition to a PhD, she holds a Master of Public Health, and a BA in Journalism, but firmly believes in the acquisition of knowledge through chance encounters as well as scientific investigation.

Theresa L Scott is a senior lecturer in Clinical Geropsychology and National Health and Medical Research Council and Australian Research Council (NH&MRC-ARC) funded Dementia Research Development Fellow, with broad research expertise in the area of healthy ageing, evidence-based therapies in later life, and supporting quality of life of people living dementia and their families. As an applied researcher, she uses mixed methods and co-design to develop and implement interventions targeted at supporting the functionality and independence of older persons and older persons living with dementia and their care partners. She has published book chapters, journal articles, and presented to international and national conferences on topics related to therapeutic horticulture.

Ben Sellar is a lecturer in Occupational Therapy and member of the IIMPACT Child Health Research Concentration at the University of South Australia. His research examines the effect of physical, social, cultural, and institutional environments on child health, disability, and development, from public playgrounds to correctional settings. A key element of this research has involved investigating the way Australian Nature-Play promotional materials reproduce culturally specific notions of nature, childhood, and parenting to the exclusion of already marginalised communities. Ben's teaching involves supervision of student community development projects and allows him to work with schools and communities to develop to improve community supports for family relationships and child development.

Josephine Spring read nutrition at Queen Elizabeth College, London University, and did research in government and industry interspersed with research management at the Science and Engineering Research Council and the Royal Academy of Engineering. She mapped research at the University College London Clinical Research Network and managed clinical research at the Royal Surrey County Hospital, and studied Horticulture at the University of Reading. At the Royal Hospital for Neuro-disability, London, she undertook research into horticultural therapy for Huntington's disease. She was awarded a Churchill Fellowship to study garden therapy for neuro-disability in Scandinavia. She is now retired, and assesses for South West in Bloom.

Takemi Sugiyama's research explores the nexus between design and health. With his interdisciplinary background in architecture, urban design, and behavioural epidemiology, Takemi currently works on the following topics:

urban design attributes facilitating adults' active living; health impact of active and sedentary transport; and office design factors related to workers' movement and interactions.

Jessica Thompson is a practitioner-academic currently combining a PhD at the University of Salford with her day job as Director of City of Trees, UK. Jess' PhD is a Realist Evaluation of the wellbeing impacts of civic environmental activity focusing on the experiences of people involved in improving public realm greenspace provision at a neighbourhood level. Jess has previously undertaken an MSc in Dementia, applying the context of public greenspace as an enabling environment for people affected by dementia which has led to the development and delivery of dementia-friendly nature-based activities. Jess aims to achieve positive social outcomes and brings practitioner perspective to academic work.

Bethaney Turner is an associate professor in Global Studies at the University of Canberra and member of the Centre for Creative and Cultural Research in the Faculty of Arts and Design. In a significant departure from her doctoral work on social revolutionary movements in Mexico, her current research explores the variety and complexity of the relationships between people and the food they grow, buy, and consume. From local community gardens to global debates about food security, this research analyses the role food plays in the formation of subjectivities, practices of meaning-making and understandings of place.

Esther Veen is a reader in Urban Food Issues at Aeres University of Applied Sciences in Almere, the Netherlands. Her work focuses on understanding food routines and creating more sustainable, healthy, and inclusive foodscapes that invite healthier and more sustainable routines. She is particularly interested in everyday urban life and the benefits of green environments for health and wellbeing.

Allison Williams completed her PhD at York University (Toronto, Canada) and, after holding permanent positions at both Brock University (St. Catharines, Ontario) and University of Saskatchewan (Saskatoon, Saskatchewan), is now Professor in McMaster University's School of Geography and Earth Sciences (Hamilton, Ontario). Her background in social health and geography continue to inform policy development. Dr Williams teaches health geography, public health and research methods and has numerous graduate student trainees.

Introduction

Pauline Marsh and Allison Williams

The healing role of gardens and gardening has been long celebrated – across mythology, literature, art, and religion. Often, these healing gardens are lush and green, and people appear to live in peace and contentment; these are images of gentle people tending to plants amid beauty and delight. In contemporary research we still find evidence for the therapeutic value of peace and contentment, but a range of disciplines now demonstrate a growing assemblage of diverse mechanisms are at work enabling health and wellbeing. The role of gardens and gardening in enhancing health and wellbeing is of increasing interest to the spectrum of human health and social sciences: medical, social, geographical, and psychological research has added greatly to our understandings of the attributes and impacts of the cultivated therapeutic landscape.

Cultivation refers to the preparation of and/or use for the raising of crops, or to foster growth in a garden, rather than in the wild. For millennia, people have engaged in various forms of cultivation for the health benefits to themselves and to the environment, as well as for practical purposes of growing food to feed themselves and others. The scholarship shows that tending to gardens helps us to relax, and to eat well, but also to socialise, to exercise and improve our mental health, and to contribute and feel a sense of purpose and meaning. Gardens are where many of us find solace in times of bereavement and grief, and where we turn for restoration from life stressors. In a time of climate change and biodiversity loss impacts on health, cultivation is under increased scrutiny.

The health-enabling functions of cultivation are due, in part, to the garden environment, but also to the embodied acts it entails. In one of the most beautiful books written about gardens and gardening, *Gardens: An essay on the human condition*, Robert Pogue Harrison (2008) invites us to consider cultivation as a form of caring. A gardener's vocation for caring, he suggests, is the ultimate expression of the human condition, because we have an innate need to care – for the earth and for each other. These are the spaces in which we cultivate not only the soil, but also nature-connectedness and hope.

During the early months of the COVID-19 pandemic, as public health directives were rolled out across the globe and people left their workplaces and public spaces, we stayed safe in the private realms of our homes and – for those

DOI: 10.4324/9781003355731-1

lucky enough – our garden sanctuaries. Researchers were quick to explore this unique situation, perhaps with the aim of demonstrating what some of us had known to be true – not only from previous studies, but also our own personal experiences: in times of trouble and tumult, gardening is especially beneficial.

But the cultivated therapeutic landscape is not only quiet and restorative; it is also intensely political. Gardening allows us to circumvent commercial pathways to feed ourselves and others. We can reclaim degraded spaces and create beauty in the wake of horror. In these ways, gardening can be a social leveller, mitigating against structural inequities caused by socio-economic disparity. In some contexts, gardening is the means by which we assert our human rights to civic contribution, protest, and cultural continuation.

In *Cultivated therapeutic landscapes: Gardening for prevention, restoration, and equity*, we explore gardening for all of the ways it contributes to improving health, wellbeing, and equity – for humans and ecological landscapes both. We build on this range of understandings, knowledges, and experiences as we explore the breadth and depth of cultivated therapeutic landscapes.

Before we go any further, we would like to point out that we did not choose the term 'cultivation' lightly. As settler-nation academics, we are acutely aware that it is a problematic and contentious term. On the surface, this practical, 'horticultural' definition is not too alarming. However, 'to cultivate' also means to *improve* in general – by labour, care, or study; 'to be cultivated' is to have '*refined* tastes and manners' (Oxford dictionary; our emphases). Consequently, in the contexts of gardening and cultural and social norms, cultivation is firmly associated with the human endeavour to 'improve' natural landscapes, and with colonial imperatives to 'civilise'. As such, it continues to play its part in furthering social and ecological hierarchies along 'sophisticated' and 'wild' spatial continuums.

Rather than turning our backs and minds away from these outdated and harmful notions of superiority, in this edition we put on our boots and dig deeply into these cultivated conundrums. This is not a horticultural therapy text; the 'therapeutic' in our cultivated landscapes refers to the multiplicity of 'health-enabling' elements present in garden spaces, as well as those generated by the practices of cultivation. We consider the eco-socio-spatial drivers and mitigators of social and health (in)equities, and approach gardeners of all abilities as potential agents of change. In this collection we challenge, invert, and renegotiate all of the contested and messy weeds, pests, blooms, and bounties of the landscape.

Therapeutic landscapes: a short history

Known as one of the most significant contributions health geographers have made to the comprehensive study of health, therapeutic landscapes theory is a conceptual framework for the analysis of physical (natural and built), social, and symbolic environments, as they contribute to physical and mental health and wellbeing in places (Williams 2009; Bell et al. 2018; Gesler 1992). Conventionally, applications of the theory have been realised in four areas: (i) traditional landscapes (i.e., reputed for health and/or healing, such as shrines, pilgrimages, etc.); (ii) natural/pristine/wild areas; (iii) landscapes for the marginalised (i.e., individuals with mental health issues, autism, etc.); and (iv) applications to health care sites (i.e., hospitals, long-term care facilities, etc.) (Williams 2007, 1999). A recent scoping review of the literature noted that, as a health geographical concept, therapeutic landscapes theory has wide application, offering in-depth insight into experiential, embodied, and emotional geographies, and promoting awareness of place as both therapeutic and exclusionary; such applications signify that the concept continues to be a relevant and lively field of enquiry across health geographies (Bell et al. 2018). Within the ongoing evolution of therapeutic landscapes theory, scholarship continues to develop more nuanced understandings of the characteristics of place that contribute to the symbolic and material processes at play in the making of health-enabling places. Some of the latest foci include the examination of how the coloured elements of healing places, particularly blue and green, promote health and wellbeing (Brooke and Williams 2021; Foley et al. 2019). In this evolving field of cultivated therapeutic landscape scholarship we also see new, exciting, critical exploration of the 'therapeutic' impacts of and by gardeners on the health gaps perpetuated by disadvantage.

The garden represents the space where natural and built environments coalesce. Natural, given that the central foci are botanical plants, and simultaneously built, in that human hands seed, plant, nurture, and shape the growth and success of the planted space. Gardens were introduced as therapeutic landscapes by Milligan, Gatrell, and Bingley (2004) in their studies of community gardens for older people in northern England. Effects included positive physical, mental, and social wellbeing, where relaxation, inclusion, purpose, safety, and restoration were experienced by participants. Health geographers have more recently begun to explore gardens as therapeutic landscapes for meeting the particular needs of a variety of population groups. For example, work has been undertaken to demonstrate the positive impacts on people with life-limiting illness and at the end of life, and those experiencing grief. Gardens function not only as restful sites during these difficult times, but also as places where people experience continuing spiritual relationships with those who have died (Marsh et al. 2017). For people living with dementia, gardens and farms are emerging as vital places in which to find both peace and company, as well as offer the chance to take risks and continue to be dignified, active citizens (Mmako et al. 2020). School gardens for young people and post-secondary students have therapeutic functions related to learning, development, and social function (Austin 2021; Marsh et al. 2020). For those who gardened during periods of social isolation during the COVID-19 pandemic, the therapeutic benefits have been multifarious (Egerer et al. 2022; Kingsley et al. 2022; Marsh et al. 2021).

This collection

This edited collection demonstrates the array of therapeutic landscapes research and practice that continues to grow, through and around the processes of cultivation. Key scholars take a close and critical look at the impacts of diverse forms of engagement with and in a variety of cultivated green spaces on health, wellbeing, and equity. More a critical exploration of therapeutic landscapes than a practical gardening guide, our deliberately interdisciplinary approach includes academics and practitioners alike – many of us identify as both. Nonetheless, there are practical examples and suggestions for the translation of theory to practice throughout. Contributions cover the disciplines of health, cultural and urban geographies, occupational therapy, psychology, disability studies, engineering, rural health, public health, health promotion, landscape design, nature-based interventions, dementia studies, nursing, architecture, urban planning, ecological sustainability, horticulture, urban forestry, environmental psychology, and food security, among others.

We divide the chapters into three parts. Part I, *Boots on: Scoping the cultivated therapeutic landscape*, offers an excellent collection of contemporary cultivated therapeutic landscape scholarship. Reading these chapters, we get a sense not only of the depth and breadth of settings, the purposes and impacts being generated, but also the uptake of disciplines and methodologies being harnessed. We begin in the public realm, where Sara Barron and colleagues invite us to consider a typology of 'tending'. In this cross-disciplinary critical essay, they argue that cultivation is a continuum of reciprocated tending – and, although our care for nature benefits human and non-human nature alike, unresolved social, cultural, and climate tensions demand attention. Allison Williams then draws on her own experiences as a volunteer in her local 'Victory Garden'. Through autoethnography, she provides rich insights into the health bounty provided by the combination of gardening and volunteering for the individuals involved and the wider community during the COVID-19 pandemic. In Chapter 3, Sumita Ghosh combines literature review and case study approaches to explore site variances across three classic cultivated landscapes: community, school, and home gardens. Through a planning and design lens, she traverses geographical and social gradients to demonstrate how the integration of local foodscapes is transforming health and wellbeing for those involved. Next, we turn to a structured therapeutic horticulture programme in the Netherlands, designed to support people living with cancer. Despite the programme's lifestyle behavioural change aims, Ester Veen and Karolina Doughty found little evidence of changes in diet and exercise; they did, however, uncover great benefits from the social, caring, and comforting enablers of the garden and the programme structure, suggesting the value lies in a holistic, public health approach. To close Part I, Selma Lunde Fjaestad and colleagues review the literature for evidence of psychological impacts from unstructured gardening activities, rather than structured therapeutic horticulture programmes. From a small collection of studies, they identify the value this form of cultivation is shown to have for mental health, but raise the need for psychologically robust tools to improve the overall strength of the scholarship.

Part II, *Companion planting: Cultivating human wellbeing*, foregrounds the variety of health needs and target populations that cultivated therapeutic landscape research and practice reaches. These critical studies and essays slant towards questions concerning the physical, emotional, spiritual, and mental health of individuals and communities – but not in isolation from the social, political, and ecological contexts that also comprise these landscapes. The increasingly common practice of 'green social prescribing' is the starting point from which Jessica Thompson and colleagues explore the health and wellbeing benefits of cultivating and nurturing public green sites, within the pressing context of climate change and biodiversity loss. Next, Josephine Spring applies her years of clinical practice experience working in the neuro-disabilities field to walk us through the pragmatic considerations of gardening in institutional spaces specifically for people experiencing cognitive symptoms. Much of the literature focuses on the needs of people who are disconnected from nature as a result of living in large cities. Occupational therapist Amy Baker and colleagues take us out of the cities to consider the needs of rural populations and explore the health possibilities of 'red' landscapes. In the final chapter of Part II, we return to the clinical setting via Theresa Scott's in-depth look at nature connection in residential aged care settings for people living with dementia. With a critical and pragmatic lens, Scott identifies the barriers and enablers to accessing health-enabling nature for people living intensely interior lives.

In the chapters of Part III, *Dig deep: Expanding and enriching cultivated therapeutic landscape*, the authors encourage us to push the boundaries of cultivation further – as a concept and a practice. They introduce some novel methodological approaches of study, while unlocking the garden gates and enticing us further down the path into the woods, the wilds, and the weeds. Pauline Marsh and colleagues report on the findings of a conservation volunteer programme with migrants which document the programme's impacts on upstream health factors, such as employment and social connection. Bruce French and Anthea Maynard share Bruce's lifetime of collecting and collating edible plant species from across the globe to demonstrate the importance – and simplicity – of addressing malnutrition through cultivating local plants, while offering a timely caution against a focus on increasing crop yields over small-scale, culturally relevant cultivation practices. Next, Bethaney Turner digs below the surface into the soil and its microbiome. In the underground, she critiques human–soil relations via ethnographic fieldwork and commentary on social and environmental inequities and multispecies injustices. We end Part III at the outskirts of the garden – in the unruly edges where Alice McSherry and Robin Kearns explore the implication of kinship relations with those plants we have hitherto been discouraged from forming relations with: weeds. They challenge many of the anthropocentric notions we raised at the beginning of this Introduction – control, order, improvement – by arguing for healing based on renewed attention to human humility and a becoming at ease with our shared botanical unruliness.

Ronan Foley, a reputed scholar of therapeutic landscapes, provides a concluding commentary, which not only synthesises the primary themes within the book, but also introduces new ideas for future work 'that frame gardens and gardening as horti-cultural therapeutic assemblages of health and wellbeing' (p. XX). While highlighting the contributions of the collection within the larger literature, Foley discusses the role gardening has had during the pandemic, as well as how gardens operate as sites of maintenance and repair; not only do gardens require ongoing upkeep and mending, but in so doing, facilitate the same in the gardeners that immerse themselves in them and, more broadly, in terms of planetary ecological pressures and climate change.

References

Austin, S, 2021, 'The school garden in the primary school: Meeting the challenges and reaping the benefits', *Education*, pp. 3–13. https://doi.org/10.1080/03004279.2021.1905017

Bell, SL, Foley, R, Houghton, F, Maddrell, A and Williams, AM, 2018, 'From therapeutic landscapes to healthy spaces, places and practices: A scoping review', *Social Science & Medicine*, vol. 196, pp. 123–130. https://doi.org/10.1016/j.socscimed.2017.11.035

Brooke, K, Williams, A, 2021, 'Iceland as a therapeutic landscape: white wilderness spaces for well-being', *GeoJournal*, vol. 86, pp. 1275–1285. https://doi.org/10.1007/s10708-019-10128-9

Egerer, M, Lin, B, Kingsley, J, Marsh, P, Diekmann, L and Ossola, A, 2022, 'Gardening can relieve human stress and boost nature connection during the COVID-19 pandemic', *Urban Forestry & Urban Greening*, vol. 68, 127483. https://doi.org/10.1016/j.ufug.2022.127483

Foley, R, Kearns, R, Kistemann, T and Wheeler, B (eds) 2019, *Blue space, health and wellbeing: Hydrophilia unbounded*, Geographies of Health, Routledge, London.

Gesler, W, 1992, 'Therapeutic landscapes: Medical geographic research in light of the new cultural geography', *Social Science & Medicine*, vol. 34, pp. 735–746. https://doi.org/10.1016/0277-9536(92)90360-3.

Harrison, RP, (2008), *Gardens: An essay on the human condition*, University of Chicago, Chicago.

Kingsley, J, Diekmann, L, Egerer, MH, Lin, BB, Ossola, A and Marsh, P 2022, 'Experiences of gardening during the early stages of the COVID-19 pandemic', *Health & Place*, vol. 76, 102854. https://doi.org/10.1016/j.healthplace.2022.102854

Marsh, P, Diekmann, LO, Egerer, M, Lin, B, Ossola, A and Kingsley, J 2021, 'Where birds felt louder: The garden as a refuge during COVID-19', *Wellbeing, Space and Society*, vol. 2, article 100055. https://doi.org/10.1016/j.wss.2021.100055

Marsh, P, Gartrell, G, Egg, G, Nolan, A and Cross, M 2017, 'End-of-life care in a community garden: Findings from a participatory action research project in regional Australia', *Health & Place*, vol. 45, pp. 110–116. https://doi.org/10.1016/j.healthplace.2017.03.006

Marsh, P, Mallick, S, Flies, E, Jones, P, Pearson, S, Koolhof, I, Byrne, J and Kendal, D 2020, 'Trust, connection and equity: Can understanding context help to establish successful campus community gardens?' *International Journal of Environmental Research and Public Health*, vol. 17, no. 20, article 7476. https://doi.org/10.3390/ijerph17207476

Milligan, C, Gatrell, A and Bingley, A 2004, '"Cultivating health": Therapeutic landscapes and older people in northern England', *Social Science & Medicine*, vol. 58, no. 9, pp. 1781–1793. https://doi.org/10.1016/S0277-9536(03)00397-6

Mmako, NJ, Courtney-Pratt, H and Marsh, P 2020, 'Green spaces, dementia and a meaningful life in the community: A mixed studies review', *Health & Place*, vol. 63, 102344. https://doi.org/10.1016/j.healthplace.2020.102344

Williams, A (ed) 1999, *Therapeutic landscapes: The dynamic between place and wellness*, University Press of America, New York.

Williams, A (ed) 2007, *Therapeutic landscapes*, Routledge/Taylor & Francis, Abingdon.

Williams, A, 2009, 'Therapeutic landscapes as health promoting places: Past, present and future', *Companion to health and medical geography*, edited by Brown, T, Moon, G and McLafferty, S, Blackwell, UK, pp. 207–223.

Part I

Boots on

Scoping the cultivated therapeutic landscape

1 Tending more than gardens

Engaging residents in public landscapes to cultivate urban nature

Sara Barron, Kate Lee, Maddison Miller, and Emily Rugel

Introduction

For individuals living in urbanised environments, finding a patch of land to tend (a term defined in greater detail below) remains challenging. Increasing demands on spaces and soils in cities are reducing opportunities to nurture plants and provide suitable landscapes for fauna, further exacerbating people's disconnection from nature. In this chapter, we re-envision the ways in which urban residents can tend to nature and foster their connections to the places they tend, moving beyond the garden to engage with the landscapes that surround them more broadly. We draw on a range of approaches to tending across a variety of public landscapes, including those led by institutions and by communities, to explore the acts of tending that are evident in multiple contexts. We also recognise and call attention to the care that Indigenous peoples apply to their homelands and the ways in which these acts have been interrupted and intercepted by colonial forces. Throughout, we draw on existing research from a wide range of disciplines conducted in urban settings around the globe, while acknowledging an overrepresentation from the places in which the authors have lived and practised: Oceania, North America, and Europe.

In cities, private gardens and cultivated spaces are one of the primary places where people are able to spend time and connect with nature. Researchers and practitioners emphasise the need to support deeper engagement with nature through these existing relationships. We propose that illuminating the concept of 'tending' can uncover the potential benefits of a wider range of experiences. Broadening our notions of tending can also help provide a path towards increasing urban resilience and equity. Not all local authorities governing urban areas have the means to create and maintain accessible and quality public landscapes. International recognition of the importance of ensuring ecosystems and biodiversity are valued within municipal planning processes is reflected in the United Nations Sustainable Development Goal 15 (United Nations 2022) and is recognised by the World Health Organization (WHO) as a key pillar for COVID recovery (WHO 2020). In addition, tending by residents can support the provision and maintenance of public landscapes, if these acts are sufficiently accepted and supported by municipal and regional authorities charged with managing these areas (Cranz and Boland 2004).

DOI: 10.4324/9781003355731-3

Natural public landscapes offer essential nature-based solutions (NBS) to the array of economic, social, and ecological challenges created by the climate crisis (European Commission 2015). In addition, robust evidence from multiple systematic reviews indicates that integrating NBS can provide profound benefits to population health through reduced heat stress and improved psychological wellbeing (van den Bosch and Ode Sang 2017). Building on these direct benefits, NBS can create a self-reinforcing virtuous cycle: opportunities to engage with nearby nature strengthen our sense of connection with nature, fostering our engagement with a range of sustainable behaviours (Prévot et al. 2018) and increasing levels of public support for NBS (Steg et al. 2014), thus leading to even more local public landscapes to tend.

Our exploration of tending acknowledges it as an act of care through which human beings have actively engaged with the landscapes that surround us over tens of thousands of years. Critically, this chapter also extends this construction of tending to consider the ways in which the landscape, in turn, cares for those who engage with it. We see this view of tending as a critical expansion of the concept of 'cultivation' – defined as, 'to prepare or prepare and use for the raising of crops; to foster the growth of; culture; to improve by labour, care, or study' (Merriam-Webster 2022) – in part because of this embedded duality and also because the types of landscapes we explore consist of much more than crops. Tending is an act that brought nourishment to our cities long before they were the places we know today, when concepts such as 'urban' or 'public' took on meanings distinct from those currently understood. As colonial forces chose the places that later became 'cities', they mistook these carefully tended places for naturally occurring landscapes suited to sustaining newly arrived populations (Pascoe 2014). Indigenous peoples have cared for *place* long before present-day cities were built and will continue their custodial roles far into the future (Darug et al. 2019; Rey 2019).

In this chapter, we re-envision the ways in which modern urban residents can tend to nature and foster their connections to the places they tend, tracing current practices back to their roots in traditional knowledge systems. We investigate the benefits of such nurturing to individual wellbeing; the ways community-based organisations can help to protect and preserve nature; and the roles local authorities play in planning and designing for nature connectedness. We focus on the role of tending nature in public landscapes and how the act of tending can transform a space into a place. Borrowing Laurie Olin's suggestion that meanings 'relate to aspects of the landscape as a setting for society and have been developed as a reflection or expression of hopes and fears for survival and perpetuation' (cited in Treib 2011, p. 84), we recognise that landscapes have a diversity and history of meanings. With these thoughts in mind, we suggest that urban landscapes – as settings for society and experience, as well as a source of identity – should allow for tending by 'the people whose daily practices and perceptions shape the social and physical landscape' (Olwig 2007, p. 581). We frame tending, in all its various forms, as a daily practice that can shape public landscapes and, as such, both reflect a community's values and provide a setting for communities to connect.

Drawing on interdisciplinary perspectives from cultural heritage, Indigenous epistemologies, horticulture, environmental psychology, landscape design, and population health, we examine these broader notions of 'tending' in the context of urban environments, moving beyond the garden as the primary site for tending to engage with public landscapes more broadly. Seeking to inform current debates in urban planning and municipal policy, we also delve into the potential trade-offs of engagement, in which optimal tending for plants and people may diverge, delineating multiple paths that institutions and communities can pursue to cultivate sustained programmes of support for urban nature. Although our focus is on resolving tensions that exist at the neighbourhood, municipal, and regional levels, we also acknowledge that such conflicts exist at a much broader, and bloodier, scale, via the creation of protected areas on Indigenous ancestral lands that displace communities and result in human-rights violations and outright violence (Project Expedite Justice 2022).

Current typologies of tending

Forms of tending – acts of care extending beyond mere cultivation – are wide and varied. Tending actions can range from those designed to transform landscapes – significantly altering their composition and appearance to something new or re-establishing a previous state – to those designed to manage, maintain, or even conserve existing landscapes. Both transformational tending and ongoing management may be directed towards creating landscapes *for* humans (such as maintaining sports fields or creating new parklands) or to be protected *from* humans (such as conserving remnant vegetation or revegetating ecologically degraded landscapes such as agricultural lands). Transformation can be viewed both positively and negatively: in the ecological sense, transformation may create novel landscapes comprising new compositions of species for which human 'tenders' have little sense of their future state or appropriate management (Lindenmayer et al. 2008). In addition, transformations of this sort may be the result of deliberate or inadvertent human action (Hobbs et al. 2006), with the latter potentially resulting in more complex management issues. The actions used to tend landscapes also range in their level of formality, from formal tending of spaces that are regularly scheduled or contracted out to appropriate experts to informal tending of spaces that fall outside regulations, are conducted irregularly, or operate solely at the neighbourhood level.

To take one example of tending, environmental justice groups are positioned as an exemplar of community-led efforts that manage landscapes with human needs and benefits at the forefront. Traditionally, environmental justice has focused on protecting vulnerable communities from environmental harms, particularly sites such as coal-fired power plants, waste-transfer facilities, or diesel-bus hubs (Bullard 1996). More recently, however, the movement's guiding notion of protection *from* environmental polluters has expanded to include improved access *to* ecosystem services (Jennings, Larson, and Yun 2016a). In contrast to community-led landscape-management efforts that value the sustainability and biodiversity of nature itself as a standalone common good (for which climate-change activists are provided as an example in Figure 1.1), this access-driven framework is primarily motivated by the need to ensure distributive justice to the benefits urban landscapes offer people (Jennings, Yun and Larson 2016b).

Spaces for tending can be plentiful in the urban landscape, from those fully in the private realm to those completely in the public domain, with the formality of their tending often shifting across this spectrum as well. Although private gardens or backyards are not available to every urban resident, the benefits of tending these private landscapes can extend beyond individual owners or even those who are invited to visit these sites. Conversely, although landscapes in the public domain are theoretically more available for tending, societal and jurisdictional constructs may prevent residents from feeling welcome to carry out this act, particularly in formal parks. Many liminal and often untended spaces exist between these extremes – nature strips (verges or berms) along streets and informal greenspace, for example – that can be opened for tending with minimal change to typical urban land management. Aesthetics can play a role in public acceptance of community-led tending, with some spaces likely to

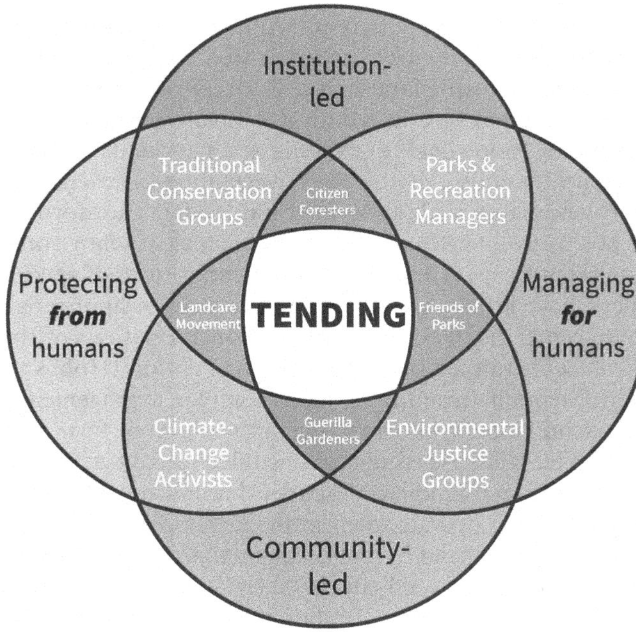

Figure 1.1 A typology of tending that specifies current actors and their actions of tending across different spaces. These actors are arrayed vertically according to their group structure and horizontally according to their aims. Those groups placed in greater proximity to one pole or another align most closely with the respective structure and aims of the pole.

require more formal or 'orderly' styles of management, while others could be tended with an eye to more ecologically diverse, or 'messy', aesthetics.

In this chapter, we seek to clarify new typologies via an extensive exploration of forces that are often in tension under current definitions of tending. We distinguish actors and their actions as the main factors to delineate the varied typologies of tending, which operate across different spaces. With respect to actors, we argue that there are two main dualities at play. The first is based upon the structure of groups that lead urban landscape-based activities, with the two poles being institutions and communities. The second is based on the overarching aims of these tenders, whether focused on managing landscapes *for* humans or protecting them *from* humans. Figure 1.1 depicts the intersection of all four poles. It provides a paradigmatic example of tending as it currently exists, and one that sits at the intersection of these distinct dualities. Although much of this chapter elucidates the current state of tending (as delineated in this diagram), as well as detailing the contested histories underlying and appearing in the present, we conclude by proposing a vision for the future.

Towards a new typology of cultivated spaces

In considering our typology, as outlined in Figure 1.1, we first acknowledge that, in many cases, urban environments in settler-colonial countries are on Indigenous lands and that Indigenous peoples have maintained their caring responsibilities for these places in the face of colonial intrusion. Despite deep ecological knowledge, Indigenous peoples have been marginalised from decision-making about the urban environment (Mata et al. 2020). The richness of Indigenous cultures, and the knowledges that are held within these communities, have often not been appreciated within such areas. This deficit reflects the colonial project of forgetting and the deliberate denigration of Indigenous knowledges, particularly in contested urban spaces (Cumpston 2020).

Nevertheless, the vital contribution of Indigenous and local knowledges to building climate-resilient cities that respond to their unique environments is now being acknowledged. In many Indigenous nations within Australia, the 'Welcome to Country' ceremony invites all to enjoy the bounty of the land and waters, but asks for care to be extended in return, thus enacting a cultural and social responsibility that is extended by Traditional Owners in ceremony to settlers and visitors to their lands (Murphy and Kennedy 2016). In this way, tending becomes an act of reconciliation and belonging for settlers and a reclamation of cultural responsibility and sovereignty for Indigenous peoples. Our case study in Box 1.1 illustrates how reconciliation can be brought into early childhood education – fostering positive relationships between settlers and Indigenous peoples from a young age. Unfortunately, examples like this are few, but there is hope for the future. Increasingly, Indigenous tending has been revitalised in movements such as cultural burning (Steffensen 2020). These acts of cultural reclamation are often hampered by enforced boundaries (such as land ownership) and are rarely practised in urban environments, but the potential exists to carry them out in a broader array of landscapes.

Box 1.1 Case Study: Tending in an early childhood centre in British Columbia, Canada

In settler-colonial societies, creating a deep connection to the land cannot happen without a deeper understanding of the practices and views of peoples who lived with the land long before colonial settlement. Indigenous principles of kinship and reciprocity that break the nature/human dichotomy can ground tending efforts (Nxumalo 2019). Establishing this understanding through early-life experiences is also a critical step in the reconciliation process, a fact recognised by a nationally renowned early childhood educator in Vancouver, British Columbia. We share the story of a childcare centre attended by one of the author's children as an example of how one individual's passion can inspire intergenerational change.

Feeling let down by her own childhood education's omission of any instruction about the history of colonial settlement and the residential school programme in Canada, an early childhood educator set about building a curriculum to advance recognition and understanding of this history among children aged 1–5 years. In a childcare facility located near a second-growth forest, students lived and learned daily within the local xʷməθkʷəy̓əm [Musqueam] culture and environment (see Figure 1.2). Daily forest walks included a land acknowledgment, play sessions centred on sharing creation stories, and details of settler history and its atrocities were introduced in an age-appropriate fashion. The children were welcomed to sit with xʷməθkʷəy̓əm [Musqueam] elders in ceremony. They learned traditional arts, knew hən̓q̓əmin̓əm̓ [Halkomelem] words for local flora and fauna, and witnessed the carving and raising of a reconciliation pole, knowing why it was important and educating their parents on the meaning of the pole's symbols. Through trial and error, and championed by a passionate educator, over the course of five years the centre cared for children who knew the stories of the land on which they played and appreciated their relationship with the forest. For these children, tending was intuitively caring.

Such an understanding cultivated in early childhood could be a path to a better, more sustainable future caring for the land. Unfortunately, this programme was driven by a single champion; when she left, the same level of connection could not be maintained. This case study underscores the importance of community champions creating a path for change, while highlighting the centrality of institutional support to maintain and expand programmes of this type.

Figure 1.2 The stories and language of the land can be a daily part of an early childhood education, fostering deeper connections for the next generations.

Photo credit: Kristin Webster.

In this section, we apply our typology of tenders to illuminate the full spectrum of tending available in urban public landscapes, detailing distinct domains of tending that are common within urban spaces. Based on Figure 1.1 these domains comprise: (i) institutions managing for people; (ii) communities managing for people; (iii) institutions managing for nature; and (iv) communities managing for nature. Within each domain, we provide a brief history, describe a range of current efforts, and delve into the future potential for tending.

Institution-led management to benefit humans

Most urban landscapes in the public domain are formally managed by institutions, such as urban parks departments, which typically have a dual focus on protecting these environments and ensuring they are accessible to residents and provide them with a range of benefits. As such, government policy and strategy documents often link land management, urban nature, and even biodiversity strategies to human health and wellbeing (State of Victoria Department of Environment, Land, Water and Planning 2017). This dual focus is perhaps most clearly reflected in municipal sustainability plans, which have cited evidence such as 'green spaces have been shown to benefit our physical and emotional health by reducing blood pressure, cholesterol, and stress' (City of Vancouver 2012) to justify preserving, managing, and cultivating urban landscapes.

The focus on managing nature *for* people has grown alongside a rapidly expanding evidence base on the health and wellbeing benefits of access to public landscapes (Elliott et al. 2020; James et al. 2015; Marselle et al. 2021). Decades of research highlight the importance of traditional spaces, such as parks, for a range of beneficial outcomes (Hartig et al. 2014). Access to nearby natural landscapes became increasingly important during the global COVID-19 pandemic, with individuals valuing them for a range of benefits, including for personal wellbeing, physical activity, and nature connection (Berdejo-Espinola et al. 2021). A mixed-methods study conducted at the height of pandemic-related movement restrictions in Australia found that many participants accessed local public landscapes to offset a feeling of being 'trapped' in their residences, and that these natural amenities were valued alongside access to more traditional services, such as supermarkets and chemists (Bower et al. 2021). Reliance on public landscapes as a release valve during the pandemic was also associated with changing patterns of use and visitation, leading to complaints of crowding (Lopez et al. 2021). Although research in this area is still underway (as at the end of 2022), it is possible that the increasing importance of local greenspaces to both community members and institutions may be associated with changes in management practices in the years to come.

Planners and managers must focus on the quality of public urban landscapes – not just their quantity – in delivering their services (Lindholst et al. 2015). Delivering and tending to landscapes to ensure they achieve the highest quality possible can take different forms, which may include transforming the environment to create new landscapes, upgrading, and revegetating existing landscapes, or funding and organising their ongoing management. One challenge for the organisations responsible for managing such places can be found in the potentially conflicting preferences and needs of distinct stakeholders and users, who may prioritise and evaluate these spaces in very different ways (Lindholst et al. 2015).

Managing public landscapes has expanded to involve deliberately and inclusively working with networks of stakeholders to address social values. As Lindholst et al. (2016, p. 167) observe: '[I]nstead of a "top-down" orientation, managers are now called to orientate themselves "outward" and "upward" in the quest for defining and providing services of public value'. Over the past few decades, management has often been seen to involve delivering new services and programmes to various parts of the community (Walker 2004). Indeed, institutions around the globe have begun supporting community involvement in tending public landscapes. For example, the City of New York (2015) has called on New Yorkers to care for their urban forest by planting new trees and maintaining them over the coming decades. Programmes like this highlight the intertwined role of institutions and the community in the actions of tending.

Recognising the need to respond to community priorities, institutions are also seeking to address longstanding inequities in greenspace distribution, access, quality, and experience. Reflecting the increasingly well-recognised role that urban natural landscapes play in supporting human health and wellbeing, inequitable access to such landscapes is now acknowledged as a form of environmental injustice – a topic that will be discussed in more detail in the following sections. In line with this framework, institutions should ensure that they do not exacerbate existing inequities when creating, transforming, and managing public landscapes. Without such attention to equity, greening efforts can inadvertently lead to the displacement of long-term residents via increased housing costs, resulting in what has been termed 'green gentrification' (Wolch et al. 2014). In addition to embedding greenspace access within the broader environmental justice paradigm, potential solutions to the problem of green gentrification include a focus on small-scale projects with extensive community input and the integration of other policies to offset the potential social and economic harms of landscape improvements (Wolch et al. 2014). In this way, communities can take advantage of the multitude of benefits urban natural landscapes offer all residents, without adversely impacting specific groups.

Reflecting research that demonstrates the importance of regular contact with nature, a broader range of natural places is being designed and managed in cities (Li et al. 2018; White et al. 2019). The potential value of 'informal' landscapes – leftover or in-between places that can occur spontaneously and receive very little formal management, including vegetated road verges and vacant lots – is also increasingly recognised in the literature (Farahani and Maller 2019). These spaces challenge Eurocentric ideas of landscapes and their purpose in urban areas, which are based upon the notion that they should be useful (Gandy 2016). It is possible that small shifts in management by institutions, such as adding simple amenities or improving access, may help improve community perceptions and experiences without transforming them into formal (and more intensively managed) spaces. This approach could be cost-effective for institutions while supporting community needs for informal spaces, thus providing 'unexpected opportunities and benefits' (Farahani and Maller 2019, p. 300) – potentially including community tending.

Community-led management to benefit humans

Community-led cultivation of landscapes with an explicit focus on maximising benefits *for* humans – those participating in such programmes and members of the broader community – is often led by environmental-advocacy organisations that sit outside, or even in opposition to, institutions such as conservation groups or parks departments. This position can offer additional flexibility when it comes to cultivating vegetated areas that may not necessarily fall under the purview of a governmental agency; it can also lead to conflict with organisations that bear formal management responsibilities and whose work is guided by legislation that may prioritise the human benefits of interacting with urban greenspaces. In recognition of the fact that the types of benefits individuals gain through contact with urban nature also vary depending on its form, researchers and urban planners have called for access to a range of natural public landscapes, from nature reserves and wildlife sanctuaries to highly designed spaces that offer sports fields, washrooms, cafés, and other amenities (Smit et al. 2011).

The first step in expanding access to urban nature generally involves clarifying the existing situation at the neighbourhood scale. This requires mapping what types of greenspace are in place, detailing the extent to which these areas meet the needs of local residents, and identifying the potential for newly developed or improved landscapes. One tool that can be used to conduct this mapping is photovoice, a qualitative method with a long history within environmental-justice efforts that is often integrated into community-based participatory research. Photovoice harnesses the unique power of images to visualise inequities and influence policymakers. A photovoice project conducted in socio-economically deprived neighbourhoods of three large cities in Canada (Toronto, Vancouver, and Winnipeg) found the criminalisation of standard behaviour, poor maintenance, and physical barriers were all common in local greenspaces (Masuda et al. 2012). Interestingly, many of the project's participant-researchers also spoke of the dual nature of exclusion from municipal planning efforts, as both limiting their ability to achieve improvements and creating a sense of common purpose and solidarity (Masuda et al. 2012).

Advances in technology mean that communities can integrate on-the-ground approaches to mapping (such as photovoice) with methods that are literally high in the sky, via satellite-based data made publicly available through platforms such as Google Earth and Street View. Efforts to directly compare systematic social observation of landscape features to satellite-based measurements have generally reported high levels of agreement between the two approaches, indicating that they are both viable methods of assessing natural landscapes (Edwards et al. 2013; Rundle et al. 2011; Taylor et al. 2011; Wu et al. 2014). That said, satellite-based methods are often ineffective at measuring certain aspects of urban greenspaces that are known to reduce the benefits humans receive from them, including high levels of traffic (Rundle et al. 2011), the presence of litter (Taylor et al. 2011) and graffiti (Edwards et al. 2013), or generally poor maintenance (Wu et al. 2014).

In addition to lacking data on such detailed aspects of cultivated landscapes, many standard approaches to measuring access to greenspace – particularly those employed in studies examining the potential health benefits of urban nature – exclude smaller parcels such as pocket parks, small-scale community gardens, or road verges (Generaal et al. 2019; Parra et al. 2010; Sugiyama et al. 2016). Studies that have included such spaces, along with larger or more formal greenspaces, have found that although smaller parcels have a relatively smaller impact on health, they can still provide mental health benefits (Wood et al. 2017) and do so in an easily accessible manner. These marginal areas may even support urban agriculture, benefitting local residents by providing access to fresh produce and fostering social ties among neighbours (Brown and Jameton 2000). In addition, a recent set of case studies from Australia highlights the example of Scrubby Hill Farm in Geeveston, Tasmania, where the benefits of carrying out agriculture on marginal lands have expanded to include training and employment via a social enterprise (Kingsley et al. 2021).

Many efforts may be community-*led*, while still profiting from policies, programmes, or structures created by institutions. In 2016, for example, the Brisbane City Council issued guidelines allowing residents to create gardens along sections of footpaths between their residences and the road, an act that would previously have resulted in fines (Brisbane City Council 2019). In Toronto, a partnership between the City and the Friends of Regent Park supported the creation of a 'bake oven', which allows members of local community groups to cook items for distribution at park events or sales at the park's farmers' market (Hassen 2016). Moving from permissibility or even partnership to outright advocacy, the regeneration plan for Salford (Greater Manchester) in the United Kingdom has built on the work of guerrilla gardening groups – who carry out urban agriculture on public lands, or even abandoned private lots, without seeking official permission – to provide dedicated space for individual gardening allotments, pop-up community gardens, and a commercial urban farm (Hardman et al. 2018).

Institution-led management to protect nature

The previous section discussed institutions managing urban landscapes to benefit humans; here, we examine the narrower historic focus of government plans and policies on the role natural landscapes can play in achieving *environmental* sustainability. For instance, the City of Madrid's nature-based climate change adaptation strategy calls for the expansion of green roofs in order to 'support water management, improve insulation and air quality, provide cooling, and create habitat for biodiversity'; the City discusses greening other forms of infrastructure primarily in the context of providing additional habitat for wildlife and improving air quality (Caro, Armour, and Fernandez 2016, p. 2).

In New York, the 'One New York' plan embeds environmental justice (a construct that is explicitly guided by human rights), but many components of the City's aim to be 'the most sustainable big city in the world' focus on landscape-based approaches to improving the quality of the air, water, and soil (New York City 2015, p. 6). Among these, the City's 'Green Infrastructure Program' integrates verge bioswales to reduce run-off flowing from impervious surfaces into the sewer system, thereby improving water quality, as well as greening improvements to the City's outdoor public spaces, which are similarly cited as leading to 'reduced pollution and improved stormwater management and flood resiliency' (New York City 2015, p. 206). This emphasis on the environmental impacts of green infrastructure exists in implementation as well as policy documentation, as captured in interviews with the municipal officials responsible for carrying out the programme and reported in a mixed-methods analysis (Meerow 2020). These experts cite stormwater management as the primary motivation for expanding green infrastructure, with any potential benefits for human health described as 'an added bonus' (Meerow 2020, p. 6).

Regardless of the stated goals or guiding viewpoints, tending public urban landscapes to achieve environmental aims can achieve a 'win-win' of providing benefits *for* humans as well. For instance, efforts that seek to increase biodiversity can prevent the predominance of a single, injury-inflicting species, such as white-tailed deer. These animals have become increasingly common in urban and suburban areas of North America due to landscape fragmentation arising from urban sprawl (Urbanek and Nielsen 2013), while their increasing population density has also been associated with a significantly higher incidence of Lyme disease in surrounding communities (Kilpatrick, Labonte, and Stafford 2014). Reduced biodiversity following the loss of apex predator species has also been associated with the over-proliferation of white-tailed deer in the eastern United States. This confluence of factors has led to large numbers of collisions between motor vehicles and deer, with one study estimating that reintroducing cougars in this region could prevent 21,400 injuries and 155 deaths over just three decades by reducing the rate of such collisions (Gilbert et al. 2017).

As with the trade-offs discussed in previous sections, however, there are instances in which managing landscapes to benefit nature can result in inadvertent harm to the human residents of urban environments. A 2021 review of the pathways linking biodiversity to human health detailed a wide range of benefits associated with higher levels of biodiversity, including providing sources for both traditional and modern medicines, reducing exposure to deadly heatwaves, and reducing levels of psychological stress (Marselle et al. 2021). Conversely, the review also noted the potential for more frequent encounters with injurious flora and fauna (such as poisonous plants), more common vector-borne diseases in urban areas due to habitat loss on their fringes, and greater exposure to a wider range of allergens.

Community-led management to protect nature

Grassroots, community-based tending of public landscapes *for* nature has mutual benefits for both people and the environment. Environmental community tending activities can include stewardship activities such as rubbish collection, removing invasive species, and introducing beneficial plants via pollinator gardens. These activities can take place across a broad range of urban landscapes – nature strips, private yards, small parks, or even major natural areas – but they are generally motivated by the desire to protect and support the environment. In addition to ecosystem impacts, this type of tending can lead to a sense of pride among participants, as well as a shared stewardship of landscapes, which can increase social connections within communities (Cranz and Boland 2004).

A common narrative suggests that residents' tending behaviours damage local ecological integrity (McWilliam et al. 2010; Roman et al. 2021). Yet, studies show that active citizen participation can actually increase biodiversity (Dennis and James 2016). In modern cities, activities that care for privately owned gardens provide a clear example of where community-led tending is known to increase biodiversity (Parris et al. 2018). Motivating individuals to create garden spaces that connect to, and support, the broader ecological system is especially important in urban areas, where privately held land accounts for the majority of urban greenspace (Parris et al. 2018). In some instances, community members may be reluctant to adopt these practices, but introducing concepts such as 'messy ecosystems, orderly frames' (Nassauer 2005) can encourage them to adopt less formal, but more ecologically rich, urban landscapes.

Broader, grassroots-organised protests, such as climate change activism, also fall within the domain of community tending for nature. Protesting climate-unfriendly policies and participating in emissions-reducing behaviours can have similar psychological benefits to the activities outlined above (Haugestad et al. 2021). Motivated by feelings of shared responsibility and a fear of future environmental threat, young people have been well represented in recent climate change activism, such as the Fridays For Future global climate strikes inspired by Greta Thunberg (Haugestad et al. 2021). Youth participating in such climate strikes have created peer-to-peer social networks and a collective identity that could lead to increased adoption of pro-environmental behaviours across their lifespans (Wallis and Loy 2021). In early childhood, carers can provide opportunities to foster tending behaviour. Our case study (see Box 1.1) demonstrates how one educator's efforts helped young children to engage in place-based environmental tending.

Although the choice to participate in environmental protests is generally motivated more by collective concerns than by personal health (Kowasch et al. 2021), contributing to such activities has been linked to a range of mental, physical, and social health benefits. Numerous studies have shown measurable increases in wellbeing among participants in environmental stewardship programmes, particularly those which provide social interaction in facilitated groups (Molsher and Townsend 2016). Stoeckl et al. (2021) note that 'giving' activities related to tending to nature can be as important for human wellbeing as the other ecosystem services humans receive from nature. Clean-up campaigns along beaches, rivers, streets, and within parks can lead to a 'helpers high' or 'warm glow' along with other positive psychological benefits among participants, as well as encouraging physical activity (Molsher and Townsend 2016). It is also important to note that pro-environmental behaviours are more enduring when they are intrinsically motivated by a sense of connectedness to nature, rather than by a desire to conform to social norms (Rosa et al. 2018). Campaigns with external motivational factors, such as financial rewards or prizes, often see relatively quick drop-offs in behaviour because they are not internally motivated (van der Linden 2015).

At the same time, conflict surrounding environmental tending can arise when there is an intersection with contested space and resources. Indigenous sovereignty and land protection movements have been met with violence by exploitative corporations and governments (Hernandez 2022). Such violence often arises in instances when Indigenous land ownership poses a threat to resource extraction – such as the 'force, surveillance, and criminalization of land defenders and peaceful protesters to intimidate, remove, and forcibly evict

Secwepemc and Wet'suwet'en Nations from their traditional lands', which the United Nations Committee on the Elimination of Racial Discrimination (CERD) called on the government of Canada to cease in May 2022 (Wilson 2022). However, similar acts of violence have occurred as part of nature-protection projects as well, with a 2022 report describing three main factors common to 'exclusionary conservation': (i) land dispossession and displacement; (ii) indirect human-rights violations occurring during such displacement; and (iii) gross abuses committed against Indigenous peoples who remain on newly 'protected' lands (Project Expedite Justice 2022).

In addition to potential conflict, other barriers to engaging in environmental tending activities may be widespread. These include the need to seek permission to access private land and dealing with toxic or dangerous items or substances during the clean-up process. Extending opportunities to undertake environmental tending in an equitable fashion is also important because not all community members will have access to natural landscapes in the form of private gardens or residential nature strips where they can carry out such tending, thus preventing them from garnering the benefits detailed above.

Tending for the future

A legacy of tending: Indigenous futures

Indigenous peoples have tended to place long before our cities were built; they continue to care for Country as custodians of the land (Murphy and Kennedy 2016). Early settler-colonists in south-east Australia recall the ground being so soft that the legs of their horses sunk deep into the earth (Pascoe 2014). This rich soil was the result of thousands of years of tending to place. The earth so lovingly turned over by the digging stick, mixing nutrition through the soil to nourish the *murrnong* that, in turn, will provide nutrition for the one tending to it. In a few short years, that soil had been compacted and rainwater was sheeting from the hills – a legacy of caring trampled underfoot by herds of newly introduced sheep. This is a story told the colonial world over, heralding the harmful and extractive colonial practices that bleed into the present (Jacobs et al. 2021).

Indigenous ontologies observe that people have a kinship relationship with Country, and that Country as a domain is dynamic and active (Darug et al. 2019). For Indigenous peoples in Australia, Country encompasses the earth, waters, skies, plants, animals, buildings, roads, and all physical aspects of the land. Country is all of these elements and the relationships held within them. Indigenous peoples conceptualise tending as Caring for Country. Care is an act of reciprocity. It dictates that, as we care for Country, Country cares for us. In tending to and for Country, the health of the community is sustained. The benefits of tending are recognised through its ability to build social cohesion, provide food security, and lower cortisol. For Indigenous peoples, tending is an act of cultural responsibility and allows for connection to Country. Where

Western neoliberal worldviews value direct exchanges between individuals, Indigenous caring is an enactment of cultural responsibilities, kinship, and reciprocity (Muller et al. 2019; Russell and Ens 2020).

Contemporary framing by settler-colonists of Country as either a wilderness or as urban areas erases the legacy of tending (Fletcher et al. 2021; Jacobs et al. 2021). Calls to rewild urban places can perpetuate the narrative that pre-colonial cities were wild places sparsely populated by a primitive people, offering a dichotomous understanding of humans as being separate from nature (Mata et al. 2020). Tending of Indigenous lands must be cognisant of, and responsive to, the deep time traditions of care. Indigenous-led and informed tending illustrates the reciprocity in tending, and how nature has the capacity to tend to us in return (Darug et al. 2019). This perspective shifts the current dynamic of humans managing for, or protecting from, other humans, while introducing Country (nature) as an additional player capable of tending.

Towards sustained tending in urban nature

The interconnection between approaches to cultivating landscapes for environmental sustainability and human wellbeing is a central component of the concept of nature-based solutions (NBS). Unfortunately, our global institutions have often focused on the potential financial implications of landscape management decisions, rather than impacts on nature and people, with economic capital per person doubling between 1992 and 2014, while the amount of natural capital declined by 40% over the same period (Dasgupta 2021). Future cities require landscapes that transcend historical ownership structures and adopt new management approaches to support long-term viability and sustainability, while providing benefits to residents of all ages, backgrounds, and interests.

Currently, cities around the globe are failing to achieve this aim. Research indicates that communities with greater levels of economic deprivation and a higher prevalence of ethnic and racial minorities often have less access to suitable natural landscapes, whether defined by quantity, distance, or quality. For instance, a study in ten large cities in the United States linked higher levels of education, income, and White residents to greater quantities of vegetation at the Census block level (Nesbitt et al. 2019). Such disparities are especially concerning given the results of a 2021 review of access to greenspace and health equity, which found that individuals with lower socio-economic status (SES) experience greater benefits from greenspace than those of higher SES. This was also true for studies conducted with residents of Europe (Rigolon et al. 2021).

Inequities also exist when looking at access by specific age groups. A pair of case studies that examined accessibility through an age-based lens found a mismatch between the distribution of existing open spaces and the areas of Catania, Italy, with the largest proportion of older adults, while the assessment of Nagoya, Japan, indicated that both older adults and children were more likely to be located near smaller parks with fewer amenities (La Rosa et al.

2018). More positive outcomes of regular connections to greenspace have also been shown to accrue to those at either end of the lifespan (Kabisch, van den Bosch and Lafortezza 2017). Engaging in a range of outdoor activities, including gardening, has also been identified as an important source of social contacts for older adults (Sugiyama and Thompson 2007), with such social engagement a critical pathway between urban design and healthy ageing (Clarke and Nieuwenhuijsen 2009). The case study included in this chapter demonstrates how early childhood education can foster connections to place. Providing support for children to engage in tending activities may also shape their lifelong relationship with nature (Gifford and Nilsson 2014), with a greater sense of connectedness to nature during this time of life more strongly associated with pro-environmental behaviours than among older adolescents or adults (Whitburn, Linklater, and Abrahamse 2020). Identifying barriers to, and enablers of, engagement, particularly for individuals who lack a strong connection to nature, is also critical to the success of urban greening initiatives extending beyond tenders to their communities more broadly (Restall and Conrad 2015).

Conclusions

The reflections in this chapter suggest a new perspective on urban tending, one that is both holistic and grounded in place. Delineating a typology of tending, we explore the ways in which landscape-based actors, actions, and spaces might sit at one end of a spectrum or, alternatively, reflect a sense of balance. We have delved into distinct domains of tending that typify dualities – institutions managing for humans, communities managing for humans, institutions managing for nature, and communities managing for nature – while also unpacking the ways in which tensions can both result in conflict and advance collaboration to achieve mutual benefits. Presenting Indigenous tending as an act of care that rests on a deep and longstanding understanding of nature's capacity to provide care, we propose that tending in urban landscapes can also be a way of tending to ourselves. In addition, we recognise the myriad mechanisms by which the actions of tenders extend beyond the spaces they tend to reshape communities.

Inherent to this perspective is the acknowledgement that, although both people and nature have immense potential to be transformed, ensuring any such transformation is enduring remains a complex challenge. Looking beyond the garden to expand the types of urban landscapes that are suitable for tending is one path towards sustainable transformation; ensuring that everyone has equitable access to the sites and services of tending is equally important, particularly in light of the centrality of exposure to nature at an early age for lifelong beliefs and behaviours, and the potential for additional benefits of tending among the most vulnerable members of our communities.

At their heart, these dualities reflect a broader reality reflected in the 2022 report of the Intergovernmental Panel on Climate Change (IPCC 2022): anthropogenic climate change has already caused widespread harm to both nature and people; urban landscapes are a crucial space to achieve climate resilience. As a result, tending has never been more urgent and its potential never more substantial.

Acknowledgements

We thank Kristen Webster for inspiring us and allowing us to share her story of embedding indigenous learning in childcare *hay č̓ xʷ q̓ə*.

References

Berdejo-Espinola, V, Suárez-Castro, AF, Amano, T, Fielding, KS, Oh, RRY and Fuller, RA 2021, 'Urban green space use during a time of stress: A case study during the COVID-19 pandemic in Brisbane, Australia', *People and Nature*, vol. 3, no. 3, pp. 597–609. https://doi.org/10.1002/pan3.10218

Bower, M, Buckle, C, Rugel, E et al. 2021, '"Trapped", "anxious" and "traumatised": COVID-19 intensified the impact of housing inequality on Australians' mental health', *International Journal of Housing Policy*, pp. 1–32. https://doi.org/10.1080/19491247.2021.1940686

Brisbane City Council 2019, 'Verge gardens'. https://www.brisbane.qld.gov.au/clean-and-green/natural-environment-and-water/plants-trees-and-gardens/verge-gardens

Brown, KH and Jameton, AL 2000, 'Public health implications of urban agriculture', *Journal of Public Health Policy*, vol. 21, no. 1, pp. 20–39. https://doi.org/10.2307/3343472

Bullard, RD 1996, 'Environmental justice: It's more than waste facility siting', *Social Science Quarterly*, vol. 77, no. 3, pp. 493–499.

Caro, JC, Armour, T and Fernandez, M 2016, *Madrid + natural: Nature-based climate change adaptation*. Madrid. http://publications.arup.com/publications/m/madrid_natural

City of Vancouver 2012, 'Greenest city 2020 action plan'. https://vancouver.ca/green-vancouver/greenest-city-2020-action-plan.aspx

Clarke, P and Nieuwenhuijsen, ER 2009, 'Environments for healthy ageing: A critical review', *Maturitas*, vol. 64, no. 1, pp. 14–19. https://doi.org/10.1016/j.maturitas.2009.07.011

Cranz, G and Boland, M 2004, 'Defining the sustainable park: A fifth model for urban parks', *Landscape Journal*, vol. 23, no. 2, pp. 102–120.

Cumpston, Z 2020, 'Cities are country: Illuminating Aboriginal perspectives of biodiversity in urban environments', *Clean Air and Urban Landscapes Hub*. https://nespurban.edu.au/publications/cities-are-country-too-illuminating-aboriginal-perspectives-of-biodiversity-in-urban-environments/

Darug, N, Dadd, L, Glass, P, Scott, R, Graham, M, Judge, S, Hodge, P and Suchet-Pearson, S, 2019, 'Yanama budyari gumada: Reframing the urban to care as darug country in western Sydney', *Australian Geographer*, vol. 50, no. 3, pp. 279–293. https://doi.org/10.1080/00049182.2019.1601150

Dasgupta, P 2021, *The economics of biodiversity: The Dasgupta review*, HM Treasury, London.

Dennis, M and James, P 2016, 'User participation in urban green commons: Exploring the links between access, voluntarism, biodiversity and well being', *Urban Forestry & Urban Greening*, vol. 15, pp. 22–31. https://doi.org/10.1016/j.ufug.2015.11.009

Edwards, N, Hooper, P, Trapp, GSA, Bull, F, Boruff, B and Giles-Corti, B 2013, 'Development of a Public Open Space Desktop Auditing Tool (POSDAT): A remote sensing approach', *Applied Geography*, vol. 38, no. 1, pp. 22–30. https://doi.org/10.1016/j.apgeog.2012.11.010

Elliott, LR, White, MP, Grellier, J et al. 2020, 'Research note: Residential distance and recreational visits to coastal and inland blue spaces in eighteen countries', *Landscape and Urban Planning*, vol. 198, article 103800. https://doi.org/10.1016/j.landurbplan.2020.103800

European Commission 2015, *Towards an EU Research and Innovation policy agenda for nature-based solutions and re-naturing cities*, European Commission, Brussels.

Farahani, LM and Maller, C 2019, 'Investigating the benefits of "leftover" places: Residents' use and perceptions of an informal greenspace in Melbourne', *Urban Forestry & Urban Greening*, vol. 41, pp. 292–302. https://doi.org/10.1016/j.ufug.2019.04.017

Fletcher, M-S, Hall, T and Alexandra, AN 2021, 'The loss of an indigenous constructed landscape following British invasion of Australia: An insight into the deep human imprint on the Australian landscape', *Ambio*, vol. 50, no. 1, pp. 138–149. https://doi.org/10.1007/s13280-020-01339-3

Gandy, M 2016, 'Unintentional landscapes', *Landscape Research*, vol. 41, no. 4, pp. 433–440. https://doi.org/10.1080/01426397.2016.1156069

Generaal, E, Timmermans, EJ, Dekkers, JEC, Smit, JH and Penninx, BWJH 2019, 'Not urbanization level but socioeconomic, physical and social neighbourhood characteristics are associated with presence and severity of depressive and anxiety disorders', *Psychological Medicine*, vol. 49, no. 1, pp. 149–161. https://doi.org/10.1017/S0033291718000612

Gifford, R and Nilsson, A 2014, 'Personal and social factors that influence pro-environmental concern and behaviour: A review', *International Journal of Psychology*, vol. 49, no. 3, pp. 141–157. https://doi.org/10.1002/ijop.12034

Gilbert, SL, Sivy, KJ, Pozzanghera, CB et al. 2017, 'Socioeconomic benefits of large carnivore recolonization through reduced wildlife-vehicle collisions', *Conservation Letters*, vol. 10, no. 4, pp. 430–438. https://doi.org/10.1111/conl.12280

Hardman, M, Chipungu, L, Magidimisha, H, Larkham, PJ, Scott, AJ and Armitage, RP 2018, 'Guerrilla gardening and green activism: Rethinking the informal urban growing movement', *Landscape and Urban Planning*, vol. 170, pp. 6–14. https://doi.org/10.1016/j.landurbplan.2017.08.015

Hartig, T, Mitchell, R, De Vries, S and Frumkin, H 2014, 'Nature and health', *Annual Review of Public Health*, vol. 35, pp. 207–228. https://doi.org/10.1146/annurev-publhealth-032013-182443

Hassen, BN 2016, *Green paths to mental health walk lab report*, Wellesley Institute, Toronto.

Haugestad, CA, Skauge, AD, Kunst, JR and Power, SA 2021, 'Why do youth participate in climate activism? A mixed-methods investigation of the #FridaysForFuture climate protests', *Journal of Environmental Psychology*, vol. 76, article 101647. https://doi.org/10.1016/j.jenvp.2021.101647

Hernandez, J 2022, *Fresh banana leaves: Healing Indigenous landscapes through Indigenous science*, North Atlantic Books, Berkeley, CA.

Hobbs, RJ, Arico, S, Aronson, J et al. 2006, 'Novel ecosystems: theoretical and management aspects of the new ecological world order', *Global Ecology and Biogeography*, vol. 15, no. 1, pp. 1–7. https://doi.org/10.1111/j.1466-822X.2006.00212.x

Intergovernmental Panel on Climate Change (IPCC) 2022, *Climate change 2022: Impacts, adaptation and vulnerability – Summary for policymakers*, IPCC, Geneva. https://report.ipcc.ch/ar6wg2/pdf/IPCC_AR6_WGII_SummaryForPolicymakers.pdf

Jacobs, LA, Hazelwood, SP, Avery, CB, and Sangster-Biye, C 2021, 'Research note: Reimagining US Federal Land Management through decolonization and Indigenous value systems', *Journal of Park and Recreation Administration*, vol. 40, no. 1. https://doi.org/10.18666/JPRA-2021-10973

James, P, Banay, RF, Hart, JE and Laden, F 2015, 'A review of the health benefits of greenness', *Current Epidemiology Reports*, vol. 2, no. 2, pp. 131–142. https://doi.org/10.1111/1750-3841.12139

Jennings, V, Larson, L and Yun, J 2016a, 'Advancing sustainability through urban green space: Cultural ecosystem services, equity, and social determinants of health', *International Journal of Environmental Research and Public Health*, vol. 13, no. 2, article 197. https://doi.org/10.3390/ijerph13020196

Jennings, V, Yun, J and Larson, L 2016b, 'Finding common ground: Environmental ethics, social justice, and a sustainable path for nature-based health promotion', *Healthcare*, vol. 4, no. 3, p. 61. https://doi.org/10.3390/healthcare4030061

Kabisch, N, van den Bosch, M and Lafortezza, R 2017, 'The health benefits of nature-based solutions to urbanization challenges for children and the elderly – A systematic review', *Environmental Research*, vol. 159, pp. 362–373. https://doi.org/10.1016/j.envres.2017.08.004

Kilpatrick, HJ, Labonte, AM and Stafford, KC 2014, 'The relationship between deer density, tick abundance, and human cases of Lyme disease in a residential community', *Journal of Medical Entomology*, vol. 51, no. 4, pp. 777–784. https://doi.org/10.1603/me13232

Kingsley, J, Egerer, M, Nuttman, S et al. 2021, 'Urban agriculture as a nature-based solution to address socio-ecological challenges in Australian cities', *Urban Forestry & Urban Greening*, vol. 60, article 127059. https://doi.org/10.1016/j.ufug.2021.127059

Kowasch, M, Cruz, JP, Reis, P, Gericke, N and Kicker, K 2021, 'Climate youth activism initiatives: Motivations and aims, and the potential to integrate climate activism into ESD and transformative learning', *Sustainability*, vol. 13, no. 21, article 11581. https://doi.org/10.3390/su132111581

La Rosa, D, Takatori, C, Shimizu, H and Privitera, R 2018, 'A planning framework to evaluate demands and preferences by different social groups for accessibility to urban greenspaces', *Sustainable Cities and Society*, vol. 36, pp. 346–362. https://doi.org/10.1016/j.scs.2017.10.026

Li, D, Deal, B, Zhou, X, Slavenas, M and Sullivan, WC 2018, 'Moving beyond the neighborhood: Daily exposure to nature and adolescents' mood', *Landscape and Urban Planning*, vol. 173, pp. 33–43. https://doi.org/10.1016/j.landurbplan.2018.01.009

Lindenmayer, DB, Fischer, J, Felton, A et al. 2008, 'Novel ecosystems resulting from landscape transformation create dilemmas for modern conservation practice', *Conservation Letters*, vol. 1, no. 3, pp. 129–135. https://doi.org/10.1111/j.1755-263X.2008.00021.x

Lindholst, AC, Sullivan, SG, van den Bosch, CCK and Fors, H 2015, 'The inherent politics of managing the quality of urban green spaces', *Planning Practice & Research*, vol. 30, no. 4, pp. 376–392. https://doi.org/10.1080/02697459.2015.1057943

Lindholst, AC, van den Bosch, CCK, Kjøller, CP, Sullivan, S, Kristoffersson, A, Fors, H and Nilsson, K 2016, 'Urban green space qualities reframed toward a public value management paradigm: The case of the Nordic Green Space Award', *Urban Forestry & Urban Greening*, vol. 17, pp. 166–176. https://doi.org/10.1016/j.ufug.2016.04.007

Lopez, B, Kennedy, C, Field, C and McPhearson, T 2021, 'Who benefits from urban green spaces during times of crisis? Perception and use of urban green spaces in New York City during the COVID-19 pandemic', *Urban Forestry & Urban Greening*, vol. 65, article 127354. https://doi.org/10.1016/j.ufug.2021.127354

Marselle, MR, Hartig, T, Cox, D et al. 2021, 'Pathways linking biodiversity to human health: A conceptual framework', *Environment International*, vol. 150, pp. 1–58. https://doi.org/10.1016/j.envint.2021.106420

Masuda, JR, Teelucksingh, C, Zupancic, T et al. 2012, 'Out of our inner city backyards: Re-scaling urban environmental health inequity assessment', *Social Science & Medicine*, vol. 75, no. 7, pp. 1244–1253. https://doi.org/10.1016/j.socscimed.2012.04.034

Mata, L, Ramalho, CE, Kennedy, J, Parris, KM, Valentine, L, Miller, M, Bekessy, S, Hurley, S and Cumpston, Z 2020, 'Bringing nature back into cities', *People and Nature*, vol. 2, pp. 350–368. https://doi.org/10.1002/pan3.10088

McWilliam, W, Eagles, P, Seasons, M and Brown, R 2010, 'Assessing the degradation effects of local residents on urban forests in Ontario, Canada', *Journal of Arboriculture*, vol. 36, no. 6, pp. 253–260. https://doi.org/10.48044/jauf.2010.033

Meerow, S 2020, 'The politics of multifunctional green infrastructure planning in New York City', *Cities*, vol. 100, article 102621. https://doi.org/10.1016/j.cities.2020.102621

Molsher, R and Townsend, M 2016, 'Improving well-being and environmental stewardship through volunteering in nature', *EcoHealth*, vol. 13, no. 1, pp. 151–155. https://doi.org/10.1007/s10393-015-1089-1

Merriam-Webster, D 2022, *Electronic resource*. https://www.merriam-webster.com/dictionary/cultivate. Accessed 14 February, 2022.

Muller, S, Hemming, S and Rigney, D 2019, 'Indigenous sovereignties: Relational ontologies and environmental management', *Geographical Research*, vol. 57, pp. 399–410. https://doi.org/10.1111/1745-5871.12362

Murphy, J and Kennedy, L 2016, *Welcome to Country*, Walker Books Australia, Newton.

Nassauer, JI 2005, 'Using cultural knowledge to make new landscape patterns' in JA Wiens and MR Moss (eds), *Issues in Landscape Ecology*, Cambridge University Press, Cambridge UK, pp. 274–280. (Revised from Culture as a Means of Experimentation and Action. In *Issues in landscape ecology*, Wiens, JA and Moss, MR, eds. International Association for Landscape Ecology.)

Nesbitt, L, Meitner, MJ, Girling, C, Sheppard, SRJJ and Lu, Y 2019, 'Who has access to urban vegetation? A spatial analysis of distributional green equity in 10 US cities', *Landscape and Urban Planning*, vol. 181, pp. 51–79. https://doi.org/10.1016/j.landurbplan.2018.08.007

New York City 2015, *One NYC: The plan for a strong and just city*, New York City. https://onenyc.cityofnewyork.us/wp-content/uploads/2019/04/OneNYC-Strategic-Plan-2015.pdf

Nxumalo, F 2019, *Decolonizing place in early childhood education*, Routledge, New York.

Olwig, KR 2007, 'The practice of landscape 'Conventions' and the just landscape: The case of the European landscape convention', *Landscape Research*, vol. 32, no. 5, pp. 579–594.

Parra, DC, Gomez, LF, Sarmiento, OL et al. 2010, 'Perceived and objective neighborhood environment attributes and health related quality of life among the elderly in Bogota, Colombia', *Social Science & Medicine*, vol. 70, no. 7, pp. 1070–1076. https://doi.org/10.1016/j.socscimed.2009.12.024

Parris, KM, Amati, M, Bekessy, SA et al. 2018, 'The seven lamps of planning for biodiversity in the city', *Cities*, vol. 83, pp. 44–53. https://doi.org/10.1016/j.cities.2018.06.007

Pascoe, B 2014. *Dark Emu: Black seeds agriculture or accident?* Magabala Books, Broome.

Prévot, AC, Cheval, H, Raymond, R and Cosquer, A 2018, 'Routine experiences of nature in cities can increase personal commitment toward biodiversity conservation', *Biological Conservation*, vol. 226, pp. 1–8. https://doi.org/10.1016/j.biocon.2018.07.008

Project Expedite Justice 2022, *Trapped outside the conservation fortress: The intersection of global conservation efforts and systematic human rights violations*, Kailua Kona. https://www.projectexpeditejustice.org/_files/ugd/b912bf_6633fdc7a7af48f3a45e81792295a68e.pdf

Restall, B and Conrad, E 2015, 'A literature review of connectedness to nature and its potential for environmental management', *Journal of Environmental Management*, vol. 159, pp. 264–278. https://doi.org/10.1016/j.jenvman.2015.05.022

Rey, JA 2019, 'Dharug custodial leadership: Uncovering country in the city', *WIN-HEC: International Journal of Indigenous Education Scholarship*, vol. 1, pp. 56–66.

Rigolon, A, Browning, MHEM, McAnirlin, O and Yoon, H 2021, 'Green space and health equity: A systematic review on the potential of green space to reduce health disparities', *International Journal of Environmental Research and Public Health*, vol. 18, no. 5, pp. 1–29. https://doi.org/10.3390/ijerph18052563

Roman, LA, Conway, TM, Eisenman, TS, Koeser, AK, Ordóñez Barona, C, Locke, DH, Jenerette, GD, Östberg, J and Vogt, J 2021, 'Beyond "trees are good": Disservices, management costs, and tradeoffs in urban forestry', *Ambio*, vol. 50, no. 3, pp. 615–630. https://doi.org/10.1007/s13280-020-01396-8

Rosa, CD, Profice, CC and Collado, S 2018, 'Nature experiences and adults' self-reported pro-environmental behaviors: The role of connectedness to nature and childhood nature experiences', *Frontiers in Psychology*, vol. 9, article 1055. https://doi.org/10.3389/fpsyg.2018.01055

Rundle, AG, Bader, MDM, Richards, CA, Neckerman, KM and Teitler, JO 2011, 'Using Google Street View to audit neighborhood environments', *American Journal of Preventive Medicine*, vol. 40, no. 1, pp. 94–100. https://doi.org/10.1016/j.amepre.2010.09.034

Russell, S and Ens, E 2020, 'Connection as country: Relational values of billabongs in Indigenous northern Australia', *Ecosystem Services*, vol. 45, article 101169. https://doi.org/10.1016/j.ecoser.2020.101169

Smit, W, Hancock, T, Kumaresen, J, Santos-Burgoa, C, Sanchez-Kobashi Meneses, R and Friel, S 2011, 'Toward a research and action agenda on urban planning/design and health equity in cities in low and middle-income countries', *Journal of Urban Health*, vol. 88, no. 5, pp. 875–885.

State of Victoria Department of Environment, Land, Water and Planning 2017, *Protecting Victoria's Environment – Biodiversity 2037*. https://www.environment.vic.gov.au/__data/assets/pdf_file/0022/51259/Protecting-Victorias-Environment-Biodiversity-2037.pdf

Steffensen, V 2020, *Fire country: How Indigenous fire management could help save Australia*, Hardie Grant, Melbourne.

Steg, L, Bolderdijk, JW, Keizer, K and Perlaviciute, G 2014, 'An integrated framework for encouraging pro-environmental behaviour: The role of values, situational factors and goals', *Journal of Environmental Psychology*, vol. 38, pp. 104–115. https://doi.org/10.1016/j.jenvp.2014.01.002

Stoeckl, N, Jarvis, D, Larson, S, Larson, A and Grainger, D 2021, 'Australian Indigenous insights into ecosystem services: Beyond services towards connectedness – People, place and time', *Ecosystem Services*, vol. 50, article 101341. https://doi.org/10.1016/j.ecoser.2021.101341

Sugiyama, T and Thompson, CW 2007, 'Outdoor environments, activity and the well-being of older people: Conceptualising environmental support', *Environment and Planning A: Economy and Space*, vol. 39, no. 8, pp. 1943–1960. https://doi.org/10.1068/a38226

Sugiyama, T, Villanueva, K, Knuiman, M, Francis, J, Foster, S, Wood, L and Giles-Corti, B 2016, 'Can neighborhood greenspace mitigate health inequalities? A study of socio-economic status and mental health', *Health & Place*, vol. 38, pp. 16–21. https://doi.org/10.1016/j.healthplace.2016.01.002

Taylor, BT, Fernando, P, Bauman, AE, Williamson, A, Craig, JC and Redman, S 2011, 'Measuring the quality of public open space using Google Earth', *American Journal of Preventive Medicine*, vol. 40, no. 2, pp. 105–112. https://doi.org/10.1016/j.amepre.2010.10.024

Treib, M (ed) 2011, *Meaning in landscape architecture and gardens*, Taylor & Francis, London.

United Nations 2022, *UN Sustainable Development Goals: 15 Life on Land*. https://www.un.org/sustainabledevelopment/sustainable-development-goals/

Urbanek, RE and Nielsen, CK 2013, 'Influence of landscape factors on density of suburban white-tailed deer', *Landscape and Urban Planning*, vol. 114, pp. 28–36. https://doi.org/10.1016/j.landurbplan.2013.02.006

van den Bosch, M and Ode Sang, Å 2017, 'Urban natural environments as nature-based solutions for improved public health – A systematic review of reviews', *Environmental Research*, vol. 158, pp. 373–384. https://doi.org/10.1016/j.envres.2017.05.040

van der Linden, S 2015, 'Intrinsic motivation and pro-environmental behaviour', *Nature Climate Change*, vol. 5, no. 7, pp. 612–613. https://doi.org/10.1038/nclimate2669

Walker, C 2004, *The public value of urban parks*, Urban Institute, Washington, DC.

Wallis, H and Loy, LS 2021, 'What drives pro-environmental activism of young people? A survey study on the Fridays For Future movement', *Journal of Environmental Psychology*, vol. 74, article 101581. https://doi.org/10.1016/j.jenvp.2021.101581

Whitburn, J, Linklater, W and Abrahamse, W 2020, 'Meta-analysis of human connection to nature and proenvironmental behavior', *Conservation Biology*, vol. 34, no. 1, pp. 180–193. https://doi.org/10.1111/cobi.13381

White, MP, Alcock, I, Grellier, J, Wheeler, BW, Hartig, T, Warber, SL, Bone, A, Depledge, MH and Fleming, LE 2019, 'Spending at least 120 minutes a week in nature is associated with good health and well-being', *Scientific Reports*, vol. 9, no. 1, pp. 1–11. https://doi.org/10.1038/s41598-019-44097-3

Wilson, L 2022, 'UN committee calls for investigation of RCMP on Secwepemc and Wet'suwet'en territories', *APTN National News*, 13 May 2022, https://www.aptnnews.ca/national-news/un-committee-calls-for-investigation-of-rcmp-on-secwepemc-and-wetsuweten-territories

Wolch, JR, Byrne, J and Newell, JP 2014, 'Urban green space, public health, and environmental justice: The challenge of making cities "just green enough"', *Landscape and Urban Planning*, vol. 125, pp. 234–244. https://doi.org/10.1016/j.landurbplan.2014.01.017

Wood, L, Hooper, P, Foster, S and Bull, F 2017, 'Public green spaces and positive mental health – investigating the relationship between access, quantity and types of parks and mental well-being', *Health & Place*, vol. 48, pp. 63–71. https://doi.org/10.1016/j.healthplace.2017.09.002

World Health Organization 2020, *WHO manifesto for a healthy recovery from COVID-19: Prescriptions and actionables for a healthy and green recovery*, World Health Organization, Geneva.

Wu, Y-T, Nash, P, Barnes, LE, Minett, T, Matthews, FE, Jones, A and Brayne, C 2014, 'Assessing environmental features related to mental health: A reliability study of visual streetscape images', *BMC Pregnancy and Childbirth*, vol. 14, no. 1, article 1094. https://doi.org/10.1186/1471-2458-14-1094

2 Gardening for good in Ontario, Canada

A case study of Hamilton's Victory Gardens

Allison Williams

Introduction

The COVID-19 pandemic created a surge of interest in gardening across the world (Mullens et al. 2021), marking a return to the austerity gardening movement which has typically emerged in times of economic depression and war (Milthorpe 2019). The Victory Garden movement, promoted in the United Kingdom, Canada, United States, Australia, and Germany in both World Wars, experienced renewed interest during the COVID-19 pandemic as it provided a way to contribute to alleviating growing food insecurity, whilst also improving the morale of those engaged in the effort. In Canada, concerns over food supply and increasing food insecurity (Music et al. 2021; Statistics Canada 2020) motivated many people to engage in victory gardening. The case study outlined in this chapter suggests that, in addition to providing fresh produce to residents of Hamilton, Ontario, who were experiencing food insecurity, victory gardening has proved to be therapeutic for the volunteers who worked in the Hamilton Victory Gardens. As other scholars have reported, gardening during the pandemic has provided a range of non-material features, such as nature connection (Egerer et al. 2022), pleasure, relief, respite, and peace of mind (Marsh et al. 2021).

Hamilton Victory Gardens (HVG) is a not-for-profit team of community volunteers dedicated to alleviating food insecurity in Hamilton, Ontario, and surrounding communities, by using urban agriculture to supply fresh produce to local food banks and social enterprises (Hamilton Victory Gardens [HVG] 2021). HVG transforms empty city lots into places of community, education, and growth. The mission of HVG is to transform the City of Hamilton by increasing the access to healthy food for marginalised populations within the local community; encouraging volunteerism; and bringing people of all ages (from elementary school age children to senior citizens) together for the common purpose of combating hunger through growing fresh produce in gardens (or raised garden beds) constructed on unused city land, without the use of pesticides or petrochemical fertilisers.

A dedicated group of volunteers known as 'Worker-Bees' plant, maintain, and harvest the fresh, nutritious produce. In the past eight years, over 230,000

DOI: 10.4324/9781003355731-4

pounds (104,000 kilograms) of fresh produce has been harvested for local food banks and hot meal programmes. In 2020, HVG grew to a total of 12 garden sites and 661 raised beds (equivalent to 4 ft × 16 ft each). The COVID-19 pandemic created many challenges throughout the 2020 growing season and into the 2021 season, resulting in the permanent closure of one of the largest gardens and the temporary closure of many other sites. Further, the expansion and development of the city has reduced the availability of city lots: in the 2021 growing season, only three active gardens (Jones Road, St Helene's, and the Venture Centre) were worked by volunteers, as shown in Figure 2.1. This has significantly decreased the amount of fresh produce supplied to the community at a time when it is needed the most. Harvested produce is either brought to one of the many food banks and/or social enterprises located throughout the city or used to stock the growing number of community fridges that target low socio-economic neighbourhoods (see Figure 2.1). Some of the produce is donated to sponsor organisations, such as Betula Restaurant, or shared at HVG-sponsored booths at local street festivals throughout the city.

HVG sponsors include Williams Dam Seeds, which donates all kinds of vegetable seeds for all the gardens, and K&K Greenhouses, which donates soil and trays for planting seed late into the winter, as well as watering and maintaining the growing seedlings in their greenhouses until the planting season in May. A wide range of other sponsors can be found on the HVG website (HVG 2021).

Figure 2.1 Hamilton Victory Gardens and their supported locations.

Having lived in Hamilton, Ontario, since the late 1990s, I have come to understand the city as a resident, cyclist, researcher, and volunteer. With a population of approximately 570,000, Hamilton is the ninth largest city in Canada, and third largest in the province of Ontario. Located between Toronto and the Niagara Falls, the city is situated west of the Niagara Peninsula and wraps around the western part of Lake Ontario. Hamilton is the geographical centre of the industrialised 'Golden Horseshoe' region of Ontario: manufacturing, the knowledge economy, transport, and steel production are the main economic drivers for this important port and transport hub, which is affectionately known as 'Steeltown'.

The experience of many Hamilton residents illustrates the relationship between poverty and food insecurity. Based on 2016 data, the poverty rate for Hamilton was 16.7%, which is higher than provincial (13.7%) and national (12.8%) averages (Hamilton Community Foundation 2018). Those most likely to live in poverty are Indigenous people, recent immigrants, and visible minorities. Compared to the general population, lone mother families experience twice the rate of poverty (Hamilton Community Foundation 2018). According to the Statistics Canada Canadian Community Health Survey (2015–2016), 14.8% of Hamiltonians reported experiencing some food insecurity in the previous year (76,580 people), while 4.2% (over 22,000 people) reported severe food insecurity, which means reduced food intake, skipping meals, and disrupted eating patterns (Hamilton Community Foundation 2018). Hamilton Food Share (HFS) is a network of all major food banks in Hamilton. Each month, HFS supports more than 5,000 households and 12,000 individuals, including the 'poorest of the poor' (Dooley et al. 2021), who are the least likely to have access to and consume fruit and vegetables – key elements in nutritional health (Walsh 2021). For many Hamiltonians, the pandemic made things worse due to economic losses, mental health stresses, and changes in eating habits (Walsh 2021). COVID-19 rates were higher in Hamilton neighbourhoods with higher rates of poverty and among residents who are racialised (Hamilton Community Foundation 2021).

Framed within the therapeutic landscape concept, the case study in this chapter employs autoethnography as a methodological approach to explore one of the HVG gardens worked in the summer of 2021. It presents evidence that the garden not only grows significant amounts of a variety of healthy produce (which reduces food insecurity for marginalised families and individuals living in low-income communities within the greater City of Hamilton), but also provides an engaging, beautiful, and accessible green space for volunteers. The therapeutic landscape concept is employed to highlight the range of therapeutic opportunities that the garden provides for the Worker-Bee volunteers who tend the garden for three-to-four-hour blocks twice weekly during the growing season (from early May through to late October). The natural, built, social, and symbolic environments and their synchronous relationships work to produce the three thematic results, which are shared here in order of importance: (1) Meaning, purpose, and community; (2) Active, outside, and dirty; and (3) Leadership and learning. Overall, the results point to the physical and psycho-social benefits experienced by the volunteers. Finally, broader applications of these learnings are shared, as are suggestions for augmenting the autoethnographic approach in future community gardening research.

Gardens as therapeutic landscapes

Gardens are represented as both natural and built landscapes: natural, in the sense that botanical plants are the central focus; built, in the sense that human hands have planted and tended to the growth and success of the garden plants. Gardens become therapeutic landscapes when gardeners work in them, reaping their many benefits.

A recent scoping review of the therapeutic landscape literature noted that, as a health geographical concept, therapeutic landscapes offer in-depth insight into experiential, embodied, and emotional geographies, promote awareness of place as both therapeutic and exclusionary, and continue to be a relevant and lively field of enquiry across health geography (Bell et al. 2018, pp. 123–130). Recognised as a significant contribution made by health geographers to the broad study of health, therapeutic landscapes theory is a conceptual framework for the analysis of physical (natural and built), social, and symbolic environments, as they contribute to physical and mental health and wellbeing in places (Gesler 1992, pp. 735–746; Bell et al. 2018, pp. 123–130).

Traditionally, therapeutic landscapes theory has been applied to four areas: (1) traditional landscapes (i.e., those reputed for health and/or healing, such as shrines, pilgrimages); (2) natural/pristine/wild areas; (3) landscapes for the marginalised (e.g., people experiencing mental health issues, with autism); and (4) health care sites (i.e., hospitals, long-term care facilities) (Williams 1999, 2007, 2019). Milligan, Gatrell and Bingley (2004, pp. 1781–1793) introduced gardens as therapeutic landscapes in their work on community gardens for older people in northern England. Outcomes included positive physical, mental, and social wellbeing, where relaxation, inclusion, purpose, safety, and restoration were experienced by the gardeners. More recently, health geographers have explored gardens as therapeutic landscapes for a range of populations, such as patients with dementia (Marsh et al. 2018, pp. 174–179), those experiencing end-of-life support (Marsh and Spinaze 2016, pp. 214–219), and post-secondary students (Marsh et al. 2020), as well as the significance of gardens as therapeutic landscapes during the COVID-19 pandemic (Egerer et al. 2022; Marsh et al. 2021). Jiang (2015) recognises medical geography as one of the four schools contributing to the research evidence regarding healing gardens, where therapeutic landscapes theory is employed (Table 2.1). In addition to psychology, horticultural therapy (HT) is noted as a recognised school.

Within the ongoing evolution of therapeutic landscapes theory, the recent focus has been the examination of how the coloured elements of therapeutic

Table 2.1 Therapeutic landscapes and healing gardens: Schools and terminology (Jiang 2015)

School	*Terminology*	*Theories*	*Research scale*	*Representative studies*
Medical Geography	Therapeutic Landscape	Sense of Place: Four dimensions of therapeutic landscapes include: Natural, built, symbolic and social environment.	Regional scale; Historic scale	Gesler (1993, 2003)
Environmental Psychology	Restorative Environment	Attention-Restoration Theory (ART); four features of restorative environment include being away, extent, fascination, and actions and compatibility.	Wild nature; Nearby nature; Urban nature	Kaplan and Kaplan (1989); Kaplan (1993, 2001)
	Healing Garden; Restorative Garden	Aesthetic-Affective Theory (AAT); Psycho-Evolution Theories; Three features of healing gardens: relief from symptoms; stress reduction; and an improvement in the overall sense of well-being.	Urban nature; Gardens or green settings in healthcare environments.	Cooper-Marcus and Barnes (1995); Cooper-Marcus (2000); Ulrich (1979, 1984, 1995); Ulrich et al. (1991)
Salutogenic Environment and Ecological Psychology	Salutogenic Environment; Therapeutic Landscape	Theories of Environmental Affordances; Ecological Psychology	Urban park system; Historic scale	Heft and McFarland (1999, Heft 2010); Grahn et al. (2010); Stigsdotter and Grahn (2003)
Horticultural Therapy	Healing Garden; Therapeutic Garden	Theory of 'flow experience'; Sensory stimulation theories	Gardens or settings for horticultural activities	Sooderback et al. (2004); Detweiler et al. (2012)

landscapes, particularly blue and green, promote health and wellbeing (Foley et al. 2019, pp. 117–131). Finlay et al. (2015, pp. 97–106) discuss the potential of 'nearby' blue and green spaces for older low-income adults in Vancouver, British Columbia, to provide spiritual restoration and improved mental health. The evidence for the importance of green spaces continues to grow in a range of disciplines, including landscape architecture, environmental psychology, urban planning, and health geography. Gardens are healthy places where activities nurture community interactions and promote physical and social wellbeing. This is well illustrated in Howarth et al.' (2020) general logic model (well-being) (Figure 2.2), where gardens and garden activities support healthier adults and children, given the positive mental, social, and physical health outcomes. Further, general outcomes include improved community health and wellbeing. These are further explored in this chapter, together with the health benefits of community gardening for the greater good.

Many researchers recognise community gardens, such as the HVG, as 'third places', which pioneering urban sociologist Ray Oldenburg (1989) defined as public spaces where informal meeting can take place outside of the home (first place) and work (second place). Given that third spaces are public, they are open to all and include a wide range of sites, including community gardens, coffee shops, bookstores, and shopping malls. Dolley's (2019, 2020) research on community gardens in Australia and Denmark finds that these third places support informal exchanges and relationship building between individuals living in proximity and, consequently, contribute to building a sense of community and social cohesion.

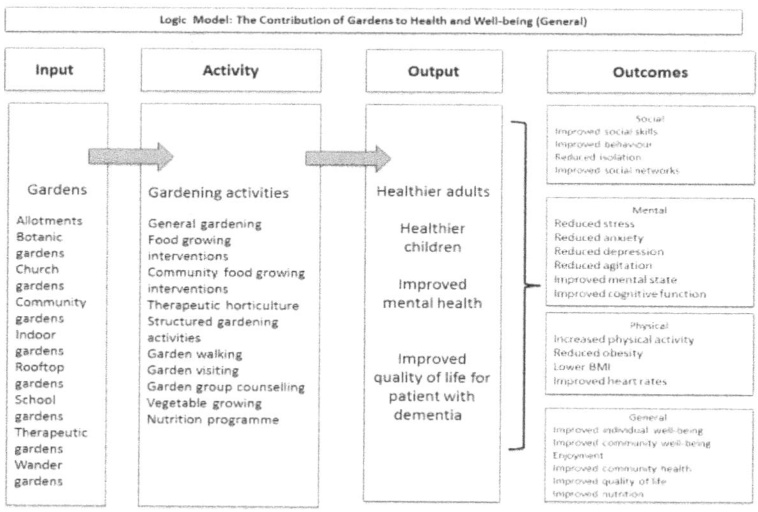

Figure 2.2 Logic model: Well-being (Howarth et al. 2020).

Methodology

The Hamilton Victory Gardens case study outlined in this chapter adopts a reflexive autoethnographic approach. Borrowing from Holman Jones (2005, p. 764), I employ autoethnographic research and writing as 'socially-just acts, where the goal is to produce analytical, accessible texts that change us and the world we live in for the better'. The HVG not only allowed individuals to volunteer their time, knowledge, and labour, but also provided an opportunity to tackle food insecurity in Hamilton during the COVID-19 pandemic, which caused many Hamiltonians to lose their employment in 2021. Autoethnography provides an approach to document the activities of the Jones Road HVG, which provided therapeutic outcomes for the Worker-Bee volunteers and contributed to improved food security for those living in Hamilton. As Ellis et al. (2010, p. 1) explain, autoethnography is a powerful methodology for capturing both personal experience (that of the researcher) and cultural experience (that of the Worker-Bee volunteer):

> Autoethnography is an approach to research and writing that seeks to describe and systematically analyze personal experience in order to understand cultural experience. This approach challenges canonical ways of doing research and representing others and treats research as a political, socially-just and socially-conscious act. A researcher uses tenets of autobiography and ethnography to do and write autoethnography. Thus, as a method, autoethnography is both process and product.

In keeping with the autoethnographic approach, I write the remainder of this chapter in first person.

Autoethnography is becoming more widely accepted and employed in human geographical research (Butz 2010; DeLyser 2015). The methodology enables researchers to engage personally in what they are studying, where the research topic becomes re-imagined via 'reflexive narrators of self' (Butz and Besio 2009, p. 1660). Health geographers apply autoethnography to describe the experience of place within the context of therapeutic landscapes and, in so doing, encourage the further use of autoethnography in therapeutic landscape enquiry (Liggins, Kearns and Adams 2013, pp. 105–109; Thomson 2021). For example, Liggins, Kearns and Adams (2013, pp. 105–109) employ an autoethnographic approach to reconsider the inpatient unit as a place of healing, to focus on the 'world within and the world without'. Another example is the work of Thompson (2021), which examines personal encounters with digital health via an autoethnographic approach, demonstrating how digital health not only upsets, but also creates new therapeutic landscapes.

Following research ethics approval, I began my field data collection at the Jones Road garden (Figure 2.1), writing copious notes, taking photographs of the produce, and documenting the various activities I was involved in, both at the garden and at the nearby food bank. These activities included cooking, cleaning, sharing, and, of course, gardening (Adams, Jones, and Ellis 2022; Chang 2008). I gathered further contextual data from the HVG website, and from literature on the Victory Gardens movement more generally. I shared with the fellow Worker-Bees and Leader my intention to conduct this research as part of a larger research project on green space and wellness. The positive response was overwhelming, as the Worker-Bees were quick to share the many benefits of volunteering and spending their time in the garden.

To understand the larger operation of the HVG, I took a day to shadow a summer contract employee, who spent his full-time hours managing the programme, including monitoring the three gardens (Figure 2.1). This allowed me to visit the two other gardens in the city (St Helene's and the Venture Centre), both of which are located in more economically challenged neighbourhoods than the Jones Road garden (Figure 2.1). Attached to a large food and clothes bank, the Venture Centre operates as the head office of the HVG; here, I met the HVG director and learned the extent of the operation. Through shadowing the contract employee, I also learned that the HVG provides produce to three community fridges located in three of the most vulnerable neighbourhoods in the city (Figure 2.1). Unlike the food banks, which require identification and documentation of need, the community fridges are freely accessible; users can take what they need and are provided with the opportunity to note which items they need for a future visit. I learnt that the produce is quickly taken, with the fridges often empty when they are restocked (twice weekly). At the end of my day of shadowing, we stocked one fridge with green beans and tomatoes.

From early May through to October 2021, I entered the primary data into my electronic field book after each Worker-Bee session in the garden/food bank. This data included both my own experience, as well as the observations of other Worker-Bees in the garden. This primary data was complemented by different data types, including contextual data. During the data analysis phase, I employed thematic analysis – a systematic method for organising and identifying insights into meaningful themes across the data (Braun and Clarke 2012). Thematic analysis procedures included creating preliminary codes, which were assigned to the data in order to describe the content. Themes were organised across the data from these preliminary codes. The themes were then reviewed, defined, and named (Braun and Clarke 2012). Following qualitative research guidelines for assuring reliability and validity (e.g., keeping an audit trail and conducting a cross-check with others working the garden), research findings became transferable, dependable, credible, and confirmable. The natural, built, social, and symbolic environments worked in synchronicity to produce the three thematic results – 'meaning, purpose, and community', 'active, outside, and dirty', and 'leadership and learning' – which are shared, below, in order of prominence. Overall, the thematic results point to the physical and psychosocial benefits I experienced (as the researcher), as well as those experienced by the other Worker-Bee volunteers, who numbered seven to eight depending on the day and week.

Thematic results

Meaning, purpose, and community

When gardening with a group of fellow gardeners, there is a shared meaning and purpose to the activity. I recall this being the case when I was a young girl. Gardening activities were done in community, whether with other family members or with multiple neighbours who shared the same garden plot. We shared the same sense of productivity, planting, and tending to the garden to reap the harvest. Rarely do I recall being in the garden by myself, except on the rare occasions when I was asked to pick lettuce, beans, or broccoli for a family meal, or find a ripe tomato for a salad. I recall the sheer boredom of weeding a neighbour's garden on my own – even though it meant having pocket money.

The shared meaning and purpose of the volunteer Worker-Bees was primarily one of contributing to the greater good – to feed those experiencing food insecurity in Hamilton. Although the primary goal of our gardening activity was altruistic, it became clearer as the summer wore on that we were all benefiting from engagement in the twice-weekly three-hour gardening sessions. For me, actively doing something to help others during the pandemic nurtured a sense of purpose, of contributing to the greater good. As with most volunteer work, it made me feel better about myself and my place in the world. By the second month into the gardening season, meeting the fellow Worker-Bees was like meeting friends for a visit; going to the garden was something I looked forward to, with anticipation. The garden became our coffee shop; our third place away from work and home. The other volunteers spoke about how they looked forward to coming to the garden, how the garden was their therapy, and discussed how much they were learning while at the garden. Intimate stories were shared, including stories of hurt and disappointment, angst and worry. These stories were rarely shared at the beginning of the three-hour sessions, but were often initiated approximately 30 minutes into a session. Not all volunteers shared, but many did. Some stories were very difficult to hear: being present to listen was important for maintaining trust and community cohesiveness. Some weeks were more talkative than others, often depending on the number and intensity of tasks being required on that particular day.

One long-time volunteer often commented that it was the garden that saved him from a range of ailments, by providing purpose, focus, and a community. Intimate stories of work, romantic relationships, familial ties, worries, and dreams were all seemingly unintentionally shared in the garden, making the activity not only beneficial for the recipients of the produce we planted, tended, and harvested, but simultaneously beneficial for each volunteer's psychological health and wellbeing. Certainly, we all shared a strong sense of belonging in our Worker-Bee community, catching up on each other's lives and troubles week after week. At first, the ability to socialise with other volunteers appeared to be a secondary activity as we planted, watered, weeded, and harvested. I soon learnt that the socialising component was just as important – if not more so – as the gardening itself. Figure 2.3 captures six Worker-Bees and a typical July harvest.

Figure 2.3 Worker-Bees with their weekly harvest. Photo by Allison Williams.

Later in the summer, the tomatoes were ripening more quickly than we Worker-Bees could harvest them, with many falling to the ground or splitting on the vine. We were all encouraged to come and pick them any time for our own use, in order that they should not go to waste. I did so on my own one Sunday morning – and felt alone, disconnected, and indifferent. I only stayed for a short 15 minutes, less than half the time I had allocated to the task. For me, this confirmed the importance of working together as a group with the same shared meaning and purpose in mind. It also confirmed the symbolism of the garden as a place that nurtured community cohesiveness, as a third place.

Active, outside, and dirty

Reflecting much of my childhood experience, my memories of gardening as a child always include being active, outside, and dirty! Dirt under my nails, mud on my shoes or boots, and dirt on my clothes were a normal part of being outside for most of the summer, and was most accentuated in the garden! I recall spending whole days planting the garden in the spring, and numerous days harvesting vegetables, pulling plants, and cleaning up the soil in preparation for spring planting. Gardening clothes were unique – often not worn elsewhere, having little aesthetic value, and fitting loosely to allow for movement.

When I reflect on my Worker-Bee experience, I realise not much has changed. The garden itself is approximately one acre in size, with 10–12 rows of various vegetables making up two halves of the garden. The perimeter of the garden is where squash, corn, and sunflowers are grown, together with crab apples and ground cherries. All the tools, seeds, and equipment for the garden are kept in a locked shed adjacent to the garden, built on the nearby parking lot (carpark) of the former municipal building. Following COVID-19 protocols, all volunteers were required to wash and sanitise their hands before entering the garden. We were also required to sign in and out in a binder that was set out on a nearby picnic table. The weekly weigh-in of produce was recorded in the same binder. A produce scale was used to weigh each crate of produce. Once it was all washed and sorted into plastic cates, the produce was brought to the nearby food bank (Stoney Creek Community Food Bank), approximately 0.5 km from the garden. Here, we cleaned (if needed), sorted, and stored the produce, dividing it into two portions for each of the two food bank days. As the first food bank day was Wednesday, we put out much of Tuesday's harvest for clients to select from, being careful to present it in the most aesthetically pleasing manner; squash, pumpkins, or sunflowers were often used to frame the produce. Produce not tabled for Wednesday's food bank was refrigerated or stored for the second food bank day (on Friday). We were coached by our Leader about how to present the vegetables, in order that they be made more attractive. The name of each vegetable was written on a cardboard nameplate and placed on the basket/bowl or vegetable to encourage the new Canadians (who were learning English as another language) to learn about the produce and the spelling of each vegetable. Time spent at the food bank always involved friendly banter, whether about the produce, what to do with the produce, or one of dozens other topics.

Each session left me dirty, physically tired, and often heat exhausted! Being comfortable with, or enjoying, being active, outside, and dirty is almost a requirement of gardening; as we garden, we will most certainly be getting dirty, while we sweat from the physical exercise, often in the heat. I found myself wearing the same lose shorts, shirt, and boots to each session, being sure not to forget my ten-year-old, wide-brimmed straw hat! One of the benefits of working the garden is that each of us can choose a handful of produce to take home. There is no question that doing this at each session improved my family's intake of vegetables: when I brought them home, they were eaten – in frittatas, omelettes, pastas, stews, and soups. Figure 2.4 provides a visual of the quantity and type of produce grown.

Sharing recipes with one another during our Worker-Bee sessions, specifically about how to prepare fresh-from-the-garden vegetables, was always a highlight. It seemed that the favourite conversation was recipes for garden-fresh salsa. Once the tomatoes ripened in the garden, we all began making a variety of salsas, which were brought to the next Worker-Bee session for sharing at the end of our gardening session. This exercise was symbolic of the mutual trust and respect we built as a gardening community. In addition to sharing salsa, one of the long-time volunteers shared sunflower seedlings with the other Worker-Bees, which were then planted in our home gardens. I received six seedlings, all of which grew to various sizes, providing me a reminder of the HVG community and the good work we were doing for the community.

Figure 2.4 Range of produce grown in the garden. Photo by Allison Williams.

Leadership and learning

Being introduced to vegetable gardening as a child, I had absorbed so much about what to do and how to do it. My parents were the 'leaders', showing and telling us four kids what and how to do things. I am sure we would have been terribly lost, were it not for their tutelage! My father was the leader, deciding when to plant and when to water, weed, and harvest. He seemed to have a sixth sense about these questions, confidently amassing the crew of family members to accomplish what was needed. Through this experience, I learned how to grow tomatoes, potatoes, onions, lettuce, cucumbers, squash, zucchini (courgettes), carrots, broccoli, and rhubarb.

The Hamilton Community Garden grew so much more than these vegetables and fruits! Given that the recipients of the produce were often new immigrants, we also grew a range of vegetables known to them, including bok choy, banana peppers, cayenne peppers, okra, and rapini. I was fortunate to learn how to grow all of these vegetables, in addition to learning about planting multiple crops in the same season. A number of wild plants that grew in the garden were also new to me, such as crab apples and ground cherries. The Leader of the Jones Road garden had grown up on a farm and had a vast and deep knowledge of gardening. She successfully led all of us Worker-Bees with kindness, being sure not to dictate to or boss us in any way; she provided the right amount of direction without being overbearing. She reminded me of my all-time favourite person growing up – my Brown Owl, who led my Brownie

Figure 2.5 HVG Mission Statement, *Planting Happiness Reducing Hunger*. Photo by Allison Williams.

pack (when I was 5–11 years of age). She could do no wrong and only planted happiness in our Worker-Bee group. I recall harvesting the first batch of green beans with her on a Tuesday harvest day, both of us revelling in the large size of the beans so early in the season. I laughed at our enthusiasm, calling us 'garden nerds'; as we shared more laughter, she wholeheartedly agreed. Leadership of the Jones Road garden became uncertain mid-season, when our Leader confirmed she would not be coming back next summer due to the growing responsibilities of caring for her ageing parents. The rest of us questioned what we would do without her guidance and her knowledge; I recall suggesting that we would need to call her regularly for step-by-step directions on what to plant, how to plant and cultivate, and when to harvest.

Discussion

In combination, the therapeutic outcomes of gardening reflect Howarth et al.'s (2020) logic model, which outlines the contribution of gardens to health and wellbeing, introduced at the start of this chapter. As the autoethnographer, I felt an improved sense of wellbeing, which was most evident in the social connection I nurtured with other team members, as well as an improved sense of general, physical, and mental health, as elaborated upon below. Further, as I was invited to be the site coordinator for the Jones Road garden in spring/summer 2022, I believe my community wellbeing improved, as noted under 'General Outcome' in the logic model. There is no question that the fresh garden produce provided to the food bank each week improved the community's nutrition, health, and quality of life, again noted as a 'General Outcome' in the logic model. Each of the three thematic results from this case study addresses various therapeutic landscape dimensions, whether it be the natural or built environment, the symbolic, or the social dimension. In keeping with the therapeutic landscape concept, the natural and built characteristics of the Jones Road garden were more obvious than the social and symbolic characteristics, which only became evident to me as I invested more time in the project.

For example, Theme 1: *Meaning, purpose, and community* addresses both the social and symbolic dimensions. The strong sense of community that was engendered over the course of the gardening season allowed each of the Worker-Bees to benefit from the many social opportunities that the garden provided. These rich social opportunities transformed the garden into a meeting place, which functioned as a third place. Theme 2: *Active, outside, and dirty* addresses both the natural and built environments. The large Jones Road garden was, for most of the season, a mix of different tones of green, reflecting the maturity and type of vegetable plants. As the season wore on, harvest days included a range of additional colours, such as red, yellow, orange, and purple. Thus, the colour-scape of the garden became more varied and vibrant as the season progressed, with produce ripening as the summer wore on. The Stoney Creek Food Bank provided a built environment that continued the social engagement of the Worker-Bees, where the garden harvest was cleaned, sorted, and presented for the food bank's clientele. Theme 3: *Leadership and learning* primarily focused on the social dimension, given that the Leader engendered safe and welcoming social engagement. This included the learning we all experienced in working the garden, whether that be seeding, weeding, harvesting, or cleaning the garden for the next season's planting. As a community, we Worker-Bees shared much by working the garden: shared physical labour, both in the natural environment of the garden and the built environment of the food banks/community fridges; shared social connection and the benefits thereof, providing psycho-social rewards; shared common altruistic goals for a better world; and shared purposiveness/meaning, which fed into our sense of place and belonging in the world. These many positive outcomes suggest that, through the community garden, we strengthened each other's health and wellbeing. Further, the findings suggest that community gardening operates as a therapeutic horticulture intervention for the Worker-Bees, given the numerous identified outcomes that reflect Howarth et al.'s (2020) logic model. Gardening in the HVG thus proved to be therapeutic – relieving stress through nature connection (Egerer et al. 2022) and social engagement, and providing a refuge during the pandemic (Marsh et al. 2021).

The autoethnographic approach used in the case study presented in this chapter was well suited to a COVID-19 summer growing season. It not only allowed me to reflect on my experience (as volunteer) in the garden, but also provided the opportunity to hear the experiences of the other Worker-Bees without being intrusive, as would have been the case had I (as researcher) elected to interview or engage them in focus group discussions. At times, I did feel like an interloper – I was not only a Worker-Bee, but also a researcher. The autoethnographic approach allowed me to reconnect with my past in memorable and real ways; as both my parents continue to keep a vegetable garden, we often exchanged produce when we saw one another every few weeks. Sharing knowledge about new fruits and vegetables and their recipes kept us all entertained; ground cherries were a favourite, as my parents had never seen those before. In short, I feel the HVG experience not only brought about personal change in the Worker-Bee volunteers, but also made a small dent in better managing food insecurity in Hamilton. In sum, our growing season made the world a bit better. Researching and writing this project encouraged my own personal sense of responsibility and agency during a pandemic, whilst also increasing consciousness and stimulating change, with respect to contributing to improved food security in Hamilton.

As we begin to move beyond COVID-19, data collection opportunities that complement autoethnography will allow for engagement in documentary film, narrative data, photovoice, and a range of other methods to capture the impact of the garden's colour-scapes, as they evolve and change throughout the gardening season.

Conclusions

Framed within the therapeutic landscape concept, this chapter uses autoethnography as a methodological approach to explore one of the Hamilton Victory Gardens cultivated in the summer of 2021 – approximately one year into the COVID-19 pandemic. It presents evidence that the garden not only grows significant and varied quantities of healthy produce (which reduces food insecurity for marginalised families and individuals living in low-income communities within the greater City of Hamilton), but also contributes to the health and wellbeing of volunteers who enjoy this engaging, beautiful, and accessible green space. The therapeutic landscape concept frames three thematic results – Theme 1: *Meaning, purpose, and community* (social and symbolic dimension); Theme 2: *Active, outside, and dirty* (natural and built dimension); and Theme 3: *Leadership and learning* (social dimension). The natural, built, social, and symbolic environments work in synchronicity to create the physical and psycho-social benefits experienced by the volunteers. Broader applications of these learnings suggest further engagement with the colour-scapes of the gardens throughout the evolving and ever-changing gardening season.

References

Adams, TE, Jones, S and Ellis, C (eds) 2022, *Handbook of autoethnography* (2nd edition), Routledge/Taylor & Francis, London.

Bell, S, Foley, R, Houghton, F, Maddrell, A and Williams, A 2018, 'From therapeutic landscapes to healthy spaces, places and practices: A scoping review', *Social Science & Medicine*, vol. 196, pp. 123–130. https://doi.org/10.1016/j.socscimed.2017.11.035

Braun, V and Clarke, V 2012, 'Thematic analysis' in H Cooper, PM Camic, DL Long, AT Panter, D Rindskopf and KJ Sher (eds), *APA handbook of research methods in psychology, vol. 2. Research designs: Quantitative, qualitative, neuropsychological, and biological*, American Psychological Association, Worcester, MA, pp. 57–71.

Butz, D 2010, 'Autoethnography as sensibility' in D DeLyser, S Herbert and S Aitken (eds), *The Sage handbook of qualitative geography*, Sage Publications, Thousand Oaks, CA, pp. 138–155. https://dx.doi.org/10.4135/9780857021090.n10

Butz, D and Besio, K 2009, 'Autoethnography', *Geography Compass*, vol. 3, no. 5, pp. 1660–1674. https://doi.org/10.1111/j.1749-8198.2009.00279.x

Chang, H 2008, *Autoethnography as method*, Left Coast Press, Walnut Creek, CA.

Cooper-Marcus, C and Barnes, M 1995, *Gardens in the healthcare facilities: Uses, therapeutic benefits, and design recommendations*. Center for Health Design, Wiley, New York.

Cooper-Marcus, C 2000, 'Gardens and health', In *Design and health: The therapeutic benefits of design*, 2nd International Congress on Design and Health, Karolinska Institute, Stockholm, pp. 461–471.

DeLyser, D 2015, 'Collecting, kitsch and the intimate geographies of social memory: A story of archival autoethnography', *Transactions of the Institute of British Geographers*, vol. 40, no. 2, pp. 209–222. https://doi.org/10.1111/tran.12070

Detweiler, MB, Sharma, T, Detweiler, JG, Murphy, PF, Lane, S, Carman, J … and Kim, KY 2012, 'What is the evidence to support the use of therapeutic gardens for the elderly?', *Psychiatry Investigation*, vol. 9, no. 2, pp. 100–110.

Dolley, J 2019, 'Third places and social capital: Case study community gardens' in J Dolley and C Bosman (eds), *Rethinking third places*, Edward Elgar, Cheltenham, pp. 136–157.

Dolley, J 2020, 'Community gardens as third places', *Geographical Research*, vol. 58, no. 2, pp. 141–153. https://doi.org/10.1111/1745-5871.12395

Dooley, M, Bokma, R and Yin, M 2021, 'The poorest of the poor use Hamilton food banks', (*Policy Options/Politiques*, 10 December 2021). https://policyoptions.irpp.org/magazines/december-2021/the-poorest-of-the-poor-use-hamilton-food-banks/

Egerer, M, Lin, B, Kingsley, J, Marsh, P, Diekmann, L and Ossola, A 2022, 'Gardening can relieve human stress and boost nature connection during the COVID-19 pandemic', *Urban Forestry & Urban Greening*, vol. 68, article 127483. https://doi.org/10.1016/j.ufug.2022.127483

Ellis, C, Adams, TE and Bochner, AP 2010, 'Autoethnography: An overview', *Forum Qualitative Sozialforschung/Forum: Qualitative Social Research*, vol. 12, no. 1, article 10. https://www.qualitative-research.net/index.php/fqs/article/view/1589/3095

Finlay, J, Franke, T, McKay, H and Sims-Gould, J 2015, 'Therapeutic landscapes and wellbeing in later life: Impacts of blue and green spaces for older adults', *Health Place*, vol. 34, pp. 97–106. https://doi.org/10.1016/j.healthplace.2015.05.001

Foley, R, Kearns, R, Kistemann, T and Wheeler, B (eds) 2019, *Blue space, health and wellbeing hydrophilia unbounded*, Routledge Geographies of Health, Routledge, Oxfordshire, UK.

Gesler, W 1992, 'Therapeutic landscapes: Medical issues in light of the new cultural geography', *Social Science & Medicine*, vol. 34, no. 7, pp. 735–746. https://doi.org/10.1016/0277-9536(92)90360-3

Gesler, WM 1993, Therapeutic landscapes: Theory and a case study of Epidauros, Greece. *Environment and Planning Design*, vol. 11, pp. 171–189.

Gesler, WM 2003, *Healing places*. Rowman & Littlefield, Lanham, MD.

Grahn, P, Ivarsson, CT, Stigsdotter, UK and Bengtsson, IL 2010, 'Using affordances as a health-promoting tool in a therapeutic garden' in CW Thompson, P Aspinall and S Bell (eds), *Innovative approaches to researching landscape and health: Open space: People space 2*, Routledge, London.

Hamilton Community Foundation 2018, 'Vital Signs', *Hamilton Community Foundation*, https://hamiltoncommunityfoundation.ca/VS2018/

Hamilton Community Foundation 2021, 'Vital Signs', *Hamilton Community Foundation*, https://hamiltoncommunityfoundation.ca/vital-signs/

Heft, H and McFarland, D 1999, Children's and Adult's Assessments of a Step Affordance for Self and Others. Poster presented at the meetings of the *Society for Research in Child Development*, Albuquerque, New Mexico.

Heft, H 2010, 'Affordances and the perception of landscape' in CW Thompson, P Aspinall and S Bell (eds), *Innovative approaches to researching landscape and health*, Routledge, London, pp. 9–32.

Holman-Jones, S 2005, 'Autoethnography: Making the personal political' in Norman K Denzin and Yvonna S Lincoln (eds), *Handbook of qualitative research*, Sage Publications, Thousand Oaks, CA, pp. 763–791.

Howarth, M, Brettle, A, Hardman, M and Maden, M 2020, 'What is the evidence for the impact of gardens and gardening on health and well-being: A scoping review and evidence-based logic model to guide healthcare strategy decision making on the use of gardening approaches as a social prescription', *BMJ Open*, vol. 10, no. 7, e036923. https://doi.org/10.1136/bmjopen-2020-036923

HVG 2021, *Hamilton Victory Gardens website*. Hamilton Victory Gardens, Hamilton, Ontario.

Jiang, S 2015, 'Encouraging engagement with therapeutic landscapes: Using transparent spaces to optimize stress reduction in urban health facilities' (PhD Thesis, Clemson University). https://tigerprints.clemson.edu/all_dissertations/1495/

Kaplan, S 1992, 'The restorative environment: Nature and human experience' in D Relf (eds) *Role of horticulture in human well-being and social development: A national symposium*, Timber Press, Arlington, VA, pp. 134–142.

Kaplan, R 1993, 'The role of nature in the context of the workplace', *Landscape and Urban Planning*, vol. 26, no. 1, pp. 193–201.

Kaplan, S 1995, 'The restorative benefits of nature: Toward an integrative framework', *Journal of Environmental Psychology*, vol. 15, no. 3, pp. 169–182.

Kaplan, R 2001, 'The nature of the view from home psychological benefits', *Environment and Behavior*, vol. 33, no. 4, pp. 507–542.

Kaplan, S, and Kaplan, R 1982, *Cognition and environment: Functioning in an uncertain world*. Praeger, New York.

Kaplan, R and Kaplan, S 1989, *The experience of nature: A psychological perspective*. CUP Archive.

Kingsley, J, Foenander, E and Bailey, A 2019, '"You feel like you are part of something bigger": Exploring motivations for community garden participation in Melbourne, Australia', *BMC Public Health*, vol. 19, article 745. https://doi.org/10.1186/s12889-019-7108-3

Liggins, J, Kearns, RA and Adams, PJ 2013, 'Using autoethnography to reclaim the place of healing in mental health care', *Social Science & Medicine*, vol. 91, pp. 105–109. https://doi.org/10.1016/j.socscimed.2012.06.013

Marsh, P, Courtney-Pratt, H and Campbell, M 2018, 'The landscape of dementia inclusivity', *Health & Place*, vol. 52, pp. 174–179. https://doi.org/10.1016/j.healthplace.2018.05.013

Marsh, P, Diekmann, LO, Egerer, M, Lin, B, Ossola, A and Kingsley, J 2021, 'Where birds felt louder: The garden as a refuge during COVID-19', *Wellbeing, Space and Society*, vol. 2, article 100055. https://doi.org/10.1016/j.wss.2021.100055

Marsh, P, Mallick, S, Flies, E, Jones, P, Pearson, S, Koolhof, I, Byrne, J and Kendal, D 2020, 'Trust, connection and equity: Can understanding context help to establish successful campus community gardens?' *International Journal of Environmental Research and Public Health*, vol. 17, no. 20, article 7476. https://doi.org/10.3390/ijerph17207476

Marsh, P and Spinaze, A 2016, 'Community gardens as sites of solace and end-of-life support: A literature review', *International Journal of Palliative Nursing*, vol. 22, no. 5 pp. 214–219. https://doi.org/10.12968/ijpn.2016.22.5.214

Milligan, C, Gatrell, A and Bingley, A 2004, '"Cultivating health": Therapeutic landscapes and older people in northern England', *Social Science & Medicine*, vol. 58, no. 9, pp. 1781–1793. https://doi.org/10.1016/S0277-9536(03)00397-6

Milthorpe, N (ed) 2019, *The poetics and politics of gardening in hard times*, Rowman and Littlefield, Washington, DC.

Mullins, L, Charlebois, S, Finch, E and Music, J 2021 'Home food gardening in Canada in response to the COVID-19 pandemic', *Sustainability*, vol. 13, no. 6, article 3056. https://doi.org/10.3390/su13063056

Music, J, Finch, E, Gone, P, Toze, S, Charlebois, S and Mullins, L 2021, 'Pandemic victory gardens: Potential for local land use policies', *Land Use Policy*, vol. 109, article 105600. https://doi.org/10.1016/j.landusepol.2021.105600

Oldenberg, R 1989, *The great good place: Cafés, coffee shops, bookstores, bars, hair salons, and other hangouts at the heart of a community*, Paragon House, New York.

Statistics Canada 2020, 'Food insecurity during the COVID-19 pandemic', *Statistics Canada*. https://www150.statcan.gc.ca/n1/pub/45-28-0001/2020001/article/00039-eng.htm

Söderback, I, Söderström, M and Schälander, E 2004, 'Horticultural therapy: The 'healing garden'and gardening in rehabilitation measures at Danderyd Hospital Rehabilitation Clinic, Sweden', *Pediatric Rehabilitation*, vol. 7, no. 4, pp. 245–260.

Stigsdotter, U and Grahn, P, 2003, 'Experiencing a garden: A healing garden for people suffering from burnout diseases', *Journal of Therapeutic Horticulture*, vol. 14, pp. 38–48.

Thompson, M 2021, 'The geographies of digital health: Digital therapeutic landscapes and mobilities', *Health & Place*, vol. 70, article 102610. https://doi.org/10.1016/j.healthplace.2021.102610

Ulrich, RS, 1979, 'Visual landscapes and psychological well-being', *Landscape Research*, vol. 4, no. 1, pp. 17–23.

Ulrich, RS 1984, 'View through a window may influence recovery', *Science*, vol. 224, no. 4647, pp. 224–225.

Ulrich, RS 1995, 'Effects of healthcare interior design on wellness: Theory and recent scientific research', *Innovations in Healthcare Design*, pp. 88–104.

Ulrich, RS, Simons, RF, Losito, BD, Fiorito, E, Miles, MA and Zelson, M 1991, 'Stress recovery during exposure to natural and urban environments', *Journal of Environmental Psychology*, vol. 11, no. 3, pp. 201–230.

Walsh, P 2021, 'Food insecurity: A preventable public health crisis', *The Hamilton Spectator*, 31 March 2021. https://www.thespec.com/opinion/contributors/2021/03/31/food-insecurity-a-preventable-public-health-crisis.html

Williams, A (ed) 1999, *Therapeutic landscapes: The dynamic between place and wellness*, University Press of America, New York.

Williams, A (ed) 2007, *Therapeutic landscapes*, Routledge/Taylor & Francis, Abingdon.
Williams, A, Patterson, A and Parnell, M 2019, 'Blue yogic culture: A case study of Sivananda Yoga Retreat, Paradise Island, Nassau, Bahamas' in R Foley, R Kearns, T Kistemann and B Wheeler (eds), *Blue space, health and wellbeing: Hydrophilia unbounded*, Routledge, Abingdon, pp. 117–131.

3 Growing health in local food gardens

Case studies of community, school, and home gardens

Sumita Ghosh

Introduction

A significant movement is underway to cultivate food locally in various productive spaces, such as home, community, rooftop, and school gardens, as well as on small and large urban and rural farms (Gaynor 2006; Thompson, Corkery and Judd 2007; Wise 2014). These typologies of urban agriculture vary in scale, function, purpose, operations, and management approaches (Mougeot 2000). There are multiple health and social benefits associated with these urban agricultural practices. Many studies have established that the process of growing vegetables, fruits, and herbs locally in food production spaces of different types can contribute to disease prevention, stress reduction, decreased obesity, and better mental health (Mead, Christiansen, Davies et al. 2021; Palar, Hufstedler, Hernandez et al. 2019; Soga, Cox, Yamaura et al. 2017). Hutchins (2020) identifies eight positive aspects of gardening, including building self-esteem, better cardiovascular health, stress relief, making a happy self, strengthening hand capability, good relationships and interactions within the family, reducing vitamin deficiency, and healthier food choices. It increases access to healthy food, allows nutritious vegetable and fruit production close to home or within the local catchment area, and includes fresh vegetables in people's daily diets (Hutchins 2020). Through gardening, physical activities increase, and mental health and cognitive functions improve (Chalmin-Pui, Griffiths, Roe et al. 2021; Koay and Dillion 2020; McVey, Nash and Stansbie 2018; Wilkinson and Orr 2017). Urban agricultural practices in gardens provide recreation and relaxing opportunities for people and benefit their quality of life and general wellbeing (Home and Vieli 2020; Schultz and Rosen 2022). The shorter farm-to-plate distance and closed-loop local food systems can provide nutritious and fresh food for the community, improve public health, and facilitate social networking and participation. These local foodscapes enable placemaking through growing fruit and vegetables and community engagement.

Cultural context plays an important role in growing culturally appropriate and traditional food in the gardens. The vegetable production in food gardens is linked to definite culinary habits and daily diets of different communities, where food as medicine has positive effects on the health of specific communities.

DOI: 10.4324/9781003355731-5

Food production in migrants' domestic gardens reflects connections to cultural identities, memories, traditions, and homeland practices (Graham and Connell 2006). A survey with 1,390 Australian households and interviews with experts and community gardeners concludes that one in two Australian families (equal to a total of 4.7 million households) is growing some fruits, vegetables, and herbs, either in individual home gardens or in community gardens (Wise 2014). A community garden in a high-rise public housing estate in Waterloo, in Sydney's inner west, found that significant physical, emotional, and spiritual well-being is associated with the garden for city-dwellers and public housing tenants (Thompson et al. 2007). School gardens at primary and secondary schools can act as a platform for social learning and can develop students' cognitive and emotional-affective abilities (Pollin and Retzlaff-Furst 2021). Individual qualities, commitments, and goals shape individual and household motivational factors for gardening, lived experiences, and sustainability awareness (Ghosh 2010).

During the COVID-19 pandemic, home gardens provided food and contributed to the physical and mental wellbeing of individuals and households, and the percentage of people using their home garden for vegetables and fruit production increased (Cerada, Guenat, Egerer, and Fischer 2022; Corley, Okely, Taylor et al. 2021). Corley et al. (2021) conducted an online survey during the COVID-19 lockdown in Scotland, which involved 171 participants (mean age = 84 years) who had access to home gardens. This study established that home garden usage amongst older adults was associated with higher levels of self-reported mental and physical wellbeing (Corley et al. 2021). During COVID-19 lockdown periods, it was challenging for community and allotment gardens to continue to provide health and social benefits due to restrictions on movement. An international research project, FEW-meter, covering five countries (France, Germany, Poland, the United Kingdom, and the United States) over two growing seasons, surveyed gardeners and allotment plot holders during COVID-19 (Schoen, Blythe, Caputo et al. 2021). The community and allotment garden challenges included maintaining pre-COVID-19-level accessibility, services, and continuing operations, resulting in lower participation in food gardening and reduced social interactions and food production (Schoen et al. 2021). During the pandemic lockdown, community gardens in Arizona applied specific changes (Buechler and Mansaray 2021); it was found that home gardens (Cerada et al. 2022) were better places to continue to grow food, providing health and wellbeing benefits to people.

Looking through a therapeutic lens, this chapter investigates the capabilities of local community, school, and home food gardens to provide health improvements and social opportunities. The following main research question is explored in this chapter: 'How do local foodscapes in community, school, and home gardens improve health and social exchanges across social and geographical gradients?'

The primary objectives of this research are to explore the differences in these three urban agriculture typologies (community, school, and home gardens) and to comprehend how growing produce in these gardens could bring changes in the dietary habits and health of individuals and households and improve social connections.

Methodology

This research examined the health benefits and social exchange possibilities of three common garden types and the real-world beneficial effects of these gardens on students, households, and individuals, when these spaces are cultivated for growing fruit and vegetables. This research connected multiple disciplines and recognised that the spatial locations of cultivated local productive spaces locally, and globally, under different settings, could have similar or dissimilar performances and outcomes. I have followed three key steps in conducting this research. First, I conducted a literature review for each of the three urban agriculture typologies to select papers for each garden type, and selected seven identified academic articles from each urban agriculture type for comparison (Tables 3.2–3.4). In respective sections and throughout the chapter, the remaining 34 selected articles were reviewed and analysed. Second, I selected three case studies of exemplar projects involving each of these typologies and undertook a close reading of each to identify and analyse the outcomes across health and social domains. Third, I systematically organised and synthesised both the outcomes from the broad literature review and the in-depth case studies, while identifying the similarities and differences in health and social benefits associated with the three types of gardens. Figure 3.1 presents a research methodology flow diagram.

The literature review investigated the therapeutic benefits of the selected local food gardens, such as health improvements, disease protection and prevention, and social exchange possibilities. I searched the literature using various relevant keywords for research papers in Scopus, Science Direct, Google Scholar databases, internet sources, and other reports. For example, keywords included 'nutrition intervention', 'health and wellbeing', 'physical activity', 'emotional wellbeing', 'mental health benefits', 'school gardening programme',

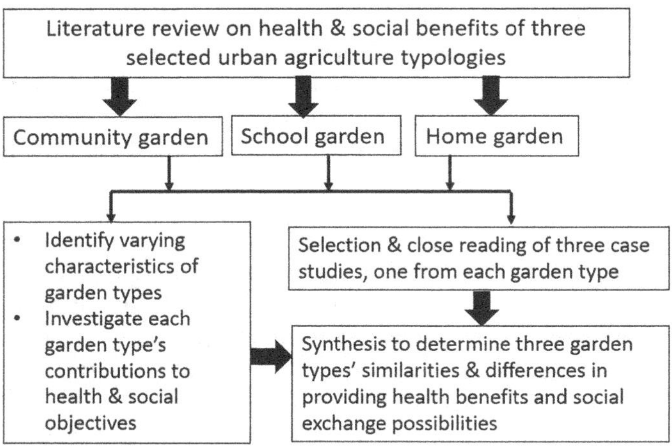

Figure 3.1 Research methodology flow diagram.

'improved diet', 'dietary intake of fruits and vegetables', and 'fruits and vegetable consumption'. I also considered different age groups for community, school, and home gardens. Search parameters limited results to studies and projects conducted between 2006 and 2022.

The objectives of this research were to comprehend the extent to which local foodscapes in community, school, and home gardens work to improve individual and household health and social exchanges, as well as the differences and similarities between these three garden types. This research builds on my previous research on urban agriculture. Relevant academic papers for a timeline from 2006 to 2022 were included through a rapid literature review process to provide a quick synthesis in a timely and resource-efficient manner (Moons, Goossens and Thompson 2021). The selected resources are in line with the foci of the objectives of this chapter.

The inclusion criteria for the literature review were studies (involving the three garden types of community, school, and home gardens) which explored the effectiveness of garden-based nutrition intervention programmes on physical health through increased fruit and vegetable intake and dietary changes of the gardeners and households, implications of the process of local food growing in the three garden types on improved mental wellbeing (e.g. stress relief, socially competent behaviour, and horticulture therapy) of different age groups and physical activities (e.g. Body Mass Index (BMI)) and role of the gardens, and operational issues faced during the COVID-19 pandemic. A total of 64 relevant resources published between 2006 and 2022 were selected across the three types of gardens. Of the 64 resources, 55 were from the academic literature. The remaining nine are grey literature, consisting of research reports, survey reports, local strategies, websites, and others. Of the 55 academic papers selected, 19 papers focused on community gardens, 15 on school gardens, and 21 on home gardens, from which seven papers were selected for each garden type (Table 3.1). A total of 21 papers covering the three garden types were compared, considering health and social objectives, research methods, and research outcomes in three tables (Tables 3.2–3.4). The remaining 34 academic papers and nine grey literature resources were reviewed and analysed. Three research studies were selected and analysed as case studies, following separate selection criteria. The results are presented in this chapter's review and case studies sections.

The case studies were selected to represent the three types of productive landscape sites and to interpret and analyse in-depth real-world situations. Selection was based on the varied geographical location of each garden; the diverse socio-economic background of participants in each case study; and the uniqueness of each case study in providing health benefits and social exchange opportunities. Case Study 1 examines the Coogee Community Garden in Sydney, where my colleagues and I conducted semi-structured interviews with the members of the Garden in 2015 (Ghosh, Wilkinson, and Adler 2015; Randwick City Council 2016). Within the local government area serviced by the Randwick City Council, this small community garden is located in Dolphin Street,

Coogee, an affluent suburb of Sydney with a pattern of high-density urban development. Case Study 2 examines a school garden in Sindhupalchok District in Nepal, where Schreinemacher and colleagues (2020) conducted a one-year cluster randomised controlled trial involving 779 school children aged 8 to 12 years in an area affected by malnutrition. Using a comprehensive intervention design (treatment group) and no intervention (control group), Schreinemacher, Baliki, Shrestha et.al. (2020) explored improvements in dietary intake. The home and school gardens of the school children were included as a connected food system. Case Study 3 examines home gardens in Santa Clara County in California (Palar et al. 2019), where Valley Verde (a non-profit organisation) and researchers from the University of California implemented a home garden programme for low-income households at risk for cardiovascular disease.

Through the review of research and analysis of case studies in this chapter, I explore the evidence base which confirms the capability of the three garden types (community, school, and home gardens) to contribute to health and social dimensions. I discuss the findings in this chapter by bringing together outcomes to determine similarities and differences and future research directions.

A review of urban agriculture in community, school, and home gardens

For three urban garden types, 64 academic and grey literature resources were selected for the literature review following selection criteria as detailed in the methodology section. Table 3.1 presents the 55 selected literature used in this research for three urban agriculture types out of the 64 resources. Grey literature, including nine sources, mainly research reports on household surveys, community food gardening, local governments' community garden strategies, and information on the health and wellbeing benefits of urban agriculture from relevant universities and a local council website and the other sources, were valuable for this research.

Community gardens

According to data from the Australian Bureau of Statistics (ABS), approximately 8.4 million Australian households live in flats, apartments, or units (Wise 2014). Community gardens are essential spaces for these households for social engagement, making friends, and growing food for better health (Ghosh et al. 2016). The City of Sydney's Community Garden Policy provides a comprehensive definition of a community garden.

> Community gardens are unique forms of public open space which are managed by the community primarily for the production of food and to contribute to the development of a sustainable urban environment. They are places for learning and sharing about sustainable living practices, and for actively building community through shared activities.
>
> (City of Sydney 2016)

Table 3.1 Selected literature resources on three urban agriculture typologies

Garden type	*Selected literature resources*	*Total no. of resources*
Community gardens	Joshi and Wende (2022); Lee, Jang, Yun et al. (2022); Buechler and Mansaray (2021); Mead et al. (2021); Schoen et al. (2021); Koay and Dillion (2020); McVey et al. (2018); Lewis, Home, and Kizos (2018); Litt, Alaimo, Buchenau et al. (2018); Soga et al. (2017); Wilkinson and Orr (2017); Ghosh, Accarigi, and Giovanangeli (2016); Ghosh et al. (2015); Firth, Maye, and Pearson (2011); Blaine, Grewal, Dawes and Snider (2010); Kingsley, Townsend, and Henderson-Wilson (2009); Alaimo, Packnett, Miles and Kruger (2008); Wakefield, Yeudall, Taron et al. (2007); Thompson et al. (2007).	19
School gardens	Schultz and Rosen (2022); Gavia, Cabingas, Rodriguez and Pallo (2021); Davis, Pérez, Asigbee et al. (2021); Pollin and Retzlaff-Furst (2021); Austin (2020); Schreinemachers et al. (2020); Kim and Park (2020); Fischer, Brinkmeyer, Karle et al. (2019); Leuven Rutenfrans, Dolfing and Leuven (2018); Wells et al. (2018); Huys, Cocker, Cocker et al. (2017); Utter, Denny, and Dyson (2016); Barry (2016); Heim, Stang, and Ireland (2009); McAleese and Rankin (2007).	15
Home gardens	Depenbusch, Schreinemachers, Brown and Roothaert (2022); Cerada et al. (2022); Corley et al. (2021); Chalmin-Pui et al. (2021); Schreinemachers et al. (2020); Gerny, Marsh, and Mwebembezi (2021); Home and Vieli (2020); Baliki et al. (2019); Rammohan, Pritchard and Dibley (2019); Palar et al. (2019); Kirkpatrick and Davison (2018); Richardson, Pearce, Shortt and Mitchell (2017); CoDyre, Fraser, and Landman (2015); Cabalda, Rayco-Solon, Solon and Solon (2011); Ghosh (2010); Blake and Cloutier-Fisher (2009); Tontisirin and Bhattacharjee (2008); Head and Muir (2007); Bhattacharjee, Saha, and Nandi (2007); Graham and Connell (2006); Gaynor (2006).	21
Total		**55**

Community gardens can be categorised as 'place-based' or 'interest-based' (Firth et al. 2011). In 'place-based' typologies, 'territoriality' and a local community's rights are important (Firth et al. 2011) and can be located in various places, such as neighbourhoods, public housing estates, refugee centres, schools, universities, and hospitals. 'Interest-based' community gardens can focus on diverse and wider communities, and social capital generated through food gardening in the community gardens remains within the specific interested community (Firth et al. 2011). Table 3.2 summarises seven selected research studies on community gardens' contributions to improving health and social connections.

Community gardens are valuable for promoting better health and wellbeing, and for better access to food, education, and social networking (Wilkinson and Orr, 2017; Ghosh et al. 2016; Wise 2014; Thompson et al. 2007), as well as for initiating a successful place-making or urban regeneration process. These gardens provide prospects for engaging in physical activities, improving nutrition, and reducing stress; participating in gardening improves overall wellbeing (Mead et al. 2021; Soga et al. 2017). In the United States, community gardeners in Denver, Colorado, consumed a higher average number of servings of fruit and vegetables daily (5.7 servings) than non-gardeners, who consumed only

Table 3.2 Summary of selected research studies: Contributions of community gardens

References and community garden location	*Health and social objectives*	*Research methods*	*Research outcomes*
Koay and Dillion (2020) Singapore	• To determine the relationship between community gardening and mental health and potential.	• A cross-sectional online participant survey. • Used Perceived Stress Scale, Personal Wellbeing Index, statistical data analysis.	• Clinical validation of positive contributions of community gardens to subjective wellbeing, higher stress resilience, and higher self-esteem.
McVey et al. (2018) Edinburgh, Scotland, UK	• To comprehend participants' viewpoints in three types of community gardens. • To examine the influences of food growing on recreation and leisure choices and children's and personal developments.	• Applied semi-structured interviews and participant observations in case studies. • 38 participants interviewed.	• Gardening creates engagement and cohesion, influences better mental health and cultural diversity promotion. • Reported physical challenges of gardening.

(Continued)

Table 3.2 (Continued)

References and community garden location	*Health and social objectives*	*Research methods*	*Research outcomes*
Lewis et al. (2018) Lausanne, Switzerland	• To determine reasons for people's involvement in gardening and choosing environmentally friendly gardening practices.	• 23 semi-structured gardener interviews.	• People garden for wellbeing, social aspects, tangible (food) and intangible (feeling of producing food) experiences.
Wilkinson and Orr (2017) Kings Cross, Sydney, Australia.	• To understand the perceived and reported mental health benefits of a Horticulture Therapy (HT) programme for hospital inpatient participants.	• A literature review and semi-structured focus group interviews with the HT programme GROW participants.	• Reported emotional, mental, and physical health wellbeing benefits. • Garden was a relaxing and socialising place; developed interpersonal links and gardening skills.
Blaine, et al. (2010) Cleveland, Ohio, USA.	• To investigate associations of dietary changes with community gardening.	• Questionnaire survey (124 surveys received, with 32% response rate).	• 71% supported the changes in diet; 29% accepted no dietary changes with gardening.
Kingsley et al. (2009) Port Melbourne, Australia	• To investigate perceptions of 'Dig In' community garden members.	Ten semi-structured garden member interviews.	• Better nutritious and organic food. • Delivered health, wellbeing, physical fitness and a sense of achievement and joy. • Garden – a place to connect to nature and spirituality. • Social connectedness, appreciation for place qualities and a supportive environment.

(Continued)

Table 3.2 (Continued)

References and community garden location	*Health and social objectives*	*Research methods*	*Research outcomes*
Wakefield et al. (2007) Toronto, Canada.	• To examine perceived health impacts of community gardening in 15 gardens.	• Participant observation. • Participation of 55 people in the focus group. • 13 in-depth interviews.	• Improved nutrition, food access, mental health, and increased physical activity. • Promoted social health and community cohesion.

3.9 servings (University of Colorado Boulder 2011). Alaimo et al. (2008) surveyed 766 adults and concluded that gardeners who participated in community gardens were 3.5 times more likely to consume fruit and vegetables daily than those who did not participate in community gardens. Blaine et al.'s (2010) survey of gardeners involved in a long-term food gardening project (over one year) in the community garden in Cuyahoga County Extension, in Cleveland, Ohio, demonstrated improvements in their dietary habits. In Australia, Wise (2014) conducted research to comprehend the gardeners who are currently growing their own food in home and community gardens, their motivations and barriers, and the potential social and health benefits associated with these food gardening practices. This study on residential and community food gardening found that 61% of food gardening households had children aged 11 years or younger, which established that the presence of children in the households is an important demographic characteristic for engaging in food gardening (Wise 2014). Over 50% of gardeners in this study consumed 2–3 servings of fruit and vegetables daily; 33% consumed four or more daily servings of both fruit and vegetables; 71% believed in their dietary changes; 44% thought that one year of gardening had created significant dietary changes (Wise 2014). A Korean study on appropriate intervention methods and programmes for an apartment community garden focusing on a sense of community established that this community garden can provide mental health benefits of reduced stress levels (Lee et al. 2022). During the COVID-19 pandemic, it was difficult to operate community gardens as usual (Schoen et al. 2021) and necessary changes were implemented in some gardens to enable participation (Buechler and Mansaray, 2021).

Located on the roof of the Refugee Centre at Kings Cross, Sydney, the St Canice Rooftop Garden grows vegetables and herbs in an area of approximately 55 m^2 on a high-density inner-city site (Figure 3.2). This garden is an initiative to provide horticulture therapy, health, wellbeing, and social connectedness to patients and surrounding communities (Ghosh et al. 2016). It is a shared community meeting place for getting together, making friends, and growing food. Participants can take produce home from the garden.

Figure 3.2 St Canice Rooftop Garden, Kings Cross, Sydney. Photo by Sumita Ghosh.

The flowers, plants, and activities in the garden provoke participants' associations with joyful memories, initiate learning new skills, and create a tranquil environment for healing (Kaziro 2015).

Wilkinson and colleagues (2017) conducted semi-structured focus group interviews with four participants in the GROW Project, a Horticultural Therapy (HT) Programme at the St Canice Rooftop Garden in 2016. The HT programme was led by a trained horticultural therapist and a professional gardener and comprised weekly, two-hour gardening sessions for eight consecutive weeks. All four participants had previously received treatment at an inpatient mental health service at a nearby hospital; the HT programme supported their recovery through engaging with food gardening. Participants reported that the HT programme developed their sensory awareness, provided physical health benefits through weekly activities, planting and potting, lifting and hauling materials, and garden bed preparation, and contributed to their mental health and wellbeing (Wilkinson and Orr, 2017).

Community gardens represent one neighbourhood-based strategy through which we can examine the ways in which behaviourally based interventions impact health (Litt et al. 2018, p. 10).

School gardens

School gardens are functional, productive spaces that can be used as a classroom to reconnect students with nature. They can also serve as a platform for developing gardening skills, imparting knowledge about nutrition, and teaching values of local food production and social responsibility in various creative ways to support children's development (Schultz and Rosen, 2022; Huys et al. 2017; Austin 2020). The learnings from school gardens can connect to other subjects taught at school (Green Heart Education 2022) and improve school children's eating behaviour (Henderson 2021). Table 3.3 summarises the selected research studies on the contributions of school gardens in enhancing health and social qualities.

Table 3.3 Summary of selected research studies: Contributions of school gardens

References and location	*Health and social objectives*	*Research methods*	*Research outcomes*
Schultz and Rosen (2022) Chicago, USA	• To evaluate the social and emotional wellbeing of primary, middle, and high-school students in school garden-based education programmes.	• Two surveys (3rd–8th grades) before and after the education programme for fruit and vegetable consumption. • Third Hello Insight online platform survey (9th–12th grades) assessed five socio-emotional learning capacities. • Surveyed 15,460 students.	• Improved access to fruit and vegetables. • Developed food appreciation by cultivating, harvesting, and cooking school garden produce. • Increased positive experience, plus social and emotional wellbeing.
Pollin and Retzlaff-Furst (2021) Germany	• To investigate development and potential of students' cognitive and emotional-affective abilities in a garden as a social learning site.	• Observations and emotion diaries in school gardens and classes and comparisons. • 31 girls and 22 boys aged 11–12 years self-reported. • 150 observations.	• Gardens provided positive emotions. • More socially and emotionally competent student behaviour in school garden lessons than in classes.
Davis et al. (2021) Texas, USA	• To examine annual impacts of school-based nutrition, and intervention gardening programme on students' obesity, dietary intake, and blood pressure.	• School-based cluster-randomised controlled trial (RCT). • 16 elementary schools with 3,135 children participated. • Intervention: culture-specific vegetables growing in the school garden.	• Did not reduce obesity or blood pressure. A longer timeframe is essential. • Significantly increased children's vegetable intake. • Guided children's eating behaviour change.

(Continued)

Table 3.3 (Continued)

References and location	*Health and social objectives*	*Research methods*	*Research outcomes*
Kim and Park (2020) Seoul, South Korea	• To validate improvement in children's vegetable eating practices by school garden-based intervention programme.	• Questionnaire survey. • A 12-week school garden-based programme with 202 elementary school students (6–11 years).	• Significantly increased children's vegetable consumption, dietary self-efficacy, gardening, and nutrition knowledge; improved eating behaviours.
Leuven et al. (2018) Nijmegen, Netherlands	• To investigate the long-term effects of school garden interventions on students' vegetable intake, preferences, and knowledge.	• Intervention programme (17 lessons over 7 months for students 10–12 years). • Students survey before and after the intervention. • Analysis of students' self-reported vegetable preferences, change in the capabilities to recognise vegetables, and attitudes.	• Increased students' knowledge of vegetables and preferences in the short term. • In the long term, increased students' willingness to taste and daily intake; intervention did not affect attitudes towards gardening.
Huys et al. (2017) Ghent, Belgium	• To explore school gardening implementation practices, and key members' and children's perceptions of a school garden.	• In four primary schools with 38 children (10–13 years). • Five focus groups and 12 interviews with key members.	• Reported students' positive perceptions and increased interests in school garden vegetables. • Importance of parents' involvement. • Shared needs for better facilitation, garden maintenance, and curriculum integration.

(Continued)

Table 3.3 (Continued)

References and location	*Health and social objectives*	*Research methods*	*Research outcomes*
Utter, Denny and Dyson (2016) New Zealand	• To assess the influences of school gardens on secondary students' eating behaviours, physical activity patterns, and Body Mass Index (BMI).	• Analysis of 2012 national study on youth survey for health and wellbeing. • 8,500 selected students from 91 randomly selected secondary schools (55% of schools had vegetable/fruit gardens).	• A lower BMI for students was associated with school gardens. • No associations between with school gardens and increased fruit or vegetable intake or physical activity by students. • The greatest benefit was experienced by students from poor households.

In Australia, Stephanie Alexander's Kitchen Garden Foundation applied the Kitchen Garden Program Model to engage students in primary schools: students attend garden classes regularly in a school garden, and learning is embedded within the curriculum (Barry 2016). The joint participation of educators and students provides successful learning, leading to improved healthy eating diets and behaviours in the students. Initially, 267 schools which implemented the Kitchen Garden Program have continued to run this programme; a new flexible programme model was implemented in another 550 schools nationwide in Australia (Barry 2016). In the United States, McAlesse and colleagues (2007) surveyed 122 (sixth grade) students in three similar elementary schools in Idaho who participated in garden-based nutrition education. The combined results showed that students' consumption of fruit and vegetables more than doubled, from an average of 1.93 to 4.50 servings per day.

In Germany, Pollin and colleagues (2021) conducted 150 observations to investigate school students' (aged 11–12 years) emotions in class and in the school garden and socially competent behaviour (including direct verbal communication and cooperation in the class). Participants included 31 girls and 22 boys, and self-reported on the school garden and classroom using 'emotional diaries' completed after each lesson in class and in the garden on the subject of biology. The research design covered three basic psychological needs: competence, autonomy, and relatedness. Results showed that positive impacts constituted 85% of the total high-intensity emotions consisting of 'happiness', 'pride', and 'surprise/wonder' in the school garden. Four categories of justification (garden work, plants, gardening, and animals) were the same for emotions of 'happiness', 'pride', and 'surprise/wonder' in the school garden; only 'classmates' was an additional category for 'surprise/wonder'. These emotions were 43% higher in the school garden than in the classroom (Pollin and Retzlaff-Furst, 2021). In self-reported emotion diaries, school students as participants contributed informally to their perceptions of emotions associated with the school garden and in the classroom. According to the emotional diaries, the highest values of school 'garden work' were related to 42% of positive emotions of 'happiness', and 48% of perceptions of 'pride'. The highest value of 'surprise/wonder' emotion was 59% for the 'plants' in the school garden. The school garden as a learning and experience space provided opportunities for social interactions and positive emotions in students which help to promote social and emotional competence (Pollin and Retzlaff-Furst, 2021).

A school garden can also positively benefit the health and wellbeing of teachers. Gavia et al. (2021) investigated the effectiveness of hands-on involvement in school-based gardening as an intervention on school teachers' wellbeing during the COVID-19 pandemic. They interviewed 14 teachers (response rate 50%) and conducted a focus group discussion. The teacher-participants supported the idea that school gardening activities increased their social interaction opportunities, communications and motivations, engagement, reduced stress, and improved overall health and wellbeing (Gavia et al. 2021).

School garden-based nutrition education and interventions can positively change students' dietary intake of fruit and vegetables. Garden-based nutritional education programmes, including gardening activities twice per week, fruit and vegetable tasting, and preparing recipes as a behaviour change intervention, can improve fruit and vegetable intake in children (Heim, Stang and Ireland 2009). Wells et al. (2018) investigated the flow-on effects of elementary school garden interventions on the availability of fruit and vegetables in children's homes in Arkansas, Iowa, New York, and Washington State in the United States. Fruits and vegetables (FV) availability at home was measured using a modified version of the GEMS Fruit, Juice and Vegetable (FJV) Availability Questionnaire survey in 24 intervention groups with school gardens and 22 control groups with no school gardens; 2,768 children participated in this study. Outcomes showed that schools with gardens overall increased the availability of fruits, vegetables, and low-fat vegetables in the homes of children (Wells et al. 2018).

School gardens are critically important for the environmental education of school students. Fischer et al. (2019, p. 35) argue for a concept of 'biodiverse edible schools', where students connect with nature, learn, and grow food in school gardens. Using a school case study from Berlin, this German study established an implementation pathway and components for a biodiverse edible school. These components include a school food growing garden, a school kitchen for cooking the produce, wild urban habitats for edible plants located near the school, and collaborative partnerships with various organisations (Fischer et al. 2019).

Utter et al. (2016) investigated 8,500 randomly selected students and 91 randomly selected secondary schools across New Zealand. They found that 55% of all schools with a fruit and/or vegetable garden where students could participate in food gardening were associated with lower BMI and improved health. In the United States, the one-year 'Texas Sprouts' project examined the health outcomes, integrated nutrition, and cooking programme intervention for school gardens at 16 elementary schools in the State of Texas (Davis et al. 2021). Schools were randomly assigned to the 'Texas Sprouts' intervention (n = 8 schools) or to a control (delayed intervention, n = 8 schools). Students' dietary intake, obesity outcomes, and blood pressure were monitored by measuring the effects of the gardening intervention on each student's vegetable intake (serving/day), BMI, and waist circumference. This case study provides information on which health benefits can and cannot be achieved in a year-long school gardening programme (Davis et al. 2021). Many studies similarly support school-based nutritional improvement through intervention programmes that offer access to multiple sources of fruit and vegetables, such as home gardens and school gardens, and which can significantly increase fruit and vegetable intake among students (Kim and Park 2020).

Home gardens

A home (or domestic) garden can be defined as a space other than a dwelling in a residential parcel or subdivision that is privately owned (Ghosh 2010). Food production in home gardens depends on household participation and the motivations of individuals and household members, household incomes, and cultural backgrounds. Home gardens differ from community and school gardens in their ownership pattern, social interaction opportunities, and household capabilities to adapt and sustain food gardening (Ghosh 2010). Food production in home gardens is a complex process involving economic aspects, the availability of productive spaces, and resources (Gaynor 2006). Table 3.4 summarises research studies on the contributions of home gardens to improving health and social participation.

A survey conducted by the National Gardening Association (2009) indicates that, from a total of 33 million households surveyed in the United States, approximately 91% had food gardens at home; 5% shared gardens with friends, neighbours, and family; and 3% had gardening spaces in community gardens. A study by Graham and Connell (2006) showed that the production of traditional fruits, vegetables, herbs, and flowers in the home gardens of Vietnamese, Greek, and Italian migrants to Australia was linked to specific cultural identities, memories, and traditional practices. These cultural identities were dynamic and continuously evolving within a multicultural landscape in Australia to generate new preferences and opportunities for culturally appropriate food gardening in home and community gardens. These new identities were reflected in second-generation food production practices in the home gardens (Graham and Connell 2006). In house-lot gardens in Santarém, Pará, Brazil, exchanges of home garden produce fostered social networks and recreated urban–rural linkages (Winklerprins 2002). The 'backyard sharing' concept was well supported by home gardeners (as seen in Coles 2019; CBC News 2022) as it

Table 3.4 Summary of selected research studies: Contributions of home gardens

References and location	*Health and social objectives*	*Research methods*	*Research outcomes*
Depenbusch et al. (2022) Cambodia	• To evaluate the distributional effects of an intervention combining gardening training and nutrition.	• A cluster randomised controlled trial before and after the intervention. • 500 rural households with children under 5 and women 16–49 years of age participated.	• Increase in the amount of vegetables harvested. • Production period extended by five months on average. • 61 gram/pp/day increase in vegetable consumption. • Increased micronutrients supply for families.
Corley et al. (2021) Scotland, UK	• To investigate older adults' mental and physical wellbeing during COVID-19 in home gardens.	• Survey to measure self-rated physical, emotional, and mental health; COVID-19 anxiety and sleep quality. • 171 individuals (mean age = 84 years) self-reported.	• Respondents: 70% gardeners, 30% non-gardeners. • Older adults spent more time in the gardens during COVID-19. • Significantly better physical, emotional, and mental health and wellbeing reported.
Chalmin-Pui et al. (2021) Seoul, South Korea	• To determine the effects of six home gardening activities on cognitive function and brain nerve growth factors in older adults.	• A 20-minute gardening practice as a physical activity intervention. • 41 participants (mean age = 77 years); blood sample analysis.	• Low-to-moderate intensity gardening activities improved memory and cognitive functions.
Home and Vieli (2020) Switzerland and Chile	• To evaluate and compare different gardeners' motivations for spending gardening time are contextual or have values beyond.	• Questionnaire survey with 409 Swiss and 169 Chilean gardeners. • 14 items in 'motivations for gardening' scale evaluated the gardeners from diverse cultural contexts.	• Significant differences between Swiss and Chilean participants at an individual scale. • Three key motivational reasons: restoration from stress; socialisation; and food production.

(Continued)

Table 3.4 (Continued)

References and location	*Health and social objectives*	*Research methods*	*Research outcomes*
Baliki et al. (2019) Four districts of Bangladesh	• To investigate the long-term (three years) behavioural impact of a vegetable production intervention programme.	• Compared three rounds of survey data from 224 control and 395 intervention households.	• Intervention encouraged and maintained annually 43 kg varieties of vegetable growing per household. • Changing vegetables' composition had a small impact on micro-nutrient supplies.
Kirkpatrick and Davison (2018) Hobart, Tasmania, Australia	• To explore variations in vegetable gardening practices, types, motivations, self-identification, and relationships between these aspects.	• A questionnaire survey on societal and personal benefits. • 101 home gardeners (65% > 50 years) in 15 suburbs in Hobart with varied socio-economic statuses, locations from CBD and physical geographies.	• Identified five gardener groups: activists, spenders, health-providers, stewards, and economisers. • Four groups assigned a high value to vegetable gardening; 86% agreed that vegetable gardening improved gardeners' health. • 'Activists' as a distinct group, supported spiritual wellbeing, environment, and positive social change benefits of home gardens.
Cabalda et al. (2011) Rizal, Philippines	• To determine dietary diversity in preschool-aged children (2–5 years) in households with or without home gardens.	• Using questionnaire survey format from US Department of Agriculture Food Security. • Dietary Diversity Score on consumption of unique food groups in the previous 24 hours.	• 52.5% of children came from households that had home gardens growing fruit and/or vegetables. • They had higher dietary diversity scores compared to the children without home gardens.

promoted better health and established critical social connections through urban gardening (Blake and Cloutier-Fisher, 2009; Yaworski 2015). A survey of 265 home gardens in Australia found three main gardener types: committed native (growing 81–100% native plants); general native (growing 41–60% native plants); and non-native gardeners (growing less than 20% native plants) (Head and Muir, 2007).

Chalmin-Pui et al. (2021) explored residents' perceptions around which aspects of food gardening activities could promote health benefits, as well as reasons for engaging in these activities. A total of 5,766 gardeners and 249 non-gardeners in the United Kingdom were surveyed using a questionnaire. Data was collected on garden types, frequency of gardening, and individual observations of health and wellbeing. The participants recognised significant relationships between gardening and health improvements, in wellbeing, perceived stress reduction, and physical activity enhancement. Frequent gardening (at least 2–3 times a week) was perceived as providing the greatest health benefits. Stress reduction was an important motivating factor for growing food and for social engagement through gardening (Home and Vieli, 2020). During COVID-19, home gardens significantly supported households' physical and mental wellbeing, as well as local food production (Cerada et al. 2022).

Research on the health benefits of home gardens establishes that gardening activities can contribute to specific health issues for different age groups. Regular gardeners are more physically active than gardeners who garden less regularly. Chalmin-Pui et al. (2021) found that wellbeing scores, as measured by the Shortened Warwick and Edinburgh Mental Well-Being Scale (SWEMWBS), were 1.84 times higher; stress scores, as measured by the Perceived Stress Scale (PSS), were 1.68 times lower; and physically active scores were higher by 1.42 days per week for regular gardeners than for those who did not garden at all.

Richardson et al.' (2017) study established that good access to home or domestic gardens can reduce social, emotional, and behavioural difficulties in children aged between four and six years. Amongst senior adults, six gardening activities – cleaning a garden plot, digging, fertilising, raking, planting/transplanting, and watering – benefited cognitive functions such as memory. This has also been demonstrated through medical evidence (Park et al. 2019). Cabalda et al.'s (2011) study in an urban setting showed that home gardening was positively associated with dietary diversity, reduced micronutrient deficiencies, and an increase in the frequency of fruit and vegetable consumption amongst children from households with home gardens. Evidence shows that children who have home gardens eat vitamin-A-rich vegetables and fruits more frequently. Encouraging households in nutritional programmes to establish a home garden can improve the health and nutrition of children and adult members of the household (Cabalda et al. 2011). In Uganda, a questionnaire and 24-hour recall nutrition surveys of 50 households participating in various nutritional programmes showed increased consumption of leafy green vegetables with micronutrients, vitamin A, and iron (Gerny et al. 2021). Using Canada's Food Guide, it was found that the most productive gardens could supply 74.2

servings per square metre from the most fertile land, and 1.08 servings per square metre from the least productive gardens (CoDyre, Fraser and Landman 2015). Rammohan et al.'s (2019) survey of 3,230 rural households in Myanmar (Burma) established that food security and dietary diversity were positively associated with home gardens.

Community-based nutrition programmes can build capacities and empower people in low-income communities to improve their nutrition (Tontisirin and Bhattacharjee, 2008). In the Government of Bangladesh's Integrated Horticulture and Nutrition Development Project (IHNDP), households with home gardens consumed more leafy green vegetables, yellow and orange vegetables, and vitamin-C-rich fruits (Bhattacharjee et al. 2007). Women had significantly higher iron, vitamin A, and vitamin C intake. Calcium and vitamin A intake were considerably increased in adolescent girls, while children had a higher intake of vitamin-A-rich fruits and vegetables (Bhattacharjee et al. 2007). Households with home gardens consumed 46–64% of the vegetables and fruits grown, sold 9–20% of the produce, and exchanged produce with their friends and families (Bhattacharjee et al. 2007). Food production in the home garden was associated with better nutritional health (Kirkpatrick and Davison, 2018) and affordable food supply.

Urban agriculture case studies of community, school, and home gardens

Case Study 1: Coogee Community Garden, Sydney, Australia

The small community garden in Dolphin Street, Coogee, within the Randwick City Council area, in Sydney in Australia, is of considerable interest. From the initial set-up process in 2013, it has grown into a functioning community garden in the neighbourhood, and the resident community participates regularly. The garden is located on a brownfield site, thus reclaiming land for urban food through an important urban renewal process in a higher-density setting, where the predominant built form is residential apartments. With colleagues, I conducted seven semi-structured interviews with the members of the Coogee Community Garden in 2015 to understand social participation and community engagement in the garden. The questions in five key areas explored participants' expectations, motivations, influences, satisfaction with different types of urban food growing activities, management, funding and time commitment, self-reported existing gardening skills, and essential training needs. These interviews provided a snapshot of people's preferences linked to the health and social benefits a community garden can provide in dense residential environments.

Most participants had a significant interest in growing food in the Coogee Community Garden due to considerable space constraints, poor quality soil, and lack of solar access for growing food in the apartment blocks where they lived. The key motivations for food gardening in the Coogee Community Garden included harvesting fresh, nutritious, and locally grown organic food, meeting people informally, making friends, and getting opportunities for physical activities through gardening, and learning new gardening methods. Participants defined better food as 'fresher, organic, more variety, and seasonal' (interview data, Ghosh et al. 2015), and noted that it contributed to better health. Gardeners preferred leafy green vegetables, artichokes, tomatoes, eggplants (aubergines), capsicums (bell peppers), potatoes, zucchini (courgettes), carrots, root vegetables, beetroots, and chard. Preferred fruits were lemons, strawberries, pomegranates, and figs; preferred herbs included basil, parsley, and coriander. Participants reported that implementing organic gardening practices was important for growing food to achieve better health. Participants expressed experiences of immense joy and happiness associated with growing food and becoming a part of a community (Ghosh et al. 2015). For example:

> [E]xpectations would be definitely [be] developing knowledge on how to actually grow things … and actually take something tangible, physical from the garden. Also creating social contact with the people that live in the area.
>
> (Interview data, Ghosh et al. 2015)

These interviews demonstrate that a community garden in this high-density environment dominated by apartments is a social-change catalyst to foster social connections, better mental health, and opportunities for growing healthy food, and can contribute to creating a healthy community.

Case Study 2: School gardening integrated with home gardening in Nepal

Nepal's Multi-sector Nutrition Plan supports school gardens as they comply with the need for health, agriculture, and education to reduce malnutrition (Government of Nepal, as quoted in Schreinemachers et al. 2020). A study of school gardens in Sindhupalchok District of Nepal, where 25% of the population lives below the national poverty line, employed a one-year cluster randomised controlled trial of 779 school children aged 8–12 years (mean age = 10 years; 55% girls). The study included a comprehensive intervention design for the treatment group with school gardens and complementary home gardens, and a control group with no intervention programme (Schreinemachers et al. 2020).

The school garden intervention included practical training for students in vegetable growing and nutritional education with 23 weekly learning modules delivered in onsite school gardens which ranged in size from 32 m^2 to 240 m^2 (average size = 90 m^2). The students' caregivers were included in the trial, and support was provided to improve their home gardens, including garden-based food growing training and nutrition training to improve household health. Nine varieties of local winter vegetables and ten types of vegetable seeds for the summer season were distributed for planting in school and home gardens (Schreinemachers et al. 2020).

Evidence shows that a comprehensive intervention programme for the treatment group provided good nutrition and increased children's vegetable intake (as a proportion of meals); 15–26% of children in the treatment group had a stronger liking of vegetables than in the control group (Schreinemachers et al. 2020). Moreover, the flow-on effect of the intervention was a 15.4% increase in food production in the home gardens for households whose children were in the treatment group (Schreinemachers et al. 2020).

> [S]chool garden interventions need to be designed in such way that they do not only stimulate children's knowledge of and preferences for vegetables, but also increase children's access to vegetables at home as well as stimulate parents to prepare and eat more vegetables.
>
> (Schreinemachers et al. 2020, p. 7)

This case study is unique because it approaches school and home gardens as connected food production spaces. This system provides a continuation of applications of knowledge and skills learned by students and caregivers in these two types of gardens. It also aims to improve health and change behaviours towards better food practices for children and households over the longer term.

Case Study 3: Home Garden Program, Santa Clara County, California

In 2019, Valley Verde (a non-profit organisation) and researchers from the University of California collaborated to implement a community-based home gardening programme in Santa Clara County (Palar et al. 2019). The primary purpose of this research was to investigate the health benefits of urban gardening with a nutritional education programme for a low-income population at high risk of cardiovascular disease. Social inequalities can negatively affect cardiovascular health. Home gardening can regenerate an environment where fresh and nutritious food is more easily accessible, thus reducing stress and providing physical activity opportunities through gardening and behaviour change.

Households participating in the programme were provided with the necessary supplies, raised beds (4 ft × 4 ft or 4 ft × 8 ft, based on available space) and materials (organic soil, compost, seedlings, etc.). The seedlings were consistent with participants' culture-specific vegetable choices: bell peppers (capsicums), spinach, tomatoes, cucumbers, carrots, cilantro (coriander), jalapeños, cabbages, radishes, cauliflowers, onions, and garlic. Participating households were required to maintain home gardens at their residences. In-depth interviews were conducted with 32 participants who were involved in the programme for a year at the household level, with an average household size of 5–6 people (Palar et al. 2019). Three main areas were explored: perceptions of improved diet and nutrition; improved physical activity; and stress reduction. Most participants reported that the programme increased their fruit and vegetable intake. They also reported that changes in diets which were enhanced by

> affordability, accessibility, freshness, flavor, and convenience of the garden produce, as well as self-efficacy for improving their health by eating garden produce.
>
> (Palar et al. 2019, p. 1040)

Case Study 3 demonstrates that produce grown in home gardens provides affordable, better-quality fresh food, as well as the convenience of bringing food directly from the garden to the plate. These two aspects are the main drivers for increasing household vegetable consumption and reducing unhealthy food intake. Participants spent considerable time in the garden, and the vegetable growing process motivated them to switch to a healthy diet with a higher vegetable intake. Some participants reported weight loss; diabetic participants maintained their diabetic diets more effectively. Regular gardening increased exercise and reduced sedentary behaviours in both adults and children. Some participants reported that they 'perceived the act of gardening to be a form of therapy for coping with stressful life situations' (Palar et al. 2019, p. 1042).

The Home Garden Program in Santa Clara County demonstrates multiple positive health impacts of home gardening. It establishes a functional urban home gardening model that integrates home gardening with gardening education and culturally appropriate nutrition. This model can significantly encourage and improve healthy diets and eating behaviours in adults and children and could provide effective disease prevention and protection and help in the management of chronic diseases.

Discussion

This chapter explores the nutritional, mental and physical health, and social benefits of three urban agricultural typologies – community, school, and home gardens – as productive foodscapes in cities and towns in different geographic locations.

Participation in, and the processes of, growing fruits and vegetables in community, school, and home gardens can provide multiple health and wellbeing benefits to the gardeners and their households. As outlined in this chapter, the diverse objectives of existing research strongly support positive health and wellbeing outcomes from food gardening through evaluation of various health aspects, such as reducing stress (Home and Vieli 2020; Koay and Dillion, 2020) and obesity (Davis et al. 2021; Utter, Denny and Dyson 2016), improving memory (Chalmin-Pui et al. 2021), mental health (Wilkinson and Orr, 2017; McVey et al. 2018; Koay and Dillion, 2020), physical fitness (Corley et al. 2021; Kingsley et al. 2009), and nutrition through increased fruit and vegetable intake (Depenbusch et al. 2022; Koay and Dillion, 2020; Davis et al. 2021; Kirkpatrick and Davison, 2018; Cabalda et al. 2011; Alaimo et al. 2008). The social and emotional contributions and cultural contexts of local food gardens as cultivated landscapes have emerged through the research (Pollin and

Retzlaff-Furst, 2021; Schultz and Rosen 2022; Graham and Connell 2006). The social and emotional benefits of growing food in urban gardens can guide significant mental wellbeing for different age groups such as seniors, middle-aged adults, and children (Chalmin-Pui et al. 2021; Corley et al. 2021; Kirkpatrick and Davison, 2018; Kim and Park, 2020). During COVID-19 isolation periods, home gardens provided households with spaces for engaging in food growing activities, as well as relaxing places that provided physical and mental health benefits (Cerada, et al. 2022). However, access to, and the functioning of, community gardens faced operational constraints during the pandemic (Schoen et al. 2021). Evidence encompasses the benefits from multiple perspectives, including clinical science, public health and wellbeing, sustainability performance, urban resilience, health geographies, cultural contexts, and social science.

Three different garden types contribute to the health and wellbeing of participants in these gardens. Community gardens can bring together diverse groups of people from varied cultural backgrounds with shared and individual growing practices and culinary habits. Community gardens are places for making friends, learning new gardening skills, and engaging in sustainable practices of growing food locally, and could be a pathway for improving community health (Mead et al. 2021; Lewis et al. 2018; McVey et al. 2018; Blaine et al. 2010; Litt et al. 2018). School gardens can act as a platform for nutritional education, gardening training, and connection to nature, as well as introducing the importance of biodiversity to students (Schultz and Rosen, 2022; Barry 2016; Fischer et al. 2019; Leuven et al. 2018; Wells et al. 2018). School gardens can also benefit students' and teachers' wellbeing (Pollin and Retzlaff-Furst, 2021; Gavia et al. 2021). Feasible solutions to overcome constraints, such as managing school gardens during school holidays, are important for the successful functioning of school gardens (Huys et al. 2017). Home gardens benefit from gardening training and nutritional programmes, which can increase household members' intake of micronutrients (such as vitamin A, zinc, and iron) through fruit and vegetable production (Depenbusch et al. 2022; Gerny et al. 2021; Baliki et al. 2019). Home and Vieli (2020, p. 5) have identified that local home gardens embody three components: 'restoration', 'food producing', and 'social'. These aspects of home gardens can connect multiple threads of health improvement and mental wellbeing for household members through growing fruit and vegetables in their home gardens.

The outcomes from the three case studies reinforce the idea that these gardens are not just spaces for growing food, but places that cultivate health, social change, knowledge sharing, and restorative and regenerative functions. The Coogee Community Garden in Sydney (Ghosh et al. 2015), school gardens in the Sindhupalchok District of Nepal (Schreinemachers et al. 2020), and a nutrition intervention programme in home gardens in Santa Clara County, California (Palar et al. 2019), provide case studies from diverse settings in which communities and individuals and households are growing vegetables and fruits. The Coogee Community Garden is located in an affluent, high-density suburb of Sydney. The community garden is the place for social interactions, growing healthy food, and making friends (Ghosh et al. 2015). The food gardening intervention programme for school children in Nepal, which combines school and home gardens, is aligned with the Nepali government's plan to reduce malnutrition (Schreinemachers et al. 2020). It is unique in that it combines two urban agricultural typologies as a connected food system for evaluating the health and social benefits of growing fruit and vegetables with children and their respective households. In a low-income population with high cardiovascular risk, the community-based home gardening programme in Santa Clara County, California, establishes the possibility of minimising the risk of diseases (such as diabetes) with a healthy diet supported by fresh fruit and vegetables from the home garden (Palar et al. 2019).

The garden types in the review and case studies presented in this chapter establish similarities and differences relating to their nature, operation, activities, ownership, and social aspects. The community, school, and home gardens have similarities in self-reported gardeners' perceptions that health and wellbeing benefits have been achieved. Growing fruit and vegetables is a common function in these garden types. Food production in these gardens is often influenced by the cultural practices, traditions, and memories of distant homelands (Graham and Connell 2006). The vegetables and fruits grown in all the gardens are informed by knowledge sharing on cross-cultural cuisines, culturally appropriate diets, and specific culinary habits of different communities. All three garden types depend on existing practices and government initiatives and programmes for developing health awareness, nutritional knowledge, and gardening skills, methods, and techniques adopted by the gardeners. These gardens have similar structures of central motivations, although the cultural contexts of home gardens differed from those of community and school gardens.

Community, school, and home gardens have differences in their organisation, management practices, motivations, gardener types, and spatial and cultural characteristics. The community gardens have a different operational framework, where households can rent plots for fruit and vegetable production. In community gardens, multiple ethnic communities share plots for growing food. These gardens act as sites of weaving cultures, giving rise to opportunities to develop new food growing practices, knowledge sharing, and social learning and participation. Community gardens can be melting pots of different cultures, an excellent platform for sharing cross-cultural knowledge, making friendships, activating social connections, and learning about and sampling multicultural cooking and associated cultures. In school gardens, the gardeners are mainly school children. These gardens are held in school ownership and, therefore, are dependent on decisions made by the school authorities for the continuity of the local food production functions of the school gardens. Teachers guide students in following a nutritional education/gardening programme included in the curriculum or not in the curriculum to develop gardening skills and nutrition awareness. In school gardens, students learn how to grow food, develop nutritional awareness and healthy eating habits, and prepare tasty recipes in the school canteen. This has flow-on effects in children's respective home gardens. For those students with home gardens, having parents who also participate (directly or indirectly) in the school garden, a continuum of continuously improving gardening skills, knowledge, and health awareness involving two generations is established. Growing food locally in home gardens is impacted by the motivations of individuals and households, and by lifestyle choices, awareness, knowledge, and income (Ghosh 2010). Home gardens are privately owned. Various age groups can participate in individual home gardens, shaped by the related household characteristics. For example, age group and motivation of the gardener, gardener types (Head and Muir, 2007) and cultural backgrounds (Graham and Connell 2006) impact the food types grown in home gardens. Produce exchanges can happen with neighbours and friends in the surroundings of the home gardens, or from the rural areas to urban locations, to create social networks along an urban–rural continuum (Winklerprins 2002).

Limitations of this research include the vast academic literature on these three garden types, which could be explored individually to a greater extent across multiple fields. Summarising the field is thus beyond the scope of this chapter. Instead, this chapter aims to provide a snapshot by investigating relevant research literature on these urban productive space categories to comprehend and compare their health and wellbeing potential and characteristics. Existing research has established an evidence base for the health, wellbeing, and social contributions of individual, household, and community engagement across three types of local food gardens. The gardening activity of producing fruit and vegetables provides physical activity, engagement, and social connectivity through working together, exchanging products, sharing skills and knowledge, making friends, and better mental wellbeing for most age groups. The significance of inclusive intervention designs to improve participants' fruit- and vegetable-eating behaviour is meaningful and has positive impacts at the individual, household, and community levels. Analysis and synthesis of the positive associations with gardening demonstrate many health, healing, and social implications. All these aspects of the three gardens can assist in efficient health and urban planning policy formulation and informed decision-making.

Given the evidence, governments and private organisations can design appropriate gardening intervention programmes and initiatives to increase people's motivation for gardening and offer funding opportunities. Low-income householders with limited funds to buy food and who are in greatest need can access fresh produce and maximise health benefits by participating in these productive gardens. The government, non-government organisations, and private organisations could provide incentives and funding to establish local food gardens. Future research should explore how different useful incentives and funding opportunities can be implemented to grow local food that can make a substantial positive difference in the health and wellbeing of individuals and households. Research on the emotional aspects of gardening – a critical part of mental health in different age groups – is limited. Future research should focus on evaluating the positive impacts of the emotional wellbeing of gardening before and after gardening, along with possible health and social benefits. Research on intervention programme design for different types of food gardens, their comparative efficiencies, and their abilities to enable long-term social change and exchange are to be explored and validated through future research. Exploring the health effects and social engagement in other garden types (such as residential rooftop gardens, small urban farms, balcony and vertical gardens, and other food production gardens) could open up new opportunities. Future research should also explore the contributions of different gardening types to providing health, social and economic benefits, and environmental performance capabilities, which could be integrated holistically as a productive network in the urban fabric.

Conclusion

This chapter investigates the benefits of local food gardens for health improvements and enhanced social exchange possibilities. Revisiting the research question – 'How do local foodscapes in community, school, and home gardens improve health and social exchanges across social and geographical gradients?' – this chapter uses three case studies to establish that local food gardens of various typologies can contribute to health, wellbeing, and social exchange opportunities for participants and households. The community, school, and home gardens are three different urban agriculture typologies embedded within the spatial fabric of cities and towns. Although these typologies have similarities in the primary function of growing fruits and vegetables, they also have differences when all other aspects are considered. Nevertheless, different urban gardens as cultivated landscapes with varied processes and mechanisms all have the potential to provide social and health benefits. These gardens with social and health functions require different circumstances to perform successfully as positive food landscapes. Research in these garden spaces can collectively build a solid evidence base for the therapeutic capacity of various gardening sites across geographical and social gradients. This research strengthens the argument that integrating local foodscapes can transform cities and towns into healthy and socially interactive places for people.

Acknowledgements

The author is very thankful to the School of Built Environment, Faculty of Design, Architecture and Building, the University of Technology, Sydney, in Australia, for providing immense support to conduct this research. The author would like to thank the reviewers for their valuable comments on this research.

References

Alaimo, K, Packnett, E, Miles, RA and Kruger, DJ 2008, 'Fruit and vegetable intake among urban community gardeners', *Journal of Nutrition Education & Behaviour*, vol. 40, no. 2, pp. 94–101. https://doi.org/10.1016/j.jneb.2006.12.003

Austin, S 2020, 'The school garden in the primary school: Meeting the challenges and reaping the benefits', *Education 3-13, International Journal of Primary, Elementary and Early Years Education*, vol. 50, no. 6, pp. 707–721. https://doi.org/10.1080/03004279.2021.1905017

Baliki, G, Brück, T, Schreinemachers, P and Nasir Uddin, M 2019, 'Long-term behavioural impact of an integrated home garden intervention: Evidence from Bangladesh', *FoodSecurity*, vol. 1, pp. 1217–1230. https://doi.org/10.1007/s12571-019-00969-0

Barry, A 2016, 'Expanding horizons: Home economics from garden to table', *Journal of the Home Economics Institute of Australia (HEIA)*, vol. 23, no. 3, pp. 10–14.

Bhattacharjee, LB, Saha, S and Nandi, BK 2007, *Food-based nutrition strategies in Bangladesh: Experiences from Integrated Horticulture and Nutrition Development Project*, RAP Publication 2007/05, Department of Agricultural Extension, Ministry of Agriculture, The People's Republic of Bangladesh/ FAO, Bangkok. https://www.fao.org/3/ag126e/AG126E00.htm

Blaine, TW, Grewal, PS, Dawes, A and Snider, D 2010, 'Profiling community gardeners', *Journal of Extension*, vol. 48, no. 6, pp. 1–12.

Blake, A and Cloutier-Fisher, D 2009, 'Backyard bounty: Exploring the benefits and challenges of backyard garden sharing projects', *Local Environment, The International Journal of Justice and Sustainability*, vol. 14, no. 9, pp. 797–807.

Buechler, S and Mansaray, S 2021, *Responding to the COVID-19 pandemic through community gardens: A lifeline to food security and community. A research report*, University of Arizona, Tucson.

Cabalda, AB, Rayco-Solon, P, Solon, JAA and Solon, FS 2011, 'Home gardening is associated with Filipino preschool children's dietary diversity', *American Dietetic Association*, vol. 111, no. 5, pp. 711–715.

CBC News 2022, 'Why some homeowners are sharing their yards with gardeners and farmers' (*CBC News*, 30 April 2022). https://www.cbc.ca/news/science/what-on-earth-yard-sharing-1.6434216

Cerada, C, Guenat, S, Egerer, M and Fischer, LK 2022, 'Home food gardening: Benefits and barriers during the COVID-19 pandemic in Santiago, Chile', *Frontiers in Sustainable Food Systems*, vol. 6, article 841386. https://doi.org/10.3389/fsufs.2022.841386

Chalmin-Pui, LS, Griffiths, A, Roe, J, Heaton, T and Cameron, R 2021, 'Why garden? – Attitudes and the perceived health benefits of home gardening', *Cities*, vol. 112, article 103118. https://doi.org/10.1016/j.cities.2021.103118

City of Sydney 2016, *Community gardens policy*, City of Sydney. https://www.cityofsydney.nsw.gov.au/policies/community-gardens-policy

CoDyre, M, Fraser, DGM and Landman, K 2015, 'How does your garden grow? An empirical evaluation of the costs and potential of urban gardening', *Urban Forestry & Urban Greening*, vol. 14, no. 1, pp. 72–79. https://doi.org/10.1016/j.ufug.2014.11.001

Coles, S 2019, 'Shared backyards' (*Sanctuary*, 29 March 2019). https://renew.org.au/sanctuary-magazine/outdoors/shared-yards/

Corley, J, Okely, JA, Taylor, AM, Page, D, Welstead, M, Skarabela, B, Redmond, P, Cox, SR and Russ, TC 2021, 'Home garden use during COVID-19: Associations with physical and mental wellbeing in older adults', *Journal of Environmental Psychology*, vol. 73, article 101545. https://doi.org/10.1016/j.jenvp.2020.101545

Davis, JN, Pérez, A, Asigbee, FM, Landry, MJ, Vandyousefi, S, Ghaddar, R, Hoover, A, Jeans, M, Nikah, K, Fischer, B, Pont, SJ, Richards, D, Hoelscher, DM and Van Den Berg, AE 2021, 'School-based gardening, cooking and nutrition intervention increased vegetable intake but did not reduce BMI: Texas sprouts – A cluster randomized controlled trial', *International Journal of Behavioral Nutrition and Physical Activity*, vol. 18, no. 18, pp. 1–14. https://doi.org/10.1186/s12966-021-01087-x

Depenbusch, L, Schreinemachers, P, Brown, S and Roothaert, R 2022, 'Impact and distributional effects of a home garden and nutrition intervention in Cambodia', *Food Security*, vol. 14, pp. 865–881. https://doi.org/10.1007/s12571-021-01235-y

Firth, C, Maye, D and Pearson, D 2011, 'Developing "community" in community gardens', *Local Environment: The International Journal of Justice and Sustainability*, vol. 16, no. 6, pp. 555–568. https://doi.org/10.1080/13549839.2011.586025

Fischer, LK, Brinkmeyer, D, Karle, SJ, Cremer, K, Huttner, E, Seebauer, M, Nowikow, U, Schütze, B, Voig, P, Völker, S and Kowarik, I 2019, 'Biodiverse edible schools: Linking healthy food, school gardens and local urban biodiversity', *Urban Forestry & Urban Greening*, vol. 40, pp. 35–43. https://doi.org/10.1016/j.ufug.2018.02.015

Gavia, MM, Cabingas, JP, Rodriguez, NP and Pallo, JE 2021, 'Hands-on school-based gardening: An intervention for teachers' wellbeing amidst pandemic', *Journal of Innovations in Teaching and Learning*, vol. 1, no. 1, pp. 41–46. https://doi.org/10.12691/jitl-1-1-8

Gaynor, A 2006, *Harvest of the suburbs: An environmental history of growing food in Australian cities*, University of Western Australia Press, Crawley.

Gerny, R, Marsh, R and Mwebembezi, J 2021, 'The promise and challenges of vegetable home gardening for improving nutrition and household welfare: new evidence from Kasese district, Uganda', *African Journal of Food, Agriculture, Nutrition and Development*, vol. 21, no. 1, pp. 17272–17289. https://doi.org/10.18697/ajfand.96.20125

Ghosh, S. 2010, 'Sustainability potential of suburban gardens: Review and new directions', *Australasian Journal of Environmental Management*, vol. 17, no. 3, pp. 49–59. https://doi.org/10.1080/14486563.2010.9725263

Ghosh, S, Accarigi, IV and Giovanangeli, A 2016, 'Social aspects of rooftop gardens' in S Wilkinson and T Dixon (eds), *Green roof retrofit: Building urban resilience*, Willey-Blackwell, New Jersey, US, pp. 189–215.

Ghosh, S, Wilkinson, S and Adler, D 2015, 'Social aspects of urban food production: A case study of Coogee Community Garden in Sydney', *Proceedings of Eighth Making Cities Liveable Conference*, Association for Sustainability in Business Inc., Pullman Melbourne on the Park, Melbourne, pp. 71–86.

Graham, S and Connell, J 2006, 'Nurturing relationships: The gardens of Greek and Vietnamese migrants in Marrickville, Sydney', *Australian Geographer*, vol. 37, no. 3, pp. 375–393. https://doi.org/10.1080/00049180600954799

Head, LM and Muir, PA 2007, *Backyard: Nature and culture in suburban Australia*, University of Wollongong Press, Wollongong.

Heim, S, Stang, J and Ireland, M 2009, 'A garden pilot project enhances fruit and vegetable consumption among children: Perspectives in practice', *Journal of American Dietetic Association*, vol. 109, no. 7, pp. 1220–1226. https://doi.org/10.1016/j.jada.2009.04.009

Henderson, E 2021, 'School gardens may get children to eat more vegetables' (*News Medical Life Sciences*, 4 February 2021). https://www.news-medical.net/news/20210204/School-gardens-may-get-children-to-eat-more-vegetables.aspx

Home, R and Vieli, L 2020, 'Psychosocial outcomes as motivations for urban gardening: A cross-cultural comparison of Swiss and Chilean gardeners', *Urban Forestry & Urban Greening*, vol. 52, article 126703. https://doi.org/10.1016/j.ufug.2020.126703

Hutchins, R 2020, '8 surprising health benefits of gardening' (*UNC Health Talk*, 18 May 2020). https://healthtalk.unchealthcare.org/health-benefits-of-gardening/

Huys, N, Cocker, K, Cocker, KD, Craemer, MD, Rosebeke, M, Cardon, G and Lepeleere, SD 2017, 'School gardens: A qualitative study on implementation practices', *International Journal of Environmental Research & Public Health*, vol. 14, no. 12, article 1454. https://doi.org/10.3390/ijerph14121454

Joshi, N and Wende, W 2022, 'Physically apart but socially connected: Lessons in social resilience from community gardening during the COVID-19 pandemic', *Landscape and Urban Planning*, vol. 223, 104418. https://doi.org.ezproxy.utas.edu.au/10.1016/j.landurbplan.2022.104418

Kaziro, P 2015, 'GROW' horticulture therapy programme at St Canice Kitchen Garden, Kings Cross, Sydney' (personal communication).

Kim, S and Park, S 2020, 'Garden-based integrated intervention for improving children's eating behavior for vegetables', *International Journal of Environmental Research and Public Health*, vol. 17, no. 4, pp. 1–14. http://dx.doi.org/10.3390/ijerph17041257

Kingsley, JY, Townsend, M and Henderson-Wilson, C 2009, 'Cultivating health and wellbeing: Members' perceptions of the health benefits of a Port Melbourne community garden', *Leisure Studies*, vol. 28, no. 2, pp. 207–219. https://doi.org/10.1080/02614360902769894

Kirkpatrick, JB and Davison, A 2018, 'Home-grown: Gardens, practices and motivations in urban domestic vegetable production', *Landscape and Urban Planning*, vol. 170, pp. 24–33. https://doi.org/10.1016/j.landurbplan.2017.09.023

Koay, WI and Dillion, D 2020, 'Community gardening: Stress, wellbeing, and resilience potentials', *International Journal of Environmental Research and Public Health*, vol. 17, no. 18, article 6740. https://doi.org/10.3390/ijerph17186740

Lee, SM, Jang, HJ, Yun, HK, Jung, YB and Hong, IK 2022, 'Effect of apartment community garden program on sense of community and stress', *International Journal of Environmental Research & Public Health*, vol. 19, article 708. https://doi.org/10.3390/ijerph19020708

Leuven, JRFW, Rutenfrans, AHM, Dolfing, AG and Leuven, RSEW 2018, 'School gardening increases knowledge of primary school children on edible plants and preference for vegetables', *Food Science & Nutrition*, vol. 6, no. 7, pp. 1060–1967. https://doi.org.ezproxy.utas.edu.au/10.1002/fsn3.758

Lewis, O, Home, R and Kizos, T 2018, 'Digging for the roots of urban gardening behaviours', *Urban Forestry & Urban Greening*, vol. 34, pp. 105–113. https://doi.org/10.1016/j.ufug.2018.06.012

Litt, JS, Alaimo, K, Buchenau, M, Villalobos, A, Glueck, DH, Crume, T, Fahnestock, L, Hamman, RF, Hebert, JR, Hurley, TG, Leiferman, J and Li, K 2018, 'Rationale and design for the community activation for prevention study (CAPs): A randomized controlled trial of community gardening', *Contemporary Clinical Trials*, vol. 68, pp. 72–78. https://doi.org/10.1016/j.cct.2018.03.005

McAleese, JD and Rankin, LL 2007, 'Garden-based nutrition education affects fruit and vegetable consumption in sixth-grade adolescents', *Journal of the American Dietetic Association*, vol. 107, no. 4, pp. 662–665. https://doi.org/10.1016/j.jada.2007.01.015

McVey, D, Nash, R and Stansbie, P 2018, 'The motivations and experiences of community garden participants in Edinburgh, Scotland', *Regional Studies, Regional Science*, vol. 5, no.1, pp. 40–56. https://doi.org/10.1080/21681376.2017.1409650

Mead, BR, Christiansen, P, Davies, JAC, Falagán, N, Kourmpetli, S, Liu, L, Walsh, L and Hardman, CA 2021, 'Is urban growing of fruit and vegetables associated with better diet quality and what mediates this relationship? Evidence from a cross-sectional survey', *Appetite*, vol. 163, article 105218. https://doi.org/10.1016/j.appet.2021.105218

Moons, P, Goossens, E and Thompson, DR 2021, 'Rapid reviews: The pros and cons of an accelerated review process', *European Journal of Cardiovascular Nursing*, vol. 20, no. 1, pp. 515–519. https://doi.org/10.1093/eurjcn/zvab041

Mougeot, LJA 2000, 'Urban agriculture: Definition, presence, potential and risks' in N Bakker, M Dubbeling, S Gündel, U Sabel-Koschella and H De Zeeuw (eds), *Growing cities growing food: Urban agriculture on the policy agenda*, Deutsche Stiftung für Internationale Entwicklung, Feldafing, pp. 1–42.

National Gardening Association 2009, *The impact of home and community gardening in America*, National Gardening Association, South Burlington, VT, pp. 1–17.

Palar, K, Hufstedler, EL, Hernandez, K, Chang, A, Ferguson, L, Lozano, R and Weiser, SD 2019, 'Nutrition and health improvements after participation in an Urban Home Garden Program', *Journal of Nutrition Education and Behavior*, vol. 51, no. 9, pp. 1037–1046. https://doi.org/10.1016/j.jneb.2019.06.028

Park, SA, Lee, AY, Park, HG and Lee, WL 2019, 'Benefits of gardening activities for cognitive function according to measurement of brain nerve growth factor levels', *International Journal of Environmental Research and Public Health*, vol. 16, no. 5, article 760. https://doi.org/10.3390/ijerph16050760

Pollin, S and Retzlaff-Furst, C 2021, 'The school garden: A social and emotional place', *Frontiers in Psychology*, vol. 12, article 567720. https://doi.org/10.3389/fpsyg.2021.567720

Rammohan, A, Pritchard, B and Dibley, M 2019, 'Home gardens as a predictor of enhanced dietary diversity and food security in rural Myanmar', *BMC Public Health*, vol. 19, no. 1, article 1145. https://doi.org/10.1186/s12889-019-7440-7

Randwick City Council 2016, 'Social gardening at Coogee', *Randwick City Council*. https://www.randwick.nsw.gov.au/about-council/news/news-items/home-page-news/gardening-gets-social-in-coogee

Richardson, EA, Pearce, J, Shortt, NK and Mitchell, R 2017, 'The role of public and private natural space in children's social, emotional and behavioural development in Scotland: A longitudinal study', *Environmental Research*, vol. 158, pp. 729–736. https://doi.org/10.1016/j.envres.2017.07.038

Schoen, V, Blythe, C, Caputo, S, Fox-Kämper, R, Specht, K, Fargue-Lelièvre, A, Cohen, N, Poniży, L and Fedeńczak, K 2021, '"We have been part of the response": The effects of COVID-19 on community and allotment gardens in the Global North', *Frontiers in Sustainable Food Systems*, vol. 5, article 732641. https://doi.org/10.3389/fsufs.2021.732641

Schreinemachers, P, Baliki, G, Shrestha, RM Bhattarai, DR, Gautam, IP, Ghimire, PL, Subedi, BP and Brück, T 2020, 'Nudging children toward healthier food choices: An experiment combining school and home gardens', *Global Food Security*, vol. 26, article 100454. https://doi.org/10.1016/j.gfs.2020.100454

Schultz, C and Rosen, AE 2022, 'School gardens' impact on students' health outcomes in low-income Midwest schools', *Journal of School Nursing*, vol. 38, no. 5, pp. 486–493. https://doi.org/10.1177/10598405221080970

Soga, M, Cox, DTC, Yamaura, Y, Gaston, KJ, Kurisu, K and Hanaki, K 2017, 'Health benefits of urban allotment gardening: Improved physical and psychological well-being and social integration', *International Journal of Environmental Research and Public Health*, vol. 14, no. 1, article 71. https://doi.org/10.3390/ijerph14010071

Thompson, S, Corkery, L and Judd, B 2007, 'The role of community gardens in sustaining healthy communities', *Proceedings of Third State of Australian Cities Conference (SOAC)*, Adelaide, pp. 161–171.

Tontisirin, K and Bhattacharjee, L 2008, 'Community based nutrition programs', *International Encyclopedia of Public Health*, Elsevier Science, pp. 791–799. https://doi.org.ezproxy.utas.edu.au/10.1016/B978-0-12-803678-5.00084-9

University of Colorado Boulder 2011, 'Community gardens improve personal and neighborhood health, CU-led research finds' (*CU Boulder Today*, 22 June 2011). https://www.colorado.edu/today/2011/06/22/community-gardens-improve-personal-and-neighborhood-health-cu-led-research-finds

Utter, J, Denny, S and Dyson, B 2016, 'School gardens and adolescent nutrition and BMI: Results from a national, multilevel study', *Preventive Medicine*, vol. 83, pp. 1–4. https://doi.org/10.1016/j.ypmed.2015.11.022

Wakefield, S, Yeudall, F, Taron, C, Reynolds, J and Skinner, A 2007, 'Growing urban health: Community gardening in South-East Toronto', *Health Promotion International*, vol. 22, no. 2, pp. 92–101. https://doi.org/10.1093/heapro/dam001

Wells, NM, Meyers, BM, Todd, LE, Henderson, CR, Barale, K, Gaolach, B, Ferenz, G, Aitken, M, Tse, CC, Pattison, KO, Hendrix, L, Carson, JB, Taylor, C and Franz, NK 2018, 'The carry-over effects of school gardens on fruit and vegetable availability at home: A randomized controlled trial with low-income elementary schools', *Preventive Medicine*, vol. 112, pp. 152–159. https://doi.org/10.1016/j.ypmed.2018.03.022

Wilkinson, S and Orr, F 2017, 'The impact of horticulture therapy on mental health care consumers on a retrofitted roof', *Proceedings of 23rd Annual Pacific-Rim Real Estate Society Conference*, Sydney.

Winklerprins, MGAA 2002, 'House-lot gardens in Santarém, Pará, Brazil: Linking rural with urban', *Urban Ecosystems*, vol. 6, pp. 43–65. https://doi.org/10.1023/A:1025914629492

Wise, P 2014, *Grow your own: The potential value and impacts of residential and community food gardening*, Policy Brief No. 59, The Australia Institute, Canberra. http://www.tai.org.au/content/grow-your-own

Yaworski, B 2015, 'Sharing backyard gardens: A community project for urban dwellers', (*Alive*, 24 April 2015). https://www.alive.com/lifestyle/sharing-backyard-gardens/

4 The cultivated 'healing garden'

Respite and support, or lifestyle change?

Esther Veen and Karolina Doughty

Introduction

Since the 1990s, there has been increasing acknowledgement in broader public health debates of the importance of everyday, non-medical settings for the purpose of health promotion (Walker et al. 2017). There is a growing understanding that various community settings can be experienced as positive for health and wellbeing, and that much crucial 'recovery', restoration, and prevention of ill health takes place outside the parameters of the biomedical field (Walker et al. 2017). So-called 'green care' settings, such as communal gardens, have received much attention as ideal sites for health promotion. A growing literature on therapeutic horticulture across several disciplines shows the potential of the communal garden as a therapeutic landscape that can elicit a set of interrelated benefits, such as improved mental health, greater physical activity, and increased social support (Veen et al. 2016), while also providing a site for more structured interventions for groups with particular health-related needs.

Research applying the perspective of therapeutic landscape (e.g., Biglin 2020; Marsh et al. 2017; Milligan et al. 2004; Pitt 2014; Sanchez and Liamputtong 2017) primarily discusses the value of communal gardens and gardening in terms of holistic, emplaced, therapeutic qualities that may help people feel well, even if they are not well (Marsh et al. 2017). This characterises what could broadly be described as a 'positive' approach to health. However, in life science fields, such as nutrition, studies focus on more direct health benefits, and often employ a narrower conception of health as the absence of disease. Such studies associate gardening with diets of higher nutritional value (Alaimo et al. 2008), healthier eating behaviour (Blair et al. 1991), and increased access to higher-quality foods. This acknowledgement of such benefits has led the community garden to be considered as a place where lifestyle changes can be stimulated.

The concept of a 'healthy lifestyle' has been central to the discourse of health promotion and the 'new public health' for several decades, resulting in a history of individual-level interventions. In relation to health, lifestyle is 'a loose aggregation of behaviours and conditions, encompassing body size, body shape, diet, exercise and the use of drugs both legal and illegal' (Davison and

DOI: 10.4324/9781003355731-6

Davey Smith 1995, p. 92). In sociological literatures on health, this individualist paradigm has been criticised as unacceptably reductive, in that it ignores the complexities of social action and environmental and cultural influences on health. Health interventions targeting 'lifestyle' have also been criticised for their narrow construction of health (absence of disease), which results in promoting health primarily through active risk-avoidance behaviours (Korp 2010). More holistic conceptions of health are gaining influence in wider health discourses, shifting the focus from risk avoidance to a more dynamic understanding of health that prioritises wellbeing and quality of life. Within this broader context, we examine the communal garden as a cultivated therapeutic landscape that also provides a setting for targeted health interventions. In doing so, we open a discussion about critical practice issues that arise from situating the communal garden, as an example of 'green care', within a landscape of (neoliberal) public health that focuses largely on individual behaviour change – through the notion of lifestyle – as a route to better health.

In the next part of this chapter, we provide an overview of therapeutic gardens in health care and beyond. We discuss the communal garden both as a supportive setting for a holistic and dynamic sense of 'everyday wellbeing', and also, potentially, as a setting for targeted interventions that risk reducing health to practices of risk-avoidance. Then, we examine how these approaches played out in the Healing Garden pilot project conducted in the Netherlands in 2017 to explore the use of therapeutic horticulture to support physical activity and healthy eating during and after cancer treatment. The project involved recovering cancer patients taking part in vegetable gardening with a gardening coach. The underlying hypothesis was that this intervention would help cancer patients adhere to lifestyle guidelines related to diet and physical exercise, and that the garden would serve as a place for social peer support. The garden proved successful in offering an opportunity to spend time in a setting that holistically supported health and wellbeing, and the gardening activity was specifically useful in creating social peer support. However, it was less successful as a diet and exercise lifestyle intervention. In the final section of this chapter, we reflect on what the Healing Garden experience tells us about the limitations of a risk-focused approach to health in the context of cancer support and therapeutic horticulture.

Therapeutic gardens in health care and beyond

It is an old idea that a garden can be a healing place, designed to please the senses, lift the spirits, and enhance overall wellbeing, both through the relatively passive enjoyment of sitting or strolling around and through the more active physical undertaking of tending the garden. Age-old stories show that human beings and gardens belong together: 'Through the millennia [we find] similar descriptions of beautiful, enclosed gardens in myths all over the world' (Stigsdotter and Grahn 2002, p. 67). Examples include the Garden of Eden of the Abrahamic religions, Arcadia of the ancient Greeks, and the heavenly paradise in Elysium (Stigsdotter and Grahn 2002). In 2000 BCE, the Mesopotamians were already using plants as mood refreshers, and the Persians built gardens with colourful and fragrant flowers to delight the senses (Jellicoe and Jellicoe 1975 cited in Thaneshwari et al. 2018). In the twelfth century, St Bernard attributed the therapeutic benefits of a hospice garden at a monastery in Clairvaux, France, to its green plants, fragrances, privacy, and birdsong (Horowitz 2012, p. 78).

Gardening activities, such as planting, weeding, watering, and harvesting, have also been part of rehabilitation and recovery treatment programmes for many years. In the United States, for instance, plant-based activities were introduced into veterans' hospitals in the 1940s and 1950s (Horowitz 2012). In Europe, horticulture has been used since the turn of the twenty-first century as a therapeutic intervention or adjunct to therapy for a range of groups, such as those recovering from major illnesses, people recovering from injury, and those living with physical disabilities, learning disabilities, or mental health issues, as well as older people, offenders, and individuals with drug or alcohol dependencies (CCFR 2003).

Although there has been a broad shift in the arena of health policy and politics towards valorising and investing in informal and non-medical settings and activities for health, wellbeing, and care – such as gardens – there are vastly different ways in which gardens can be utilised as the setting for health promotion. Consequently, there are differences in the types of outcomes and evidence that practitioners and policymakers find acceptable and actionable.

Below, we discuss two broad perspectives on health and how they translate into the context of the cultivated therapeutic landscape of the communal garden: (i) a holistic, or positive, perspective on health; and (ii) a more risk-focused perspective on health which translates into individual-level interventions targeting 'lifestyle' – exemplifying what critical health promotion literatures term 'lifestyle drift'.

The garden as therapeutic space: Holistic and positive approaches to health

Everyday environments and situated forms of physical practice can be beneficial for wellbeing. Practices such as gardening, walking, swimming, and yoga have been shown to have a positive effect on both physical and mental health (Frensham et al. 2018; Pisu et al. 2017; Stan et al. 2016). The literature on 'nature' as a space for health promotion overlaps with efforts to understand the potentially therapeutic aspects of places more broadly. Concepts such as 'therapeutic landscapes' (Gesler 1992), 'enabling settings' (Walker et al. 2017), 'restorative environments' (Hartig 2004), 'enabling environments' (Steinfeld and Danford 1999), or 'enabling places' (Duff 2012) have been used to theorise how some places can facilitate 'access to an array of social, material and/or affective resources' (Duff 2012, p. 1388) that assist in the process of recovery or promote better health and wellbeing more generally. The underlying premise of therapeutic horticulture draws on the broader idea that nature and landscape can influence emotions, health, and behaviour (CCFR 2003). This is supported by a large body of literature, across medical and social sciences, which acknowledges this potential of the natural environment. Research from several perspectives, including health geography (Gesler 2003), environmental psychology (Ulrich 1981), and ecological psychology (Roszak et al. 1995), has sought to explain how natural views and nature experiences can promote health and wellbeing. Here, health is often understood as emerging through relationships and embedded in places (Conradson 2005). The translation of this research knowledge into community health practices can be captured by the term 'green care': an umbrella term for a broad spectrum of health-promoting interventions that utilise elements of nature to foster social, physical, mental, and even educational wellbeing (Haubenhofer et al. 2010). Community gardens, alongside care farms, green exercise, and wilderness therapy, are all examples of green care.

Stigsdotter and Grahn (2002) identify three schools of thought that have developed around different explanatory models for the healing effects of gardens: (i) experiential; (ii) activity-based; and (iii) cognitive. According to the experiential school – the 'healing garden' school – health effects are derived from the *experiences* of being in the garden, depending on its design and contents. These effects are understood according to different theories from the fields of environmental psychology and landscape architecture: health effects are thought to derive from the restorative influence on emotional centres in the limbic system of the brain; from the restorative influence of green on cognitive functions; or from the fact that the garden makes demands that gently balance the gardener's own ability and control (Stigsdotter and Grahn 2002, p. 62). The activity-based school – the 'horticultural therapy' school – holds that the health effects of gardening result mainly from the *activities* people perform in the garden; for instance, because these activities are experienced as meaningful and enjoyable. In other words, where the 'healing garden' school highlights the passive experience of being in a garden, the 'horticultural therapy' school focuses on the activities performed in a garden. Finally, the 'cognitive school' sits between the experiential and the activity-based poles of the spectrum. This school contends that the health effects of gardening relate both to the experiences of, and the activities performed in, the garden and to the gardener's background and character. It argues that the garden space, in combination with the activities performed, can restore a person's positive view of themselves (Stigsdotter and Grahn 2002).

Following this approach, gardens can be experienced passively (simply being in the garden) and actively (performing activities in the garden), but whether and how a garden is experienced as healing also depends on whether or not therapy is actively offered. The difference between 'horticultural therapy' and 'therapeutic horticulture' is useful in this respect. The terms are sometimes used interchangeably and the literature is inconclusive about their differences; however, 'horticultural therapy' is generally linked to setting client goals and is often implemented by trained therapists, while 'therapeutic horticulture' is less goal-driven, more easily implemented, and not necessarily implemented by therapists (Gonzalez et al. 2009). Therapeutic horticulture, then, can be defined as 'a nature-based intervention that involves social and behavioural activation, participation in enjoyable activities, and moderate levels of physical activity in pleasant surroundings' (Gonzalez et al. 2011b, pp. 119–120). Therapeutic horticulture is less explicitly a form of therapy, while horticultural therapy mostly refers to using green space to mediate mental healing (Söderback et al. 2004, p. 245).

Research that applies a therapeutic landscape lens to gardens and gardening has tended to focus on community gardens, as settings for therapeutic horticultural activities for a local community (Pitt 2014; Sanchez and Liamputtong 2017), or as a place of support for specific groups, such as the elderly (Milligan et al. 2004), refugees (Biglin 2020), or the terminally ill (Marsh et al. 2017). These studies offer qualitative insights into restorative experiences, practices, and social relations in the setting of such communal gardens, and focus primarily on their value for 'everyday wellbeing' (Mossabir et al. 2021), that is, encompassing mental, physical, nutritional, social, and place-related aspects. What sets this literature apart from the three schools of thought outlined by Stigsdotter and Grahn (2002) is its explicitly holistic perspective on the relationship between a physical setting and its effects on wellbeing and health, with an understanding that 'collectively the environmental, cultural, social and individual constituents of a space form a complex landscape of health and healing – a therapeutic landscape' (Marsh et al. 2017, p. 110).

Gardening as lifestyle intervention: Risk-focused approaches to health

A focus on preventing ill health forms part of a broader agenda of decentralisation and 'personalisation' of health services across Europe, which is heavily influenced by neoliberal forms of health governance and austerity measures (Karanikolos et al. 2013). Through the concept of 'lifestyle', everyday life behaviours and habits have become the target of health interventions. Lifestyle factors are addressed in public health policy, primarily through individual behavioural interventions, which target diet and physical activity. Examples of exercise and healthy eating guidelines that have been promoted in national public health campaigns across many countries include the World Health Organization (WHO) guidelines on 'health-enhancing physical activity', which recommends 150 minutes of moderate physical activity per week (WHO 2010), and the widely adopted '5 a day' recommendation in relation to fruit and vegetables. While such recommendations may seem reasonable in themselves, the broad policy focus on issuing such behavioural advice has been discussed in critical public health literatures as a problem of 'lifestyle drift' (Powell et al. 2017) – a phenomenon whereby health policymakers begin with a recognition of the social, political, and economic determinants of health, only to drift back into designing policies largely targeted at modifying individual behaviour (Godziewski 2020, p. 1).

Lifestyle drift arguably constitutes the most significant barrier for public health in addressing health disparities because it overlooks abundant evidence on the social determinants of health (Fisher et al. 2015): the 'conditions in which people are born, grow, work, live, and age, and the wider set of forces and systems shaping the conditions of daily life' (WHO n.d.). Ignoring the social determinants of health is problematic because the interactive characteristics of individuals and environments that underlie health outcomes should guide public health practice, as argued by Golden and Earp (2012). Nevertheless, most health interventions are still siloed into areas such as physical activity and nutrition, which are, in turn, framed through individualised behavioural approaches. Such interventions do nothing to address the social structures and place effects that generate health inequities. It is primarily the structural perpetuation of poverty, poor education, and poor housing that keeps people with lower socio-economic status less well than wealthier populations.

The developing phenomenon of 'green care' provides one avenue through which structural health disparities may be practically addressed. As Marsh (2020, p. 146) writes, for green care interventions, such as community gardening, to truly address public health goals, '[w]e need to move beyond a discussion contained within the established domains of exercise, social engagement and eating'. That said, within the broader discourses on nature-based solutions for health, there is a strong focus on how to motivate people to be physically active in green space (Drakou et al. 2011). This is also reflected in therapeutic gardening programmes designed as lifestyle interventions. Such behavioural approaches also drive demand for evidence-based outcomes aligned with a risk-focused understanding of health. Focusing on individual health behaviours, however, obscures the 'ways in which gardening [can address] the range of factors that contribute to health inequities: social, physical, cultural, ecological and emotional' (Marsh 2020, p. 146).

The potential of cultivated 'healing gardens' to address the multiple causes and impacts of health inequity depends to a great extent on their operationalisation and the principles of evaluation they apply. A focus on lifestyle changes related to physical activity and 'healthy eating' will entail a different set of criteria for evaluation from a focus on the more holistic and relational wellbeing benefits that come from offering participants a space for respite and mindful engagement with 'nature' within a supportive social setting. We use the Healing Garden pilot project to further discuss this difference and its consequences.

The Healing Garden pilot project

The Healing Garden pilot project was conducted in Almere, a medium-sized city in the Netherlands (population approximately 200,000 in 2020) in 2017. Its primary aim was to study whether gardening would help recovering cancer patients adhere to suggested lifestyle guidelines around eating and exercise. A secondary aim was to understand the extent to which the project would enhance social peer support – considering the importance of social support for cancer patients (Taylor et al. 1986) and that gardening is often found to boost social relations (Wang and MacMillan 2013). The project was designed by an interdisciplinary research group consisting of sociologists, human nutritionists, and plant scientists, supported by an interdisciplinary supervisory team representing various institutes. Inspired by a similar project conducted in Birmingham, Alabama, in the United States (Cases et al. 2016), our pilot project invited participants to garden in a group (rather than individually, as in the US study).

The objectives of our pilot were threefold: (i) to test the research design; (ii) to discover how the activity could be organised to meet participants' needs; and (iii) to gain a first impression of the feasibility of our hypotheses. The first author (a sociologist) was part of the research group; with others, she was responsible for establishing the garden and was involved in interviewing the participants (see below). After the completion of this phase of the project, she invited the second author (a cultural geographer) to help analyse the sociological data and contextualise it within public health and 'green care' literatures.

The Healing Garden pilot was conducted over one growing season, from April to October 2017. Participants were recovering cancer patients who were invited to garden in wooden gardening containers placed in the grounds of a cancer support centre (see Figure 4.1). Gardening was offered as one of several activities, such as yoga, peer support groups, or mandala drawing. Participants gardened as a group in weekly sessions of 90 minutes each, guided by two experienced gardeners. Seeds, seedlings, and tools were offered free of charge. There was a total of ten containers, of which six were at the ground level, and four at the table level. Six people agreed to participate in the gardening activity. Sadly, one participant left the project after the first measurement week due to returning cancer. Gardeners each gardened in 'their own' container; the remaining containers were maintained communally. Figures 4.1–4.4 give an impression of the containers and their placement next to the support centre.

Figure 4.1 The gardening containers as laid out next to the cancer support centre.

Figure 4.2 A container at the ground level.

Figure 4.3 The gardening containers growing produce.

Figure 4.4 Some of the harvest.

Methodology and methods

Cancer survivors have an increased risk of chronic illnesses, such as cardiovascular disease, diabetes, osteoporosis, and depression (Wu and Harden 2015; Sarfati et al. 2016), which is believed to be a result of both cancer treatments and lifestyle factors. These illnesses are associated with lower quality of life, higher mortality rates, and higher health care costs (Demark-Wahnefried et al. 2005; Fu et al. 2015). The working hypothesis for the Healing Garden pilot project was that offering an attractive leisure activity (i.e., gardening) would prove more successful in terms of long-term participation than receiving suggestions on lifestyle: diet and exercise interventions for cancer survivors often have only short-term results (Cases et al. 2016). This hypothesis was based on Social Practice Theory, a sociological theory which argues that human beings participate in various 'practices' in daily life and that we rely on routines. The project team hypothesised that if participants became engaged with a practice they enjoyed, they would be able to perform this practice more easily and for a longer period. This practice would then become a routine in daily life. As the practice of gardening involves both physical activity and freshly harvested produce, stimulating people to become engaged with gardening might be more beneficial than telling them to exercise more or eat more healthily.

While the Healing Garden research group approached the project as a shared endeavour, its methodological set-up reflected two main disciplines: human nutrition and sociology. Each discipline used its own data collection methods. Data were collected in three data collection weeks across the half-year period of the gardening activity: at the beginning, midpoint, and the end. The main question of the three nutritionists in the research team was whether the garden would be able to help people adhere to lifestyle guidelines around healthy eating and exercise. To study this, a set of similar measurement tools was used during each of the three data collection weeks: participants were asked to fill in a number of standardised validated questionnaires on their physical health (diet, physical activity, tiredness, and pain in hands and feet); they performed a few light physical tests; wore an accelerometer for a week; kept an activity diary; and their height, weight, and Body Mass Index (BMI) were measured. These measurements took no more than 30 minutes in total and were planned around the gardening activity. (We do not discuss the results of these measurements here but do report on some outcomes to help put the rest of our findings in perspective.) Due to the small number of participants in the pilot project, and because the research team did not manage to secure funding for a larger study, the data from the validated questionnaires and physical tests have not been published.

With a background in community gardening and social cohesion, the sociologist and her assistant aimed to understand the extent to which the gardening project would help create social bonds between participants, and whether gardening could be considered as a form of social peer support. Semi-structured interviews were conducted during each measurement week; each round had a specific focus. The first round of interviews centred around expectations and wishes, and the practical issues needed to plan the activity. Ample attention was given to these issues as the project team aimed to create a gardening activity that was as pleasant and suitable as possible. The second round of interviews focused on the experience of the gardening activity, and what respondents did or did not enjoy. In the final round, respondents reflected on the activity as a whole – the extent to which their expectations had been met, and whether the activity had led to social support (by asking this explicitly and discussing participants' views). We also asked respondents to reflect on whether the physical goals had been met. Additionally, participants were asked to rate certain elements of the activity in order of importance using cards (i.e., participants chose six cards that illustrated what they enjoyed best, and rated cards in categories to express how they would imagine an ideal gardening activity (see Figure 4.5)). Interviews were conducted at a location chosen by the participant

Figure 4.5 Scoring cards.

(often the cancer support centre; sometimes the participant's house or a café). Interviews lasted between 30 and 45 minutes and were recorded and transcribed. The data was manually coded in an iterative process.

Five women participated in the gardening project. Their average age was 50 years (ranging from 34 to 64 years); four of the five were of Dutch nationality. Three of the women were married, three had lower education levels, and three were employed. All participants were recovering from cancer, all had undergone surgery, and all had received treatments for cancer, such as chemotherapy or radiotherapy. Two of the women were still receiving treatment. None of the participants smoked (although two had previously smoked) and their mean BMI was relatively high (29.1 kg/m^2).

Results: Garden design and experience

The garden designed for the Healing Garden project consisted of ten gardening containers of varying sizes, which were placed in the grounds of a garden support centre (see Figure 4.1). The choice for this garden design was pragmatic. The research team discussed various potential gardening locations. Almere is not a very dense city (especially by Dutch standards), so almost any location would be far away for some of the potential participants. We argued that the cancer support centre would at least be accessible and reachable for a group of potential participants. Moreover, the cancer support centre has facilities, such as a toilet, storage space for tools and seeds, a coffee machine and tables and chairs to sit on/around. Additionally, we considered the option that the activity could turn out to be emotionally draining. At the cancer support centre, volunteers who could offer support would always be close by.

Whether to use containers or garden in open ground was discussed with the participants who initially signed up for the gardening activity. Participants had no clear preference. Containers, however, were much easier to realise than a garden in open ground: the design team included a landscaper who could sponsor and deliver the containers. Moreover, using containers made it possible to place the garden immediately adjacent to the cancer support centre, where there was a gravel pitch. This was considered important, as this piece of land belonged to the support centre itself. Creating a garden in the grass field behind the support centre, although tempting, would not only take much more time and energy (and the gardening season was fast approaching), but would also mean requesting a permit from the municipality, which owns the land. Additionally, the chosen location was guarded by a camera, which was considered important as local youth sometimes vandalised the location at night. Hence, the choice of containers was based on several practical realities and confirmed by the fact that none of the participants had a clear preference for gardening in open ground. The containers were placed close to the entrance of the support centre but were not very visible to passers-by, as they were shielded from the pathway to the entrance and the carpark by a row of bushes. Nevertheless, some of the support centre's offices looked right onto the garden.

Interviewees indicated that the gardening experience was highly satisfying. One of the most important reasons for this was that participants simply enjoyed the activity. The first round of interviews showed that the activity itself was an important motivation for all participants to join the project (besides other motivations, such as the harvest or the social aspect). Participants differed, however, in what they specifically enjoyed about gardening: the healthy harvest *(A blissful feeling or so because you think, my, I now have unsprayed vegetables. R1)*; the pleasure of doing manual labour *(I like working with my hands (…). I like seeing plants grow and then you can even eat them. R2)*; mental rest (*A little rest in your head. R5)* – and mostly a combination of these. One participant specifically mentioned the effect of working with nature, and how she related that to her disease and the struggles she had with realising she had survived cancer while others would not.

> That man then said to me some [seeds] come up and some don't. Then I thought, hey, in fact that's true. Some survive and some do not. Don't think about that too much. And that connected me even more with sowing and seeing it come up.
>
> (R6)

These responses show that the garden itself and the gardening activity can be understood as providing a setting that supported the everyday 'work' of recovery for the participants.

Whereas all participants agreed that the atmosphere in the group was pleasant, both the pilot project as a whole and the length of the weekly activities were too short to build deep relationships. Indeed, when prompted, respondents stated that the group in itself was 'interchangeable' (i.e., the experience of the activity would have been the same with another group of people). Moreover, over time, commitment weakened and participation decreased, so that during some gardening afternoons group size was very small. This was considered an important drawback. Nevertheless, participants enjoyed working with the group and mentioned this as another important reason to enjoy the project. Working in a group was considered more fun (*Of course it is pleasurable. Look, if you have a garden by yourself, well, to whom should you be talking? R5)*; people also felt understood, appreciated not having to explain or defend themselves if they were tired, in pain, or unable to work, and just felt involved with each other.

> You didn't have to – what do you call that? Defend. If you say, well, my hands are painful now, I need a break, or I cannot do this… you can say all these things.
>
> (R6)

> You are spending time together, you share something, which is the disease in this case, but also something very nice. Like, 'Oh, yours is growing like that. Why isn't mine growing?'
>
> (R4)

Hence, there was a shared feeling of being in a supportive environment. Moreover, people not only discussed the gardening activity itself, but also more personal issues, often in relation to their disease. They shared 'our diseases, our fights with the company doctor, our fights with supervisors' (R6) and 'everyone got a chance to share her story or tell how she was doing' (R1).

Despite participants agreeing on the supportive environment, they did not agree on whether or not the gardening activity could be seen as social peer support. Only one respondent clearly perceived the gardening activity as such, because she did not have to 'prove herself'. She had tried a talking group but did not feel comfortable in that setting – talking in the garden came more naturally to her (R6). Conversely, another respondent argued that support is better received in a talking group. Others stated that they were not looking for peer support and that it was more of an added benefit than an explicit goal. Hence, while respondents valued the option to talk to others in the garden and to be supported and understood by other participants in a similar, positive way, they did not agree on whether this supportive environment 'counted' as social peer support. This relates to their differing ideas about wanting or needing support, their differing experiences with talking groups and, simply, their differing understanding of what social peer support entails. Moreover, the score cards used to evaluate the project showed that participants considered gardening and physical exercise to be more important goals than social peer support. In other words, most respondents were not specifically looking for social peer support and therefore did not value the garden in such terms, but they did appreciate the supportive environment they encountered. Indeed, participants agreed that a project like this would not need any specific attention to social peer support because it happens 'naturally'.

Although the project was generally well received, the pilot was not followed up. The research team who initiated the pilot did not secure funding for a larger study that would take the lessons learnt into account. Although the cancer support centre happily and actively hosted the pilot, the centre chose not to continue the project because of its clientele's limited interest in gardening, in combination with the efforts that the project would require in terms of supervision. The gardening containers were removed from the cancer support centre's grounds and gifted to interested organisations in Almere.

Discussion

The pilot project had some limitations – most notably, the limited number of participants, which was low even for a qualitative enquiry. Second, while aiming for interdisciplinarity is often considered worthwhile, this became a complicating factor in our pilot study, as the project struggled with two sets of aims and methods that were hard to fully and meaningfully integrate. Efforts directed towards creating a multidisciplinary project came, unfortunately, at the expense of a solid conceptual foundation. Third, several (design) choices were based on practical realities rather than on clear research hypotheses. That said, these issues are apparent consequences of working in a real-life, non-controlled situation, in which people with different disciplinary backgrounds collaborate. After all, the project was a pilot. Finally, the research team did not include any landscape architects or garden designers. We do recognise that what makes a garden is a matter of context, perception, and experience (St-Denis 2007); however, we also perceive that the garden as we designed it was fairly limited: there was no grass, no trees, and not much green besides the ten containers. Although the cancer support centre itself is considered a safe place by the participants and our garden was somewhat private, with the building on one side and a hedge on the other, it would be difficult to argue that the garden would 'offer the visitor a rich variety of experiences' or '[activate] all the senses' (Stigsdotter and Grahn 2002, p. 61). The garden we created was alarmingly close to Stigsdotter and Grahn's description of utilitarian, but somewhat uninspiring, cultivation beds.

> Many so-called gardens focusing on horticultural therapy have rational cultivation beds adjacent to care institutions. These cultivation beds, often raised, are not planned or designed to be part of the construction and composition of a garden room. On the other hand, the cultivation beds may be planned so that the cultivation functions well, the accessibility for the patients is excellent, water is within easy reach, convenient storehouses are nearby, etc. But there is hardly a garden room as such.
>
> (2002, pp. 63–64)

From this viewpoint, our gardening project was not very much like a garden room. Nevertheless, in designing the garden, we took the participants and their ideas as a starting point. We offered a group activity which started with a shared coffee and was guided by experienced gardeners, and which involved activities like planting, some weeding, and harvesting. We did not offer therapy or therapists, but we hoped to create a pleasant atmosphere that might entice social peer support and physically activate people. In that sense, the garden aligns with the definition of therapeutic horticulture, as offered by Gonzalez et al. (2011b), or of a restorative garden aimed at reducing stress and regaining energy (Thaneshwari et al. 2018).

Despite the project's aim to influence patients' lifestyles in terms of diet and physical exercise, the physical effects of the gardening activity were minimal. (Please recall that, due to the small number of participants, these findings have not been published.) Although the questionnaires revealed that participants ate slightly more vegetables over the course of the project, the physical tests showed that the gardening activity had limited, if any, physical impact. The accelerometers indicated that the activity should be considered light or even sedentary. Weight and BMI did not change during the course of the pilot project. Interviewees themselves did not consider the gardening experience as physically challenging either. That said, although all participants stated that a future gardening project should be larger than 10 m^2 and four out of five felt that gardening should take longer than 90 minutes, respondents disagreed about whether the gardening activity should have been more physically demanding. Some participants were content with the project not being physically challenging; others were disappointed by the lack of exercise. Similarly, some participants would have preferred to garden in the ground; others enjoyed the benefits of container gardening. Indeed, some participants felt that it was fine for the activity to be relatively easy physically.

Overall, at face value, the project did not result in adherence to lifestyle guidelines – despite the intention that gardening would lead to routinised engagement in a healthy activity. The garden was simply too small and the harvest too limited to have much effect in terms of these lifestyle guidelines. That said, most participants continued gardening when the project had finished; two of them in a large allotment. Moreover, the project managed to ensure that participants left the house and that they engaged in meaningful activity. Gardening, therefore, appears to have served as a 'cue to action', drawing participants outside (Cases et al. 2016, p. 202). However, promoting lifestyle change amongst cancer patients is problematic. On an individual level, there are significant barriers to behavioural change, as many patients lack the aspiration, capacity, or energy to make lifestyle changes while dealing with the physical and mental effects of a cancer diagnosis and treatment (Courneya et al. 2003). Individual lifestyle guidelines may, therefore, not be the best route to supporting improvements in health and wellbeing for this vulnerable group. Although those guidelines did not play a very large role in our project – considering that more attention was given to creating a pleasant place to garden than to meeting these guidelines – targeting lifestyle aspects such as diet and exercise led to the setting up of 'measures of success' that aligned with those goals. The fact that measurable improvements in physical stamina or a reduction in BMI was not found to be an outcome of the pilot meant that projected results were not met, which affected perceptions of the project's success. An important insight stemming from this project, therefore, is that, despite good intentions about interdisciplinary work, our ideas about how to support health differed. More importantly, perhaps, the differences lay at the foundations of our perspectives and often remained unspoken, implicitly creating starting points and guiding decisions that, only in retrospect, revealed the tension between positive and risk-focused approaches to health, as discussed in the introductory part of this chapter.

How the garden and gardening activities are designed and, even more crucially, how 'success' is measured depend on an underlying conception and model of health. As argued, therapeutic horticulture can be mobilised in different ways, depending on whether projects are underpinned by a positive or risk-focused approach to health. Our Healing Garden project embodied elements of both but fit neither category perfectly: while lifestyle guidelines played a role in the set-up of the garden, they did not enjoy full focus. Thus, although there were no landscape architects involved in our project team and we had limited attention for creating an actual garden that would be pleasant to spend time in, we did try to take the participants' wishes and needs into account. We thought about people who might want to garden at table height and involved a trained gardener for support as we wanted to ensure a positive experience, but we did not think about scents or colours, or the feeling of the ground under participants' feet. In sum, although our Healing Garden pilot project did not 'answer to the traditional definition of a Western garden' and was, instead, 'entirely focused ... on the activity' (Stigsdotter and Grahn 2002, p. 66), we did endeavour to create a pleasant environment in which to participate in the activity of gardening.

Importantly, gardening was perceived as a valuable activity. Not only did the participants really enjoy it, but they also had the opportunity to spend time in a quietly supportive environment that offered a moment of respite. This is mirrored in other work. Heliker et al. (2001, p. 41) argue that the community aspect of gardening is crucial, as it 'occurs within a restful, non-competitive environment and allows the development of a community of mutual caring and being cared-for'. Similarly, Gonzalez et al. (2011b) have reported on the importance of the social component of the intervention. Participants in that study affirmed that they enjoyed the group atmosphere and composition, in part because it consisted of people in the same situation. Another important finding is that participants simply enjoyed the gardening activity, which may be particularly important in the difficult times of illness. This aspect of therapeutic horticulture is also in line with other literature, which finds that participants evaluated working in a garden as meaningful in various settings (Brown et al. 2016; Gonzalez et al. 2011a; Park et al. 2016).

Conclusion

Our findings demonstrate that framing gardening (only) as a lifestyle intervention limits the view on its broader impacts, thereby overlooking more holistic or relational aspects of wellbeing, as well as aspects of resilience in coping with illness and recovery, or whatever life throws at you. Here, approaches applying a therapeutic landscape framework are able to show how cultivated landscapes, such as gardens, can offer opportunities for benefits to physical, mental/emotional, and social wellbeing. Even when it is a bit of a stretch to speak of a true garden – for instance, because the garden is not much more than a collection of gardening containers next to a treatment facility, as in our case – participants can consider them a valuable, comforting space. Such approaches hold great value in advancing our understanding of how these benefits arise (for example, in gardens), for different people, at different times, and in a variety of ways.

Declaration of interest

The authors report no potential conflict of interest.

Acknowledgements

The authors would like to thank the participants of the Healing Gardens project for their involvement and trust. We also thank the staff of Parkhuys Almere for welcoming our pilot project, and the Idemas for so enthusiastically guiding the gardeners in the weekly sessions. Finally, a word of thanks to Carleen Laloli, Iris Rijnaarts, Nicole de Roos, Jan Eelco Jansma, Ellen Kampman, and the other members of the team which made this project happen.

References

Alaimo, KE, Packnett, E, Miles, RA and Kruger, DJ 2008, 'Fruit and vegetable intake among urban community gardeners', *Journal of Nutrition Education and Behavior*, vol. 40, no. 2, pp. 94–101. https://doi.org/10.1016/j.jneb.2006.12.003

Biglin, J 2020, 'Embodied and sensory experiences of therapeutic space: Refugee place-making within an urban allotment', *Health & Place*, vol. 62. https://doi.org/10.1016/j.healthplace.2020.102309

Blair, D, Giesecke, CC and Sherman, S 1991, 'A dietary, social and economic evaluation of the Philadelphia Urban Gardening Project', *Journal of Nutrition Education*, vol. 23, no. 4, pp. 161–167.

Brown, GC, Bos, E, Brady, G and Kneafsey, M 2016, 'An evaluation of the Master Gardener programme at HMP Rye Hill: A horticultural intervention with substance misusing offenders', *Prison Service Journal*, no. 225, pp. 45–51.

Cases, MG, Frugé, AD, De Los Santos, JF, Locher, JL, Cantor, AB, Smith, KP, Glover, TA, Cohen, HJ, Daniel, M, Morrow, CD, Moellering, DR and Demark-Wahnefried, W 2016, 'Detailed methods of two home-based vegetable gardening intervention trials to improve diet, physical activity, and quality of life in two different populations of cancer survivors', *Contemporary Clinical Trials*, vol. 50, pp. 201–212. https://doi.org/10.1016/j.cct.2016.08.014

CCFR (Centre for Child and Family Research) 2003, 'Social and therapeutic horticulture: Evidence and messages from research. Project report', *CCFR Evidence*, vol. 6. https://dspace.lboro.ac.uk/dspace-jspui/bitstream/2134/2928/1/Evidence6.pdf

Conradson, D 2005, 'Landscape, care and the relational self: Therapeutic encounters in rural England', *Health & Place*, vol. 11, no. 4, pp. 337–348. https://doi.org/10.1016/j.healthplace.2005.02.004

Courneya, KS, Mackey, JR, Jones, LW, Field, CJ and Fairey, AS 2003, 'Randomized controlled trial of exercise training in postmenopausal breast cancer survivors: Cardiopulmonary and quality of life outcomes', *Journal of Clinical Oncology*, vol. 21, no. 9, pp. 1660–1668. https://doi.org/10.1200/JCO.2003.04.093

Davison, C and Davey Smith, G 1995, 'The baby and the bath water. Examining socio-cultural and free-market critiques of health promotion' in R Bunton, S Nettleton and R Burrows (eds), *The sociology of health promotion. Critical analyses of consumption, lifestyle and risk*, Routledge, London.

Demark-Wahnefried, W, Aziz, NM, Rowland, JH and Pinto, BM 2005, 'Riding the crest of the teachable moment: Promoting long-term health after the diagnosis of cancer', *Journal of Clinical Oncology*, vol. 23, no. 24, pp. 5814–5830. https://doi.org/10.1200/JCO.2005.01.230

Drakou, A, de Vreese, R, Lofthus, T and Muscat, J 2011, 'Motivating people to be active in green spaces' in K Nilsson, M Sangster, C Gallis, T Hartig, S de Vries, K Seeland and J Schipperijn (eds), *Forests, trees and human health*, Springer, London, pp. 283–308.

Duff, C 2012, 'Exploring the role of "enabling places" in promoting recovery from mental illness: A qualitative test of a relational model', *Health & Place*, vol. 18, no. 6, pp. 1388–1395. https://doi.org/10.1016/j.healthplace.2012.07.003

Embuldeniya, G, Veinot, P, Bell, E, Bell, M, Nyhof-Young, J, Sale, JE, and Britten, N 2013, 'The experience and impact of chronic disease peer support interventions: A qualitative synthesis', *Patient Education and Counseling*, vol. 92, no. 1, pp. 3–12. https://doi.org/10.1016/j.pec.2013.02.002

Eom, CS, Shin, DW, Kim, SY, Yang, HK, Jo, HS, Kweon, SS, Kang, YS, Kim, JH, Cho, BL and Park, JH 2013, 'Impact of perceived social support on the mental health and health-related quality of life in cancer patients: Results from a nationwide, multicenter survey in South Korea', *Psycho-Oncology*, vol. 22, no. 6, pp. 1283–1290. https://doi.org/10.1002/pon.3133

Firth, C, Maye, D and Pearson, D 2011, 'Developing "community" in community gardens', *Local Environment*, vol. 16, no. 6, pp. 555–568. https://doi.org/10.1080/13549839.2011.586025

Fisher, M, Baum, F, MacDougall, C, Newman, L and McDermott, D 2015, 'A qualitative methodological framework to assess uptake of evidence on social determinants of health in health policy', *Evidence &Policy: A Journal of Research, Debate and Practice*, vol. 11, no. 4, pp. 491–507. https://doi.org/10.1332/174426414X14170264741073

Frensham LJ, Parfitt, G and Dollman, J 2018, 'Effect of a 12-week online walking intervention on health and quality of life in cancer survivors: A quasi-randomized controlled trial', *International Journal of Environmental Research and Public Health*, vol. 15, no. 10, article 2081. https://doi.org/10.3390/ijerph15102081

Fu, MR, Axelrod, D, Guth, AA, Cleland, CM, Ryan, CE, Weaver, KR, Qiu, JM, Kleinman, R, Scagliola, J, Palamar, JJ and Melkus, GD 2015, 'Comorbidities and quality of life among breast cancer survivors: A prospective study', *Journal of Personalized Medicine*, vol. 5, no. 3, pp. 229–242. https://doi.org/10.3390/jpm5030229

Gesler, W 1992, 'Therapeutic landscapes: Medical issues in light of the new cultural geography', *Social Science & Medicine*, vol. 34, no. 7, pp. 735–746.

Gesler, W 2003, *Healing places*, Rowman & Littlefield, Oxford.

Giddens, A 1991, *Modernity and self-identity: Self and society in the late modern age*, Stanford University Press, Stanford, CA.

Godziewski, C 2020, 'Evidence and power in EU governance of health promotion: Discursive obstacles to a "health in all policies" approach', *Journal of Common Market Studies*, vol. 58, no. 5, pp. 1307–1324. https://doi.org/10.1111/jcms.13042

Golden, S and Earp, J 2012, 'Social ecological approaches to individuals and their contexts: Twenty years of health education and behaviour health promotion interventions', *Health Education & Behavior*, vol. 39, no. 3, pp. 364–372. https://doi.org/10.1177/1090198111418634

Gonzalez, MT, Hartig, T, Patil, GG, Martinsen, EW and Kirkevold, M 2009, 'Therapeutic horticulture in clinical depression: A prospective study', *Research and Theory for Nursing Practice*, vol. 23, no. 4, pp. 312–328. https://doi.org/10.1891/1541-6577.23.4.312

Gonzalez, MT, Hartig, T, Patil, GG, Martinsen, EW and Kirkevold, M 2010, 'Therapeutic horticulture in clinical depression: A prospective study of active components', *Journal of Advanced Nursing*, vol. 66, no. 9, pp. 2002–2013. https://doi.org/10.1111/j.1365-2648.2010.05383.x

Gonzalez, MT, Hartig, T, Patil, GG, Martinsen, EW and Kirkevold, M 2011a, 'A prospective study of existential issues in therapeutic horticulture for clinical depression', *Issues in Mental Health Nursing*, vol. 32, no. 1, pp. 73–81. https://doi.org/10.3109/01612840.2010.528168

Gonzalez, MT, Hartig, T, Patil, GG, Martinsen, EW and Kirkevold, M 2011b, 'A prospective study of group cohesiveness in therapeutic horticulture for clinical depression', *International Journal of Mental Health Nursing*, vol. 20, no. 2, pp. 119–129. https://doi.org/10.1111/j.1447-0349.2010.00689.x

Hartig, T 2004, 'Restorative environments' in C Spielberger (ed), *Encyclopedia of Applied Psychology, vol. 3*, Elsevier Academic Press, San Diego, CA, pp. 273–279.

Haubenhofer, DK, Elings, M, Hassink, J and Hine, RE 2010, 'The development of green care in Western European countries', *Explore*, vol. 6, no. 2, pp. 106–111. https://doi.org/10.1016/j.explore.2009.12.002

Helgeson, VS and Cohen, S 1996, 'Social support and adjustment to cancer: Reconciling descriptive, correlational, and intervention research', *Health Psychology*, vol. 15, no. 2, pp. 135–148. https://doi.org/10.1037//0278-6133.15.2.135

Heliker, D, Chadwick, A and O'Connell, T 2001, 'The meaning of gardening and the effects on perceived well being of a gardening project on diverse populations of elders', *Activities, Adaptation & Aging*, vol. 24, no. 3, pp. 35–56. https://doi.org/10.1300/J016v24n03_03

Horowitz, S 2012, 'Therapeutic gardens and horticultural therapy: Growing roles in health care', *Alternative and Complementary Therapies*, vol. 18, no. 2, pp. 78–83. https://doi.org/10.1089/act.2012.18205

Karanikolos, M, Mladovsky, P, Cylus, J, Thomson, S, Basu, S, Stuckler, D, Mackenbach, J and McKee, M 2013, 'Financial crisis, austerity, and health in Europe', *The Lancet*, vol. 381, no. 9874, pp. 1323–1331. https://doi.org/10.1016/S0140-6736(13)60102-6

Korp, P 2010, 'Problems of the healthy lifestyle discourse', *Sociology Compass*, vol. 4, no. 9, pp. 800–810. https://doi.org/10.1111/j.1751-9020.2010.00313.x

Marsh, P 2020, 'Agroecology as public health: The island example of Tasmania' in M Egerer and H Cohen (eds), *Urban agroecology: Interdisciplinary research and future directions*, Taylor & Francis, London, pp. 143–154.

Marsh, P, Gartrell, G, Egg, G, Nolan, A and Cross, M 2017, 'End-of-Life care in a community garden: Findings from a participatory action research project in regional Australia', *Health & Place*, vol. 45, pp. 110–116. https://doi.org/10.1016/j.healthplace.2017.03.006

Milligan, C, Gatrell, A and Bingley, A 2004, '"Cultivating health": Ttherapeutic landscapes and older people in northern England', *Social Science & Medicine*, vol. 58, no. 9, pp. 1781–1793. https://doi.org/10.1016/S0277-9536(03)00397-6

Mossabir, DR, Froggatt, K and Milligan, C 2021, 'Therapeutic landscape experiences of everyday geographies within the wider community: A scoping review', *Social Science & Medicine*, vol. 279. https://doi.org/10.1016/j.socscimed.2021.113980

Park, SA, Lee, AY, Son, KC, Lee, WL and Kim, DS 2016, 'Gardening intervention for physical and psychological health benefits in elderly women at community centers', *HortTechnology*, vol. 26, no. 4, pp. 474–483. https://doi.org/10.21273/HORTTECH.26.4.474

Pisu, M, Demark-Wahnefried, W, Kenzik, KM, Oster, RA, Lin, CP, Manne, S, Alvarez, R and Martin, MY 2017, 'A dance intervention for cancer survivors and their partners (RHYTHM)', *Journal of Cancer Survivorship: Research and Practice*, vol. 11, no. 3, pp. 350–359. https://doi.org/10.1007/s11764-016-0593-9

Pitt, H 2014, 'Therapeutic experiences of community gardens: Putting flow in its place', *Health & Place*, vol. 27, pp. 84–91. https://doi.org/10.1016/j.healthplace.2014.02.006

Powell, K, Thurston, M and Bloyce, D 2017, 'Theorising lifestyle drift in health promotion: Explaining community and voluntary sector engagement practices in disadvantaged areas', *Critical Public Health*, vol. 27, no. 5, pp. 554–565. https://doi.org/10.1080/09581596.2017.1356909

Roszak, T, Gomes, M and Kanner, A (eds) 1995, *Ecopsychology: Restoring the Earth, healing the mind*, Sierra Club Books, San Francisco.

Sanchez, L and Liamputtong, P 2017, 'Community gardening and health-related benefits for a rural Victorian town', *Leisure Studies*, vol. 36, no. 2, pp. 269–281. https://doi.org/10.1080/02614367.2016.1250805

Sarfati, D, Koczwara, B and Jackson, C 2016, 'The impact of comorbidity on cancer and its treatment', *CA: A Cancer Journal for Clinicians*, vol. 66, no. 4, pp. 337–350. https://doi.org/10.3322/caac.21342

Söderback, I, Söderström, M and Schälander, E 2004, 'Horticultural therapy: The "healing garden" and gardening in rehabilitation measures at Danderyd hospital rehabilitation clinic, Sweden', *Pediatric Rehabilitation*, vol. 7, no. 4, pp. 245–260. https://doi.org/10.1080/13638490410001711416

Stan, DL, Croghan, KA, Croghan, IT, Jenkins, SM, Sutherland, SJ, Cheville, AL and Pruthi, S 2016, 'Randomized pilot trial of yoga versus strengthening exercises in breast cancer survivors with cancer-related fatigue', *Support Care Cancer*, vol. 24, no. 9, pp. 4005–4015. https://doi.org/10.1007/s00520-016-3233-z

St-Denis, B 2007, 'Just what is a garden?', *Studies in the History of Gardens & Designed Landscapes*, vol. 27, no. 1, pp. 61–76. https://doi.org/10.1080/14601176.2007.10435457

Steinfeld, E and Danford, GS (eds) 1999, *Enabling environments: Measuring the impact of environment on disability and rehabilitation*, Kluwer Academic/Plenum Publishers, New York.

Stigsdotter, UA and Grahn, P 2002, 'What makes a garden a healing garden?', *Journal of Therapeutic Horticulture*, vol. 13, pp. 60–69.

Taylor, SE, Falke, RL, Shoptaw, SJ and Lichtman, RR 1986, 'Social support, support groups, and the cancer patient', *Journal of Consulting and Clinical Psychology*, vol. 54, no. 5, pp. 608–615.

Thaneshwari, T, Kumari, P, Sharma, R and Sahare, HA 2018, 'Therapeutic gardens in healthcare: A review', *Annals of Biology*, vol. 34, no. 2, pp. 162–166.

Ulrich, R 1981, 'Natural versus urban scenes: Some psychophysiological effects', *Environment & Behaviour*, vol. 13, no. 5, pp. 523–556.

Veen, EJ, Bock, BB, Van den Berg, W, Visser, AJ and Wiskerke, JSC 2016, 'Community gardening and social cohesion: Different designs, different motivations', *Local Environment*, vol. 21, no. 10, pp. 1271–1287. https://doi.org/10.1080/13549839.2015.1101433

Walker, C, Hart, A and Hanna, P 2017, *Building a new community psychology of mental health: Spaces, places, people and activities*, Palgrave Macmillan, London.

Wang, D and MacMillan, T 2013, 'The benefits of gardening for older adults: A systematic review of the literature', *Activities, Adaptation & Aging*, vol. 37, no. 2, pp. 153–181. https://doi.org/10.1080/01924788.2013.784942

Williams, A (ed) 2007, *Therapeutic landscapes*, Routledge, Abingdon.

World Health Organization 2010, *Global recommendations on physical activity for health*, WHO, Geneva. https://www.who.int/publications/i/item/9789241599979

World Health Organization n.d., *Social Determinants of Health*. https://www.who.int/health-topics/social-determinants-of-health#tab=tab_1

Wu, HS and Harden, JK 2015, 'Symptom burden and quality of life in survivorship: A review of the literature', *Cancer Nursing*, vol. 38, no. 1, E29–E25. https://doi.org/10.1097/NCC.0000000000000135

Zabalegui, A, Cabrera, E, Navarro, M and Cebria, MI 2011, 'Perceived social support and coping strategies in advanced cancer patients', *Journal of Research in Nursing*, vol. 18, no. 5, pp. 409–420.

5 Mental health outcomes associated with gardening

A scoping review

Selma Lunde Fjaestad, Jessica L Mackelprang, Takemi Sugiyama, and Jonathan Kingsley

Introduction

Mental ill health affects approximately 14 per cent of the world's population and is the leading cause of global disease burden (Doran and Kinchin 2019; World Health Organization [WHO] 2013). Improving mental health is a global health priority (Patel et al. 2018; WHO 2018). With more than two-thirds of the world's population projected to live in urbanised areas by 2050 (United Nations 2014), the impact of urbanisation on mental health must be considered. A meta-analysis that reviewed differences in mental disorders between rural and urban areas using data from 20 population surveys found that the prevalence of psychiatric disorders was significantly higher in urbanised areas (Peen et al. 2010). This suggests that urbanisation may pose a threat to mental health, in part due to reduced contact with nature (Ventriglio et al. 2021), as it is well documented that disconnection from nature is associated with adverse health outcomes (Callaghan et al. 2021; Cox et al. 2018; Jiang et al. 2016; Sarkar et al. 2018).

Gardening is a health-enhancing activity in which people can engage across the lifespan and in diverse contexts, such as private garden spaces, community parks, school settings, and hospitals (Kabisch et al. 2017; Kingsley et al. 2021 Spano et al. 2020). Soga et al. (2017a, p. 93) define gardening as 'an activity in which people grow, cultivate, and take care of plants (flowers or vegetables) for non-commercial use'. Gardens are commonly classified into three types of settings, as outlined in Table 5.1. Table 5.1 does not capture the diverse spectrum of garden types that exist (e.g., verge gardens, urban agriculture enterprises, market gardens, school gardens, or kitchen gardens).

Research investigating the pathways through which gardening can benefit human health demonstrates that gardening has the potential to increase human–nature interaction. The health and wellbeing benefits of increased human–nature interaction and nature exposure are supported by both theoretical and empirical evidence (Lin et al. 2018; Scott et al. 2015; Soga and Gaston 2016; Spano et al. 2020). The restorative and rehabilitative effects of gardening have been documented for different groups (e.g., psychiatric inpatients) across the lifespan (e.g., children and older adults; Callaghan et al. 2021;

DOI: 10.4324/9781003355731-7

Table 5.1 Common types of garden setting

Type of garden	*Definition*
Allotment garden	A garden that involves 'a piece of land allocated for personal use on a lease or rent basis' (Kingsley et al. 2019, p. 3).
Community garden	A garden that involves 'individual to collective plots that are diverse in size, produce, governance structures, geography and function ... community gardens require individuals to converge to share space, resources, food and knowledge in a collective and cooperative way' (Kingsley et al. 2020, p. 3)
Domestic garden	A garden that is 'adjacent to a domestic dwelling, which itself is either privately owned or rented' (Cameron et al. 2012, p. 129)

Wendelboe-Nelson et al. 2019). Theories that posit the therapeutic benefits of gardening – including the Attention Restoration Theory (Kaplan and Kaplan 1989), Stress Recovery Theory (Berto 2014), and the biophilia hypothesis (Ulrich 1993; Wilson 1984) – attempt to explain the effects of interaction with nature through a combination of neurological, cognitive, emotional, and evolutionary pathways.

Several reviews exploring structured gardening activities and how they interact with physical and psychological health outcomes have been published to date. Soga et al. (2017b) reviewed studies that examined the physical and psychological health benefits of a range of structured gardening programmes (e.g., horticultural therapy interventions) and unstructured gardening activities (e.g., daily recreational gardening). Howarth et al. (2020) conducted a review of empirical studies and systematic reviews to map the state of the literature on the impact of gardens and gardening on health and wellbeing, and to guide health practitioners and policymakers to incorporate gardening spaces and activities into healthcare systems. These reviews found consistent and significant positive associations between gardening and a range of health measures, across heterogeneous populations. These reviews included intervention studies that assessed the impacts of structured and prescribed gardening programmes, such as horticultural therapy, as well as observational studies in which gardening was not prescribed (i.e., unstructured gardening).

Structured gardening programmes can include additional curricula (Spano et al. 2020), such as information sessions (e.g., education about plants and vegetables, information on how to garden), briefing sessions (e.g., group discussions), and social support groups. Aligning with Soga et al.' (2017a) definition of gardening, these non-gardening activities may influence psychological outcomes, especially given their social component, as social interaction is associated with enhanced wellbeing (Thoits 2011). Thus, it is possible that mental health benefits observed in reviews of intervention studies may be driven, in part, by activities unrelated to gardening. For this reason, a scoping review that focuses on unstructured gardening is needed to complement existing reviews which have summarised the combined impact of structured and unstructured gardening interventions and additional support interventions (e.g., Howarth et al. 2020; Soga et al. 2017b; Spano et al. 2020). Synthesising the extant research on the relationship between unstructured gardening and psychological outcomes will enable a deeper understanding of the role gardening may play in wellbeing in day-to-day life. The aim of the scoping review presented in this chapter is to summarise the current state of knowledge on the relationship between unstructured gardening and mental health outcomes among adults. To effectively compare findings across studies, we focus on research using quantitative tools to measure psychological wellbeing.

Methods

To generate a comprehensive review of the existing research on mental health and gardening, we employed a scoping review methodology. A scoping review does not include assessments of study quality as a basis for inclusion; rather, it aims to synthesise a wider breadth of research (Munn et al. 2018). Arksey and O'Malley's (2005) validated five-stage framework was applied to (i) determine the research question; (ii) identify and select relevant studies; (iii) chart the data; (iv) collate and summarise the data; and (v) report our findings. This enabled us to map the research landscape on the psychological outcomes associated with gardening as an unstructured and unprescribed activity; identify keywords; and finalise search terms. The review process followed the Preferred Reporting Items for Systematic Reviews and Meta-Analyses (PRISMA) model (Moher et al. 2009).

Procedure

A systematic search of peer-reviewed journal articles was conducted using four databases: EBSCO Host, PsycInfo, Scopus, and Web of Science. Three sets of key words were used for database searches: gardening ('garden*'), outcome ('mental*' OR 'well*' OR 'psychological health'), and age group ('adult'). The search was conducted by the first author (SLF) and concluded on 26 May 2021. Two authors (SLF and JK) independently screened articles based on their title and abstract; full-text articles were subsequently screened. Discrepancies in the selection process were resolved by discussion between the two authors. EndNote 20 was used to manage articles throughout the screening and full-text review process.

Inclusion and exclusion criteria

Articles were eligible if they met the following inclusion criteria: (i) peer-reviewed journal article; (ii) published between 2000 and 2021; (iii) full-text available in English; (iv) presented quantitative measures of the associations between gardening as an unstructured, stand-alone activity and one or more mental health outcomes; and (v) included adult samples. Studies were excluded if they were (i) intervention studies that examined the effect of gardening-related programmes; (ii) conference proceeding papers or book chapters; or (iii) did not report empirical findings (e.g., reviews, meta-analysis, perspectives/opinions).

Data extraction

The following information was extracted from the included articles: location of study; study aim(s); study design; data collection method (e.g., questionnaires); participant characteristics (e.g., age, gender); garden type (e.g., community, domestic); psychological outcome measures; and key psychological findings.

Results

Characteristics of included studies

The initial database searches identified 753 articles, from which 138 duplicates were removed, leaving 615 articles to be screened. Screening based on title and abstract reduced the pool of articles to 39; of these, 13 met the inclusion criteria after full-text review. Figure 5.1 shows the PRISMA flow diagram, adapted from Moher et al. (2009). Sixty-two percent of the studies (n = 8) were conducted in the United Kingdom and the United States; the remaining studies were conducted in Portugal, the Netherlands, Japan, Singapore, and Australia (see Figure 5.2). The sample sizes ranged from 28 to 7,814 participants. Participants' ages ranged from 16 to over 95 years. Characteristics of included studies are summarised in Table 5.2.

Of the 13 included studies, 92 per cent (n = 12) were cross-sectional design (Corley et al. 2021; de Bell et al. 2020; Hawkins et al. 2011; Koay and Dillon 2020; Mourão et al. 2019; Park et al. 2009; Scott et al. 2020; Soga et al. 2017a; Sommerfeld et al. 2010; Van den Berg et al. 2010; Waliczek et al. 2005; Webber et al. 2015). One study employed a case-control design (Wood et al. 2016).

The 13 studies included in this scoping review varied in the type of garden setting in which gardening was studied. Six articles focused on gardening within allotment spaces (Hawkins et al. 2011; Mourão et al. 2019; Soga et al. 2017a; Van den Berg et al. 2010; Webber et al. 2015; Wood et al. 2016), three focused on gardening within any type of garden space (i.e., unspecified; Park et al. 2009; Sommerfeld et al. 2010; Waliczek et al. 2005), two articles focused on gardening activities within domestic gardens (Corley et al. 2021; de Bell et al. 2020), and two focused on gardening in both community and domestic garden settings (Koay and Dillon 2020; Scott et al. 2020).

Ten studies included a combination of psychological and physiological outcome measures (Corley et al. 2021; de Bell et al. 2020; Hawkins et al. 2011; Koay and Dillon 2020; Park et al. 2009; Scott et al. 2020; Soga et al. 2017a; Sommerfeld et al. 2010; Van den Berg et al. 2010; Waliczek et al. 2005), while

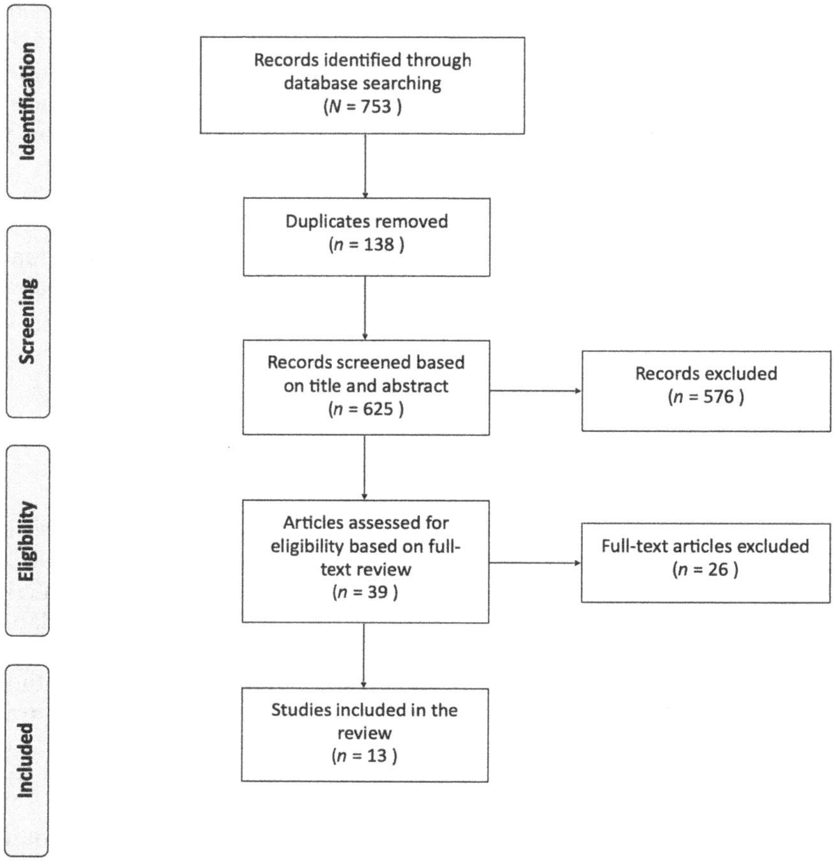

Figure 5.1 PRISMA flowchart diagram.

Note. PRISMA = Preferred Reporting Items for Systematic Reviews and Meta-Analyses.

three studies collected data from psychological measures exclusively (Mourão et al. 2019; Webber et al. 2015; Wood et al. 2016). The following sections describe outcomes and instruments used in the included studies. Table 5.3 presents a summary of the validated measures used across the included studies.

Self-reported measures of mental health

All included studies employed self-reported measures of mental health: 12 administered self-reported questionnaires (Corley et al. 2021; Hawkins et al. 2011; Koay and Dillon 2020; Mourão et al. 2019; Park et al. 2009; Scott et al. 2020; Soga et al. 2017a; Sommerfeld et al. 2010; Van den Berg et al. 2010; Waliczek et al. 2005; Webber et al. 2015; Wood et al. 2016) and one used structured, face-to-face questionnaires (de Bell et al. 2020).

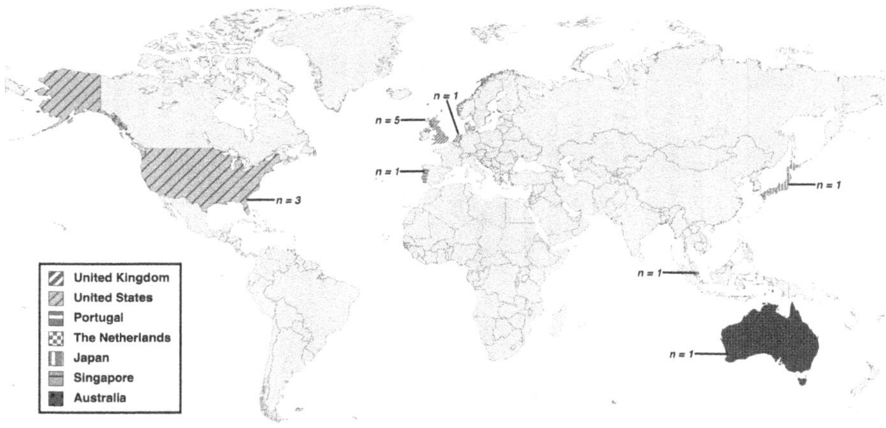

Figure 5.2 Locations of included studies.

Note. Created with mapchart.net.

Wellbeing

General wellbeing was assessed in nine studies. The 36-item Short Form Health Survey (SF-36) was utilised in three (Hawkins et al. 2011; Park et al. 2009; Scott et al. 2020) to measure eight domains of physical and mental health. One study used a single item from the SF-36 (Scott et al. 2020) to assess quality of life. The seven-item Personal Wellbeing Index was employed in two studies (Koay and Dillon 2020; Mourão et al. 2019) to measure subjective wellbeing and quality of life; one utilised the Portuguese version (i.e., Bem-Estar Pessoal Scale; Mourão et al. 2019). Evaluative and eudaimonic wellbeing were assessed in two studies using two questions on subjective wellbeing generated by the UK Office for National Statistics (Corley et al. 2021; de Bell et al. 2020). The 21-item Questionnaire for Eudaimonic Well-Being was employed in one study (Webber et al. 2015). This questionnaire measures six aspects of eudaimonic wellbeing (i.e., self-discovery, perceived development of one's best potentials, a sense of purpose and meaning, investment of significant effort in pursuit of excellence, intense involvement in activities, enjoyment of activities as personally expressive). One study administered the 24-item WHOQOL-BREF (Webber et al. 2015) to measure quality of life in four domains (physical, psychological, social, and environmental). Two studies assessed general mental health using the 12-item General Health Questionnaire (Soga et al. 2017a; Wood et al. 2016). Scott et al. (2020) evaluated wellbeing using the 24-item Attitudes to Ageing Questionnaire, which measures quality of life and subjective experiences of ageing through psychological loss, physical change, and psychological growth.

Table 5.2 Key characteristics and findings of included studies ($n = 13$)

Citation	Location	Study design	Data collection	Sample	Garden type	Psychological measurement tool(s)	Key findings
1 Corley et al. (2021)	UK	Cross-sectional	Participants from the Lothian Birth Cohort 1936 completed an online survey	Older adults ($n = 171$); 84 ± 0.5 years; 47% women	Domestic garden	Single-item self-rated questions on subjective, evaluative, and eudaimonic wellbeing; anxiety about COVID-19; anxiety and depressive symptoms were measured using the 21-item HADS.	Gardeners were significantly less likely to have a history of depressive symptoms compared to non-gardeners ($p = 0.015$), on the HADS. No significant differences between anxiety symptoms were present between gardeners/ non-gardeners. Those who gardened more from pre- to post-lockdown, experienced better emotional mental health ($p = 0.005$).
2 de Bell et al. (2020)	UK	Cross-sectional	Secondary data from English Monitor of Engagement with the Natural Environment, a structured survey (from the first seven waves 2009/10 – 2015/16)	English population ($n = 7,814$); 16–34 years, 35–64 years, 65+ years; 53% women	Domestic garden	Evaluative and eudaimonic wellbeing was measured using two questions on subjective wellbeing from the UK Office for National Statistics.	Using a garden was associated with better outcomes compared to not using a garden. Using the garden for gardening showed the highest evaluative wellbeing ($p < 0.001$), compared to those who relaxed in the garden and those did not use the garden (e.g., did not spend time in the garden). Gardening was significantly associated with eudaimonic wellbeing ($p < 0.001$).

3	Hawkins et al. (2011)	UK	Cross-sectional	Participants were grouped into: (1) allotment gardeners, (2) home gardeners, (3) outdoor walking group members, or (4) indoor exercise group participants. Participants completed the self-administered questionnaire.	Older adults (n = 94); 50–88 years (M = 67 ± 8.5 years); 68% women	Allotment and home garden	Stress levels were measured using the PSS-10; social support was assessed with the 10-item SPS; health-related quality of life the SF-36.	Only perceived stress was significantly lower in the allotment gardening group compared to the indoor exercise group ($p < 0.05$).
4	Koay and Dillon (2020)	Singapore	Cross-sectional	Participants were divided into (1) community gardeners, (2) home gardeners, and (3) those who engaged in non-gardening outdoor activities (control). All participants completed a self-administered questionnaire.	Adults who take part in outdoor activities (n = 111); 25–77 years (M = 53.40 ± 14.58 years); 61.3% female	Community and domestic garden	Resilience was measured using the 6-item BRS; stress the PSS-10; subjective wellbeing the 7-item PWI-A; self-esteem the 10-item RSE; optimism the 10-item Chinese Revised LOT-R.	Compared to outdoor activities, gardening had significant effects on PWI-A scores (p = 0.005) and LOT-R scores (p = 0,016); however, there was no significant effect on PSS-10, BRS, or RSE scores. Overall, community gardeners showed better mental health benefits compared to individual gardening or outdoor activities. Community gardeners reported higher levels of subjective wellbeing, resilience, and optimism.

(Continued)

Table 5.2 (Continued)

Citation	*Location*	*Study design*	*Data collection*	*Sample*	*Garden type*	*Psychological measurement tool(s)*	*Key findings*
5 Mourão et al. (2019)	Portugal	Cross-sectional	Participants were recruited from an urban organic allotment gardening community. All participants completed a self-administered questionnaire in a single assessment.	Adult allotment gardeners ($n = 65$); 46–65 years; 43.1% women	Allotment garden	Personal wellbeing was measured using the 7-item Bem-Estar Pessoal scale; happiness the 4-item FS scale.	Overall, wellbeing scores among gardeners were high; happiness scores averaged 7 ('a happier person'); one item, personal perspective of pessimism in relation to life, had a mean value of 3.26 which was below the neutral scale of 4; frequency of visit was significantly positively correlated with happiness scores on the SHS ($p < 0.000$).
6 Park et al. (2009)	USA	Cross-sectional	Community members, completed the Community Health Activities Model Program for Seniors questionnaire.	Older adults ($n = 53$); 58–86 years; 60.4% women	Any form of garden	Mental health was assessed using the SF-36.	There were no significant differences in mental health scores between gardeners and non-gardeners; the sample had better mental health compared to the general US population.

7	Scott et al. (2020)	Australia	Cross-sectional	Participants completed a self-administered gardening survey that consisted of qualitative and quantitative questions; administered online and via mail-out.	Older adults who garden for more than 1hr/week (n = 331); 60–95 years (M = 68.9 ± 7.36 years); 83% women	Community and domestic garden	Subjective health and quality of life was measured using two SF-36 items; attitudes to ageing using the 24-item AAQ; benefits of gardening were measured using 42 bespoke items.	The SF-36 did not significantly correlate with time spent gardening. Restoration benefits correlated with positive self-perceptions of ageing. Time spent gardening was significantly associated with restoration, attachment, physical, social identity, and purpose. Time spent gardening was not associated with AAQ subscales.
8	Soga et al. (2017a)	Japan	Cross-sectional	Participants who identified as allotment gardeners and non-gardeners completed surveys; gardeners completed additional qualitative questions.	Adult allotment gardeners or non-gardeners (n = 332); 61.5 years ± 16.8 years; 43% women	Allotment garden	Mental health was measured using the 12-item GHQ.	Allotment gardeners had significantly better mental health compared to non-gardeners ($p < 0.05$). Neither frequency nor duration of allotment gardening had a significant influence on mental health outcomes.
9	Sommerfeld et al. (2010)	USA	Cross-sectional	Participants completed a self-administered survey consisting of both qualitative and quantitative questions.	Older adults (n = 261); 50–90 (or older) years; 58.2% women	Any form of garden	Perceptions of life satisfaction were measured using the 20-item LSIA.	Gardeners had significantly higher LSIA scores compared to non-gardeners ($p = 0.001$): optimism ($p = 0.000$), zest of life ($p = 0.002$), energy level ($p = 0.004$), and congruence between desired and achieved goals ($p = 0.028$).

(Continued)

Table 5.2 (Continued)

Citation	Location	Study design	Data collection	Sample	Garden type	Psychological measurement tool(s)	Key findings
10 Van den Berg et al. (2010)	The Nether-lands	Cross-sectional	Participants completed a self-administered questionnaire consisting of qualitative and quantitative questions; gardening group completed additional section specific to allotment gardening.	Allotment gardeners and neighbours ($n = 184$); 20–87 years ($M = 58.7 \pm 12.8$ years); 51% women	Allotment garden	Stress was assessed using a 2-item self-report measure; life satisfaction the 8-item LSIA; loneliness using a 2-item measure; social contacts were assessed using 2 bespoke items.	Allotment gardening had a significant positive impact on stress, life satisfaction, and loneliness, with the effects on wellbeing being significantly greater for allotment gardeners compared to the control group ($p < 0.04$). Allotment gardening benefits were more significant in older (>62 years) compared to younger gardeners (<62 years).
11 Waliczek et al. (2005)	USA	Cross-sectional	Participants completed a self-administered survey.	Adult allotment gardeners or non-gardeners ($n = 443$); under 25–over 95 years; gardeners 68.8% female; non-gardeners 76.7% women	Any form of garden	Perceptions of life satisfaction were measured using the 20-item LSIA.	Self-identified gardeners showed significantly more positive LSIA results compared to non-gardeners ($p = 0.001$). Significant differences in scores were observed in energy level, optimism, zest for life, and physical self-concept.
12 Webber et al. (2015)	UK	Cross-sectional	Participants completed an online, self-administered gardening survey consisting of qualitative and quantitative questions	Adult allotment gardeners ($n = 171$); 24–78 years ($M = 50$ years); 67.8% women	Allotment garden	Eudaimonic wellbeing was measured using the 21-item QEWB; subjective wellbeing using the 24-item WHOQOL-BREF.	Comparing mean scores to the general population, allotment gardeners had significantly higher QEWB scores. Weekly time spent gardening was significantly positively associated with eudaimonic wellbeing.

13	Wood et al. (2016)	UK	Case-control	Participants in the case group completed a quantitative and qualitative questionnaire pre- and post- an unstructured allotment gardening session. Participants in the control group completed a one-off questionnaire during a supermarket trip.	Adult allotment gardeners or non-gardeners ($n = 269$); ($M = 55.6 \pm 13.6$ years); 44% women	Allotment garden	Self-esteem was measured using the 10-item RSE; mood using the 30-item POMS questionnaire; general mental health the GHQ-12	Self-esteem ($p < 0.05$) and mood ($p < 0.001$) were significantly better from pre- to post-allotment sessions ($p < 0.05$). There were no significant associations between length of participants tenure, time spent allotment gardening in previous 7 days, or time spent allotment gardening in the current session, and scores on the RSE or POMS. The RSE, POMS, and GHQ-12 scores for allotment gardeners were better for self-esteem, moods, and psychological functioning compared to non-gardeners.

Note. AAQ = Attitudes to Ageing Questionnaire; BRS = Brief Resilience Scale; FS = Felicidade Subjectiva; COVID-19 = coronavirus disease 2019; GHQ = General Health Questionnaire; HADS = Hospital Anxiety and Depression Scale; LSIA = Life Satisfaction Index-A; LOT-R = Life Orientation Test Revised; POMS = Profile of Mood States; PSS-10 = 10-item Perceived Stress Scale; PWI-A = 7-item Personal Wellbeing Index – Adult; QEWB = Questionnaire of Eudaimonic Wellbeing; RSE = Rosenberg Self-Esteem Scale; SF-36 = 36-item Short-Form Health Survey; SPS = Social Provisions Scale; UK = United Kingdom; USA = United States of America; WHOQOL-BREF = World Health Organization Quality of Life-Brief Version.

Table 5.3 Summary of measures used to assess mental health

Measure	*Description*
AAQ	The AAQ is a 24-item self-report measure to assess quality of life and experiences of ageing for older adults, specifically their 'ageing satisfaction' (Laidlaw et al. 2007). The 24 items comprise three factors: psychosocial loss, physical change, and psychological growth.
BRS	The BRS is a six-item measure that assesses resilience as the ability to bounce back or recover from stress (Smith et al. 2008).
FS	The FS scale (Pais-Ribeiro 2012) is a four-item measure that assesses subjective happiness and was adapted for the Portuguese population based on the Subjective Happiness Scale (Lyubomirsky and Lepper 1999).
Gardening Benefits Questionnaire[a]	A 42-item questionnaire designed to measure opinions and attitudes towards gardening relative to purpose (Scott et al. 2020).
GHQ	The GHQ comprises 12 items that assess the presence or absence of distress, such as anxiety and depression (Goldberg 1992).
HADS	The 14-item HADS measures anxiety and depressive symptoms during the past week (Zigmond and Snaith 1983). The HADS includes two scales, one for anxiety (HADS–A; 7 items) and one for depression (HADS–D; 7 items).
LSIA	The LSIA comprises 20 items that assess perceptions of life satisfaction as a measure of wellbeing across five domains: zest for life; resolution and fortitude; congruence between desired and achieved goals; physical, psychological, and social self-concept; and optimism (Neugarten et al. 1961). An eight-item version of the LSIA is also available (Hoyt and Creech 1983), adapted by one study (Van den Berg et al. 2010) included in this review.
LOT-R	The ten-item LOT-R assesses an individual's level of optimism (Lai et al. 1998; Scheier et al. 1994). The Chinese Revised LOT (Lai et al. 1998) was adapted from the English version and used by one study (Koay and Dillon 2020).
ONS	Two items assessing evaluative wellbeing and eudaimonic wellbeing, as measures of subjective wellbeing (Waldron et al. 2010).
POMS	The 30-item POMS measures six domains of mood states: tension-anxiety; depression-dejection; anger–hostility; vigour–activity; fatigue–inertia; and confusion–bewilderment (McNair et al. 1992).
PSS-10	Ten items measuring perceived stress levels in the past month (Cohen 1988).

(Continued)

Table 5.3 (Continued)

Measure	*Description*
PWI-A	The PWI-A is a seven-item measure that assesses subjective wellbeing, in which respondents rate their life satisfaction across seven life domains: standard of living; personal health; achieving in life; personal relationships; personal safety; community-connectedness; and future security (Cummins and Lau 2006; International Wellbeing Group 2013). One study used the Portuguese version of the PWI-A, Bem-Estar Pessoal Scale (Pais-Ribeiro and Cummins 2008).
QEWB	The QEWB comprises 21 items that assess six aspects of eudaimonic wellbeing: self-discovery; perceived development of one's best potentials; a sense of purpose and meaning; investment of significant effort in pursuit of excellence; intense involvement in activities; and enjoyment of activities as personally expressive (Waterman et al. 2010).
RSE	The RSE consists of ten items that measure self-esteem (Rosenberg 1965).
SF-36	The SF-36 comprises 36 items that measure physical and mental health in the past four weeks across eight domains of health: physical functioning; role limitations resulting from physical health; bodily pain; general health perceptions; vitality; social functioning; role limitations resulting from emotional problems; and mental health (Ware 2004). One study (Hawkins et al. 2011) employed the SF-36 v2 using 6 items to measure physical and mental health related to quality of life (Ware and Sherbourne 1992). One study (Scott et al. 2020) used two items from the SF-36 to assess self-rated health and quality of life (Ware et al. 2001).
SPS	The SPS includes 24 items that measure availability of perceived social support (Cutrona and Russell 1987).
WHOQOL-BREF	The WHOQOL-BREF contains 24 items to assess quality of life across four domains: physical; psychological; social; and environmental (WHOQOL Group 1998).

Note. AAQ = Attitudes to Ageing Questionnaire; BRS = Brief Resilience Scale; FS = Felicidade Subjectiva; GHQ = General Health Questionnaire; HADS = Hospital Anxiety and Depression Scale; LSIA = Life Satisfaction Index-A; LOT-R = Life Orientation Test Revised; ONS = UK Office for National Statistics; POMS = Profile of Mood States; PSS-10 = 10-item Perceived Stress Scale; PWI-A = 7-item Personal Wellbeing Index – Adult; QEWB = Questionnaire of Eudaimonic Wellbeing; RSE = Rosenberg Self-Esteem Scale; SF-36 = 36-item Short-Form Health Survey; SPS = Social Provisions Scale; WHOQOL-BREF = World Health Organization Quality of Life-Brief Version;

[a] Gardening Benefits Questionnaire assesses mental health through gardening

Mood, anxiety, and stress

General mood was reported in two articles. One study employed the 30-item Profile of Mood States questionnaire to assess mood in terms of tension–anxiety, depression–dejection, anger–hostility, vigour–activity, fatigue–inertia, and confusion–bewilderment (Wood et al. 2016). Anxiety levels were assessed in one study using a single, self-rated question on feelings of anxiety related to the COVID-19 pandemic (Corley et al. 2021). Corley et al. (2021) also employed the 21-item Hospital Anxiety and Depression Scale (HADS) to measure anxiety and depressive symptoms, with the main focus on symptoms of generalised anxiety disorder and anhedonia.

Stress was measured in three studies. Two of these employed the ten-item Perceived Stress Scale (Hawkins et al. 2011; Koay and Dillon 2020), which measures the level to which certain life situations experienced over the previous four weeks are appraised as stressful. Another study assessed stress using two validated items (self-report survey) that query respondents' perceived stress and ability to cope with stress (Van den Berg et al. 2010).

Sociability and loneliness

One article utilised the ten-item version of the Social Provisions Scale to measure the availability of social support (Hawkins et al. 2011). Another measured sociality as part of psychological health by assessing the size of participants' friendship circles and querying frequency of contacts with friends (Van den Berg et al. 2010). Van den Berg et al. (2010) explored loneliness using two validated items that measured feelings of loneliness and need for social contacts.

Life satisfaction, happiness, and resilience

Life satisfaction was reported in three studies. All studies used the Life Satisfaction Index-A (LSIA). Two of these studies employed the 20-item version of the LSIA, which assesses zest for life; resolution and fortitude; congruence between desired and achieved goals; physical-, psychological-, and social self-concept; and optimism (Sommerfeld et al. 2010; Waliczek et al. 2005). Another study used a brief version of the LSIA, which includes eight items (Van den Berg et al. 2010).

One study measured happiness with the four-item *Felicidade Subjetiva* scale, the Portuguese version of the Subjective Happiness Scale (Mourão et al. 2019). Resilience was assessed in one article (Koay and Dillon 2020), using the six-item Brief Resilience Scale, which evaluates one's ability to bounce back or recover from stress. One study measured optimism (Koay and Dillon 2020) using the ten-item Chinese Revised Life Orientation Test. Self-esteem was evaluated in two papers using the ten-item Rosenberg Self-Esteem scale (Koay and Dillon 2020; Wood et al. 2016).

Type of garden setting and mental wellbeing

The relationships between gardening and mental health outcomes were examined in two ways. One approach was to compare mental health outcomes for gardeners (a) within different garden spaces and (b) with non-gardeners; the other approach was to examine associations between time spent in (or frequency of) gardening and mental health outcomes.

When compared with non-gardeners, individuals who gardened in allotments reported significantly better general mental health (Soga et al. 2017a; Wood et al. 2016), higher self-esteem (Wood et al. 2016), better mood (Wood et al. 2016), and more positive wellbeing (Van den Berg et al. 2010). Those who gardened within allotment spaces were significantly more likely to have better psychological wellbeing compared with the general population (Webber et al. 2015); however, no significant differences were evident for social or eudaimonic wellbeing (Webber et al. 2015). One study found that individuals who gardened in allotments had high wellbeing and happiness scores overall; however, that study was limited by the lack of a comparison group (Mourão et al. 2019). Another study found a significant increase in levels of self-esteem and mood from pre- to post- allotment gardening session responses (Wood et al. 2016).

When compared with non-gardeners, those who gardened in domestic settings were less likely to have a history of depression (Corley et al. 2021) and more likely to score significantly higher in evaluative and eudaimonic wellbeing (de Bell et al. 2020). No significant differences in anxiety levels were evident (Corley et al. 2021). Another study found that the restorative effects of gardening in community and domestic spaces were significantly correlated with attitudes towards ageing and psychological growth; however, no significant correlations were evident for the other constructs assessed by the Gardening Benefits Questionnaire (e.g., attachment, social, spiritual, identity, and purpose; Scott et al. 2020).

Two studies found that individuals who gardened (i.e., those who identified themselves as gardeners) were significantly more likely to report higher life satisfaction (Sommerfeld et al. 2010; Waliczek et al. 2005) compared with non-gardeners. Another study found no significant differences in levels of general mental health between gardeners and non-gardeners (Park et al. 2009).

One study found no significant differences in stress levels, social support, and quality of life between those who identified as allotment gardeners, domestic gardeners, or members of other activity groups (e.g., running group), although stress levels were significantly lower in allotment gardeners compared with indoor exercisers (Hawkins et al. 2011). Another study found that individuals who gardened in community and domestic spaces had significantly higher levels of wellbeing and optimism when compared with individuals who engaged in other outdoor activities; however, no significant differences in levels of stress, self-esteem, or resilience were evident (Koay and Dillon 2020).

Time spent gardening and mental health

Six studies assessed the associations between time spent gardening and/or gardening frequency and mental health outcomes. Corley et al. (2021) found a significant, positive association between emotional and mental health with increased frequency of gardening from pre-to-post-COVID-19 lockdown. Mourão et al. (2019) reported that frequency of allotment gardening visits was significantly and positively correlated with happiness levels. Webber et al. (2015) found significant positive associations between eudaimonic wellbeing and hours spent gardening, and a significant negative association between gardening duration and social quality of life. Scott et al. (2020) reported significant positive associations between gardening duration and restoration, attachment, social, identity, and purpose, as measured by the Benefits of Gardening Questionnaire. Soga et al. (2017a) did not find significant associations between gardening duration or frequency and mental health outcomes. Likewise, the relationship between time spent gardening (i.e., in the current gardening session, in the past week, length of allotment tenure in years) and self-esteem or overall mood was not found to be significant in a study by Wood and colleagues (2016).

Discussion

The purpose of this scoping review was to summarise the existing evidence on the relationship between gardening as an unstructured activity and mental wellbeing among adults. Thirteen studies conducted in eight countries and published between 2000 and 2021 met the inclusion criteria. Consistent with previous reviews that have summarised the results of intervention studies (i.e., structured gardening programmes) and observational studies on gardening and mental health (e.g., Genter et al. 2015; Howarth et al. 2020; Soga et al. 2017b), we found some evidence that engaging in unstructured gardening is associated with better wellbeing and life satisfaction, and reduced depression and loneliness. This provides evidence that gardening is independently associated with better mental health in adults.

Findings from this scoping review support the notion that gardening may be beneficial for adult wellbeing, although the cross-sectional nature of the included studies preludes determination of causality. Indeed, Soga et al.' (2017b) meta-analysis of structured and unstructured gardening found similar results, pointing to a causal relationship between gardening and improved health outcomes. They reported enhanced sense of community, life satisfaction and quality, and decreased levels of depression and anxiety across a range of age groups and subpopulations, although some inconsistency was evident. Similarly, Howarth et al. (2020) found positive correlations between gardening, gardens (e.g., spending time in gardens without engaging in gardening activities) and mental wellbeing. Our review adds to the current literature by assessing how gardening was measured and by documenting the wide range of mental health measures used in studies conducted in the last two decades.

The impact of gardening on mental health varied across studies that met the inclusion criteria for this review. Gardening within community spaces was found to be more beneficial to mental health in several studies, when compared with gardening in domestic settings (Koay and Dillon 2020; Scott et al. 2020). This is consistent with previous studies (Chalmin-Pui et al. 2021b; Lanier et al. 2015). These benefits are thought to derive from the social aspects of communal engagement (Lanier et al. 2015). This has been described extensively in gardening literature, but further research on gardening in domestic settings is needed (Chalmin-Pui et al. 2021a; Kingsley et al. 2009). Supporting previous research, the present review found mixed associations concerning the relationship between dose (i.e., duration and frequency) of gardening and mental health outcomes (Mourão et al. 2019; Scott et al. 2020; Soga et al. 2017b; Soga et al. 2017a; Webber et al. 2015; Wood et al. 2016).

Inconsistencies in the association between mental health outcomes and gardening may be due to several factors, ranging from sample differences (e.g., age, general population, psychiatric outpatients) or comparison groups (e.g., gardeners versus activity group members) to a lack of systematic tools to measure gardening as an unstructured activity. Previous research has found that gardening may affect middle-aged and older adults differently (Kingsley et al. 2022; Van den Berg et al. 2010), such that the positive association between gardening and wellbeing is stronger for older adults. The literature also points to type of gardening and dose (i.e., intensity, frequency, duration; Chalmin-Pui et al. 2021a; Kingsley et al. 2022; Pretty et al. 2005; Scott et al. 2020), as well as garden design (e.g., amount of greenery; Jarvis et al. 2020; Young et al. 2020), as variables that may influence physical and psychological health outcomes. Future research could clarify the influence of these factors. The immense variability in gardening activities and measurement tools utilised across studies may also have contributed to the observed inconsistencies, as different scales and questions that purport to measure the same construct make direct comparisons difficult (Spano et al. 2020). For example, gardening can vary from a communal (e.g., community gardening, verge gardening, allotment gardening) to a more solitary activity (e.g., home or domestic gardening, gardening for employment). Differences in garden quality can also impact the outcomes associated with the activity, where increased greenery and ecological diversity has been found to be more beneficial to mental health (e.g., Carrus et al. 2015; Krols et al. 2022). Gardening activities vary in type of activity and physical intensity (e.g., watering is less strenuous than mulching or turning soil), which may impact the mental health outcomes (e.g., enjoyment levels of the activity).

There was diversity in measures of mental health across the studies that met the inclusion criteria for this scoping review. This is also reflected in the broader literature on this topic (Howarth et al. 2020; Soga et al. 2017b; Spano et al. 2020). Most measures employed by the reviewed studies were validated or drawn from validated measures. Howarth et al.' (2020) systematic review identified 35 validated health and wellbeing measures across 77 articles. Thirteen of the 21 measures from the current review were not identified in Howarth and colleagues' study, which speaks to the variety of assessment tools being deployed in gardening research. It was not uncommon for studies to use only one or two items from a tool to assess certain mental health outcomes (e.g., one item measuring quality of life drawn from the SF-36; Corley et al. 2021; de Bell et al. 2020; Scott et al. 2020; Van den Berg et al. 2010), which may affect the validity and reliability of those findings. Further, there is a glaring lack of research on key mental health constructs, such as depression and anxiety, which are among the most common mental health disorders (WHO 2017).

Mental health is a multifaceted construct. Indeed, the large number of measures documented in the body of research summarised in this review may reflect both the complexity of mental health and the diverse outcomes of interest to gardening researchers. Gardening research to date has been conducted by a wide range of health disciplines, with the majority of studies conducted by researchers within public health and epidemiology sectors (as well as in horticulture). Different disciplines vary in their preferences for measures to assess wellbeing; measures commonly used in psychology to assess depression and anxiety were largely absent from the included studies. It is imperative that future studies administer psychometrically robust measures of common psychological difficulties (e.g., depression, anxiety) to obtain valid and reliable measures of wellbeing. This will have implications for enriching knowledge on gardening and mental health disorders and for developing interventions.

Observed associations between gardening and mental health outcomes present within this review can partly be explained by human–nature theoretical frameworks, such as the Attention Restoration Theory (Kaplan 1995; Kaplan and Kaplan 1989), Stress Recovery Theory (Berto 2014), and the biophilia hypothesis (Ulrich 1993; Wilson 1984). Engaging in gardening also promotes physical activity (Soga et al. 2017a), which, in turn, enhances wellbeing through mechanisms such as endorphin production (Leuenberger 2006), lymphocyte cell activity (Ideno et al. 2017), and reduced diastolic and systolic blood pressure (Li 2010). Recent studies have found that the environmental microbiome in biodiverse soil can positively impact the human immune system (Deckers et al. 2019; Sbihi et al. 2019), with at least two studies using mouse models finding anti-inflammatory and anxiety-reducing properties in environmentally acquired microbes (Ottman et al. 2019; Liddicoat et al. 2020). These findings add to existing research which suggests that gardening may offer health benefits to individuals and communities (e.g., Kingsley et al. 2020). The present review may have implications for encouraging gardening as a health-promoting intervention, particularly in urban settings.

Strengths and limitations

To the authors' knowledge, this scoping review is the first to synthesise the current state of knowledge on the mental health benefits of gardening as an unstructured activity. We have focused on unstructured gardening activities in non-experimental settings, which has enabled us to provide unique insights into how gardening, in and of itself, is associated with a variety of mental health outcomes. By contrast, previous reviews have examined structured and unstructured gardening activities collectively. This study provides evidence of positive associations between gardening and mental health in instances where gardening activities were assessed in naturalistic settings exclusively, adding further evidence that gardening may be beneficial to wellbeing, even in the absence of structure. The present review also systematically outlines how gardening is measured within the literature and what measures are used to assess the psychological outcomes associated with this activity, which has implications for future research.

This study has limitations. The included research comprises predominantly Western samples, including from the United Kingdom and United States – a common feature of the gardening literature that limits the broader generalisability of these findings (Callaghan et al. 2021; Soga et al. 2017b). The correlational nature of the reviewed studies precluded us from drawing conclusions about causality and raises the possibility of a bidirectional relationship between gardening and psychological wellbeing (i.e., it may be that people with better mental health are more inclined to engage in gardening). Another limitation relates to the bias that may be evident in the exclusion and inclusion of mental health measures in the present review. Given the multidimensionality of mental health and the diversity of measures used to assess this construct, it was necessary to establish clear parameters: we chose to include measures such as social support and loneliness, and exclude measures of physical health.

As most of the included studies relied on self-report measures of both mental health symptoms and gardening, findings are prone to reliability and validity limits, such as response bias and measurement error (Spector 2019). This is particularly relevant for gardening, as research has found that individuals tend to over-report how much physical activity they engage in (Tudor-Locke and Myers 2001); such response bias and measurement error may influence the strength and validity of the investigated relationships. Another limitation lies with the methodology used in the review. The intention of a scoping review is to synthesise a wide range of research within a specified area; it does not include quality assessment (Pham et al. 2014). Future reviews should interrogate the calibre of existing studies.

Conclusion

This scoping review evaluated more than a decade of empirical evidence from studies conducted across several countries concerning the associations between gardening as an unstructured activity and mental health outcomes among adults. The findings extend previous reviews that have documented the combined effects of structured gardening programmes and unstructured gardening activities on mental health (e.g., Howarth et al. 2020; Soga et al. 2017b; Spano et al. 2020) by demonstrating that, even when gardening is not part of a prescribed intervention, it is associated with positive wellbeing outcomes, although some mixed results were present. This study adds to the literature by assessing how unstructured gardening is measured and by outlining the tools currently used to assess mental health outcomes of unstructured gardening. Future research should aim to develop instruments that capture relevant aspects of gardening activities, including duration, frequency, intensity (e.g., moderate, vigorous), setting, season, and type (e.g., growing food, flowers, other greenery), and use psychometrically robust measures of wellbeing to collect empirically sound data. The latter, in particular, will be important for building the evidence base on common mental health symptoms, such as depression and anxiety, which have been studied minimally to date in the gardening literature. The findings from this study demonstrate the value in increased opportunities for and access to gardening activities, especially in urbanised settings.

References

Arksey, H and O'Malley, L 2005, 'Scoping studies: Towards a methodological framework', *International Journal of Social Research Methodology*, vol. 8, no. 1, pp. 19–32. https://doi.org/10.1080/1364557032000119616

Berto, R 2014, 'The role of nature in coping with psycho-physiological stress: A literature review on restorativeness', *Behavioral Sciences*, vol. 4, no. 4, pp. 394–409. https://doi.org/10.3390/bs4040394

Cameron, RWF, Blanuša, T, Taylor, JE, Salisbury, A, Halstead, AJ, Henricot, B and Thompson, K 2012, 'The domestic garden – Its contribution to urban green infrastructure', *Urban Forestry & Urban Greening*, vol. 11, no. 2, pp. 129–137. https://doi.org/10.1016/j.ufug.2012.01.002

Callaghan, A, McCombe, G, Harrold, A, McMeel, C, Mills, G, Moore-Cherry, N and Cullen, W 2021, 'The impact of green spaces on mental health in urban settings: A scoping review', *Journal of Mental Health*, vol. 30, no. 2, pp. 179–193. https://doi.org/10.1080/09638237.2020.1755027

Carrus, G, Scopelliti, M, Lafortezza, R, Colangelo, G, Ferrini, F, Salbitano, F, Agrimi, M, Portoghesi, L, Semenzato, P and Sanesi, G 2015, 'Go greener, feel better? The positive effects of biodiversity on the well-being of individuals visiting urban and peri-urban green areas', *Landscape and Urban Planning*, vol. 134, pp. 221–228. https://doi.org/10.1016/j.landurbplan.2014.10.022

Chalmin-Pui, LS, Griffiths, A, Roe, J, Heaton, T and Cameron, R 2021a, 'Why garden? Attitudes and the perceived health benefits of home gardening', *Cities*, vol. 112, article 103118. https://doi.org/10.1016/j.cities.2021.103118

Chalmin-Pui, LS, Roe, J, Griffiths, A, Smyth, N, Heaton, T, Clayden, A and Cameron, R 2021b, '"It made me feel brighter in myself" – The health and well-being impacts of a residential front garden horticultural intervention', *Landscape and Urban Planning*, vol. 205, article 103958. https://doi.org/10.1016/j.landurbplan.2020.103958

Cohen, S 1988, 'Perceived stress in a probability sample of the United States' in S Spacapan and S Oskamp (eds), *The social psychology of health*, SAGE Publications, Thousand Oaks, CA, pp. 31–67.

Corley, J, Okely, JA, Taylor, AM, Page, D, Welstead, M, Skarabela, B, Redmond, P, Cox, SR and Russ, TC 2021, 'Home garden use during COVID-19: Associations with physical and mental wellbeing in older adults', *Journal of Environmental Psychology*, vol. 73, article 101545. https://doi.org/10.1016/j.jenvp.2020.101545

Cox, DT, Shanahan, DF, Hudson, HL, Fuller, RA and Gaston, KJ 2018, 'The impact of urbanisation on nature dose and the implications for human health', *Landscape and Urban Planning*, vol. 179, pp. 72–80. https://doi.org/10.1016/j.landurbplan.2018.07.013

Cummins, RA and Lau, A 2006, *Personal well-being index–Adult. Manual*, 4th edition, Australian Centre on Quality of Life, Deakin University, Melbourne.

Cutrona, CE and Russell, DW 1987, 'The provisions of social relationships and adaptation to stress', *Advances in Personal Relationships*, vol. 1, no. 1, pp. 37–67.

de Bell, S, White, M, Griffiths, A, Darlow, A, Taylor, T, Wheeler, B and Lovell, R 2020, 'Spending time in the garden is positively associated with health and wellbeing: Results from a national survey in England', *Landscape and Urban Planning*, vol. 200, article 103836. https://doi.org/10.1016/j.landurbplan.2020.103836

Deckers, J, Lambrecht, BN and Hammad, H 2019, 'How a farming environment protects from atopy', *Current Opinion in Immunology*, vol. 60, pp. 163–169. https://doi.org/10.1016/j.coi.2019.08.001

Doran, CM and Kinchin, I 2019, 'A review of the economic impact of mental illness', *Australian Health Review*, vol. 43, no. 1, pp. 43–48. https://doi.org/10.1071/AH16115

Genter, C, Roberts, A, Richardson, J and Sheaff, M 2015, 'The contribution of allotment gardening to health and wellbeing: A systematic review of the literature', *British Journal of Occupational Therapy*, vol. 78, no. 10, pp. 593–605. https://doi.org/10.1177/0308022615599408

Goldberg, D 1992, *General Health Questionnaire (GHQ-12)*, NFER-Nelson, Windsor.

Hawkins, JL, Thirlaway, KJ, Backx, K and Clayton, DA 2011, 'Allotment gardening and other leisure activities for stress reduction and healthy aging', *HortTechnology*, vol. 21, no. 5, pp. 577–585. https://doi.org/10.21273/horttech.21.5.577

Howarth, M, Brettle, A, Hardman, M and Maden, M 2020, 'What is the evidence for the impact of gardens and gardening on health and well-being: A scoping review and evidence-based logic model to guide healthcare strategy decision making on the use of gardening approaches as a social prescription', *BMJ Open*, vol. 10, no. 7, e036923. https://doi.org/10.1136/bmjopen-2020-036923

Hoyt, DR and Creech, JC 1983, 'The Life Satisfaction Index: A methodological and theoretical critique', *Journal of Gerontology*, vol. 38, no. 1, pp. 111–116. https://doi.org/10.1093/geronj/38.1.111

Ideno, Y, Hayashi, K, Abe, Y, Ueda, K, Iso, H, Noda, M, Lee, J and Suzuki, S 2017, 'Blood pressure-lowering effect of shinrin-yoku (forest bathing): A systematic review and meta-analysis', *BMC Complementary and Alternative Medicine*, vol. 17, no. 1, pp. 1–12. https://doi.org/10.1186/s12906-017-1912-z

International Wellbeing Group 2013, *Personal Wellbeing Index* (5th edition), Australian Centre on Quality of Life, Deakin University, Melbourne. https://www.acqol.com.au/instruments#measures

Jarvis, I, Koehoorn, M, Gergel, SE and van den Bosch, M 2020, 'Different types of urban natural environments influence various dimensions of self-reported health', *Environmental Research*, vol. 186, article 109614. https://doi.org/10.1016/j.envres.2020.109614

Jiang, B, Li, D, Larsen, L and Sullivan, WC 2016, 'A dose-response curve describing the relationship between urban tree cover density and self-reported stress recovery', *Environment and Behavior*, vol. 48, no. 4, pp. 607–629. https://doi.org/10.1177/0013916514552321

Kabisch, N, van den Bosch, M and Lafortezza, R 2017, 'The health benefits of nature-based solutions to urbanization challenges for children and the elderly – A systematic review', *Environmental Research*, vol. 159, pp. 362–373. http://doi.org/10.1016/j.envres.2017.08.004

Kaplan, R and Kaplan, S 1989, *The experience of nature: A psychological perspective*. Cambridge University Press, Cambridge.

Kaplan, S 1995, ‘The restorative benefits of nature: Toward an integrative framework’, *Journal of Environmental Psychology*, vol. 15, no. 3, pp. 169–182. https://doi.org/10.1016/0272-4944(95)90001-2

Kingsley, J, Egerer, M, Nuttman, S, Keniger, L, Pettitt, P, Frantzeskaki, N, Gray, T, Ossola, A, Lin, B, Bailey, A, Tracey, D, Barron, S and Marsh, P 2021, ‘Urban agriculture as a nature-based solution to address socio-ecological challenges in Australian cities’, *Urban Forestry & Urban Greening*, vol. 60, article 127059. https://doi.org/10.1016/j.ufug.2021.127059

Kingsley, J, Foenander, E and Bailey, A 2019, ‘“You feel like you’re part of something bigger”: Exploring motivations for community garden participation in Melbourne, Australia’, *BMC Public Health*, vol. 19, no. 1, pp. 1–12. https://doi.org/10.1186/s12889-019-7108-3

Kingsley, J, Foenander, E and Bailey, A 2020, ‘“It’s about community”: Exploring social capital in community gardens across Melbourne, Australia’, *Urban Forestry & Urban Greening*, vol. 49, article 126640. https://doi.org/10.1016/j.ufug.2020.126640

Kingsley, J, Hadgraft, N, Owen, N, Sugiyama, T, Dunstan, DW, and Chandrabose, M 2021b, ‘Associations of vigorous gardening with cardiometabolic risk markers for middle-aged and older adults’, *Journal of Aging and Physical Activity*, vol. 1, pp. 1–7. https://doi.org/10.1123/japa.2021-0207

Kingsley J et al. (2022) Associations of vigorous gardening with cardiometabolic risk markers for middle-aged and older adults. *Journal of Aging and Physical Activity* 30(3), 466–472. https://doi.org/10.1123/japa.2021-0207

Kingsley, JY, Townsend, M and Henderson-Wilson, C 2009, ‘Cultivating health and wellbeing: Members’ perceptions of the health benefits of a Port Melbourne community garden’, *Leisure Studies*, vol. 28, no. 2, pp. 207–219. https://doi.org/10.1080/02614360902769894

Koay, WI and Dillon, D 2020, ‘Community gardening: Stress, well-being, and resilience potentials’, *International Journal of Environmental Research and Public Health*, vol. 17, no. 18, article 6740. https://doi.org/10.3390/ijerph17186740

Krols, J, Aerts, R, Vanlessen, N, Dewaelheyns, V, Dujardin, S and Somers, B 2022, ‘Residential green space, gardening, and subjective well-being: A cross-sectional study of garden owners in northern Belgium’, *Landscape and Urban Planning*, vol. 223, article 104414. https://doi.org/10.1016/j.landurbplan.2022.104414

Lai, JC, Cheung, H, Lee, WM and Yu, H 1998, ‘The utility of the revised Life Orientation Test to measure optimism among Hong Kong Chinese’, *International Journal of Psychology*, vol. 33, no. 1, pp. 45–56. https://doi.org/10.1080/002075998400600

Laidlaw, K, Power, MJ and Schmidt, S 2007, ‘The Attitudes to Ageing Questionnaire (AAQ): Development and psychometric properties’, *International Journal of Geriatric Psychiatry: A Journal of the Psychiatry of Late Life and Allied Sciences*, vol. 22, no. 4, pp. 367–379. https://doi.org/10.1002/gps.1683

Lanier, J, Schumacher, J and Calvert, K 2015, ‘Cultivating community collaboration and community health through community gardens’, *Journal of Community Practice*, vol. 23, no. 3–4, pp. 492–507. https://doi.org/10.1080/10705422.2015.1096316

Leuenberger, A 2006, ‘Endorphins, exercise, and addictions: A review of exercise dependence’, *Impulse: The Premier Journal for Undergraduate Publications in the Neurosciences*, vol. 3, pp. 1–9. https://impulse.appstate.edu/sites/impulse.appstate.edu/files/2006_06_05_Leuenberger.pdf

Liddicoat, C, Sydnor, H, Cando-Dumancela, C, Dresken, R, Liu, J, Gellie, NJC, Mills, JG, Young, JM, Weyrich, LS, Hutchinson, MR, Weinstein, P and Breed, MF (2020,

'Naturally-diverse airborne environmental microbial exposures modulate the gut microbiome and may provide anxiolytic benefits in mice', *Science of the Total Environment*, vol. 701, article 134684. https://doi.org/10.1016/j.scitotenv.2019.134684

Lin, BB, Egerer, MH and Ossola, A 2018, 'Urban gardens as a space to engender biophilia: Evidence and ways forward', *Frontiers in Built Environment*, vol 4, 79. https://doi.org/10.3389/fbuil.2018.00079

Lyubomirsky, S and Lepper, HS 1999, 'A measure of subjective happiness: Preliminary reliability and construct validation', *Social Indicators Research*, vol. 46, no. 2, pp. 137–155. https://doi.org/10.1023/A:1006824100041

McNair, DM, Lorr, M and Droppleman, LF 1992, *Revised manual for the Profile of Mood States (POMS)*, Educational and Industrial Testing Service, San Diego.

Moher, D, Liberati, A, Tetzlaff, J and Altman, DG 2009, 'Preferred reporting items for systematic reviews and meta-analyses: The PRISMA statement', *Annals of Internal Medicine*, vol. 151, no. 4, pp. 264–269. https://doi.org/10.7326/0003-4819-151-4-200908180-00135

Mourão, I, Moreira, MC, Almeida, TC and Brito, LM 2019, 'Perceived changes in well-being and happiness with gardening in urban organic allotments in Portugal', *International Journal of Sustainable Development and World Ecology*, vol. 26, no. 1, pp. 79–89. https://doi.org/10.1080/13504509.2018.1469550

Munn, Z, Peters, MD, Stern, C, Tufanaru, C, McArthur, A and Aromataris, E 2018, Systematic review or scoping review? Guidance for authors when choosing between a systematic or scoping review approach', *BMC Medical Research Methodology*, vol. 18, no. 1, pp. 1–7. https://doi.org/10.1186/s12874-018-0611-x

Neugarten, BL, Havighurst, RJ and Tobin, SS 1961, 'The measurement of life satisfaction', *Journal of Gerontology*, vol. 16, pp. 134–143. https://doi.org/10.1093/geronj/16.2.134

Ottman, N, Ruokolainen, L, Suomalainen, A, Sinkko, H, Karisola, P, Lehtimäki, J, Lehto, M, Hanski, I, Alenius, A and Fyhrquist, N 2019, 'Soil exposure modifies the gut microbiota and supports immune tolerance in a mouse model', *Journal of Allergy and Clinical Immunology*, vol. 143, no. 3, pp. 1198–1206. https://doi.org/10.1016/j.jaci.2018.06.024

Pais-Ribeiro, J and Cummins, R 2008, 'O bem-estar pessoal: Estudo de validação da versão portuguesa da escala' in I Leal, J Pais-Ribeiro, I Silva and S Marques (eds), *Actas do 7° Congresso Nacional de Psicologia da Saúde*, ISPA, Lisbon, pp. 505–508.

Pais-Ribeiro, JL 2012, 'Validação transcultural da escala de felicidade subjectiva de Lyubomirsky e Lepper', *Psicologia, Saúde e Doenças*, vol. 13, no. 2, pp. 157–168. https://www.redalyc.org/articulo.oa?id=36225171003

Park, S, Shoemaker, C and Haub, M 2009, 'Physical and psychological health conditions of older adults classified as gardeners or nongardeners', *HortScience*, vol. 44, no. 1, pp. 206–210. https://doi.org/10.21273/hortsci.44.1.206

Patel, V, Saxena, S, Lund, C, Thornicroft, G, Baingana, F, Bolton, P, Chisholm, D, Collins, PY, Cooper, JL, Eaton, J and Herrman, H 2018, 'The Lancet Commission on global mental health and sustainable development', *The Lancet*, vol. 392, article 10157, pp. 1553–1598. https://doi.org/10.1016/S0140-6736(18)31612-X

Peen, J, Schoevers, RA, Beekman, AT and Dekker, J 2010, 'The current status of urban-rural differences in psychiatric disorders', *Acta Psychiatrica Scandinavica*, vol. 121, no. 2, pp. 84–93. https://doi.org/10.1111/j.1600-0447.2009.01438.x

Pham, MT, Rajić, A, Greig, JD, Sargeant, JM, Papadopoulos, A and McEwen, SA 2014, 'A scoping review of scoping reviews: Advancing the approach and enhancing the consistency', *Research Synthesis Methods*, vol. 5, no. 4, pp. 371–385. https://doi.org/10.1002/jrsm.1123

Pretty, J, Peacock, J, Sellens, M and Griffin, M 2005, 'The mental and physical health outcomes of green exercise', *International Journal of Environmental Health Research*, vol. 15, no. 5, pp. 319–337. https://doi.org/10.1080/09603120500155963

Rosenberg, M 1965, 'Rosenberg self-esteem scale (RSE)', *Acceptance and commitment therapy. Measures package*, vol. 61, no. 52, p. 18. https://doi.org/10.1037/t01038-000

Sarkar, C, Webster, C and Gallacher, J 2018, 'Residential greenness and prevalence of major depressive disorders: A cross-sectional, observational, associational study of 94 879 adult UK Biobank participants', *The Lancet Planetary Health*, vol. 2, no. 4, e162–e173. https://doi.org/10.1016/S2542-5196(18)30051-2

Sbihi, H, Boutin, RC, Cutler, C, Suen, M, Finlay, BB and Turvey, SE 2019, 'Thinking bigger: How early-life environmental exposures shape the gut microbiome and influence the development of asthma and allergic disease', *Allergy*, vol. 74, no. 11, pp. 2103–2115. https://doi.org/10.1111/all.13812

Scheier, MF, Carver, CS and Bridges, MW 1994, 'Distinguishing optimism from neuroticism (and trait anxiety, self-mastery, and self-esteem): A reevaluation of the Life Orientation Test', *Journal of Personality and Social Psychology*, vol. 67, no. 6, article 1063. https://doi.org/10.1037//0022-3514.67.6.1063

Scott, TL, Masser, BM and Pachana, NA 2015, 'Exploring the health and wellbeing benefits of gardening for older adults', *Ageing and Society*, vol. 35, no. 10, article 2176. https://doi.org/10.1017/S0144686X14000865

Scott, TL, Masser, BM and Pachana, NA 2020, 'Positive aging benefits of home and community gardening activities: Older adults report enhanced self-esteem, productive endeavours, social engagement and exercise', *SAGE Open Medicine*, vol. 8, article 2050312120901732. https://doi.org/10.1177/2050312120901732

Smith, BW, Dalen, J, Wiggins, K, Tooley, E, Christopher, P and Bernard, J 2008, 'The Brief Resilience Scale: Assessing the ability to bounce back', *International Journal of Behavioral Medicine*, vol. 15, no. 3, pp. 194–200. https://doi.org/10.1080/10705500802222972

Soga, M, Cox, DTC, Yamaura, Y, Gaston, KJ, Kurisu, K and Hanaki, K 2017a, 'Health benefits of urban allotment gardening: Improved physical and psychological well-being and social integration', *International Journal of Environmental Research and Public Health*, vol. 14, no. 1, article 71. https://doi.org/10.3390/ijerph14010071

Soga, M and Gaston, KJ 2016, 'Extinction of experience: The loss of human–nature interactions', *Frontiers in Ecology and the Environment*, vol. 14, no. 2, pp. 94–101. https://doi.org/10.1002/fee.1225

Soga, M, Gaston, KJ and Yamaura, Y 2017b. 'Gardening is beneficial for health: A meta-analysis', *Preventive Medicine Reports*, vol. 5, pp. 92–99. https://doi.org/10.1016/j.pmedr.2016.11.007

Sommerfeld, AJ, Waliczek, TM and Zajicek, JM 2010, 'Growing minds: Evaluating the effect of gardening on quality of life and physical activity level of older adults', *HortTechnology*, vol. 20, no. 4, pp. 705–710. https://doi.org/10.21273/horttech.20.4.705

Spano, G, D'Este, M, Giannico, V, Carrus, G, Elia, M, Lafortezza, R, Panno, A and Sanesi, G 2020, 'Are community gardening and horticultural interventions beneficial for psychosocial well-being? A meta-analysis', *International Journal of Environmental Research and Public Health*, vol. 17, no. 10, article 3584. https://doi.org/10.3390/ijerph17103584

Spector, PE 2019, 'Do not cross me: Optimizing the use of cross-sectional designs', *Journal of Business and Psychology*, vol. 34, no. 2, pp. 125–137. https://doi.org/10.1007/s10869-018-09613-8

Thoits, PA 2011, 'Mechanisms linking social ties and support to physical and mental health', *Journal of Health and Social Behavior*, vol. 52, no. 2, pp. 145–161. https://doi.org/10.1177/0022146510395592

Tudor-Locke, CE and Myers, AM 2001, 'Challenges and opportunities for measuring physical activity in sedentary adults', *Sports Medicine*, vol. 31, no. 2, pp. 91–100. https://doi.org/10.2165/00007256-200131020-00002

Ulrich, RS 1993, 'Biophilia, biophobia, and natural landscapes' in SR Kellert and EO Wilson (eds), *The biophilia hypothesis*, Island Press, Washington, DC, pp. 73–137.

United Nations 2014, *World urbanization prospects: The 2014 revision and highlights*, United Nations Department of Economic and Social Affairs, New York. https://population.un.org/wup/Publications/Files/WUP2014-Highlights.pdf

Van den Berg, AE, Van Winsum-Westra, M, De Vries, S and Van Dillen, SM 2010, 'Allotment gardening and health: A comparative survey among allotment gardeners and their neighbors without an allotment', *Environmental Health: A Global Access Science Source*, vol. 9, no. 1, article 74. https://doi.org/10.1186/1476-069X-9-74

Ventriglio, A, Torales, J, Castaldelli-Maia, J, De Berardis, D and Bhugra, D 2021, 'Urbanization and emerging mental health issues', *CNS Spectrums*, vol. 26, no. 1, pp. 43–50. https://doi.org/10.1017/S1092852920001236

Waldron, S, Tinkler, L and Hicks, S 2010, *Measuring subjective wellbeing in the UK*, Office for National Statistics, London. https://link.springer.com/content/pdf/10.1007/s11205-013-0384-x.pdf

Waliczek, TM, Zajicek, JM and Lineberger, RD 2005, 'The influence of gardening activities on consumer perceptions of life satisfaction', *HortScience*, vol. 40, no. 5, pp. 1360–1365. https://doi.org/10.21273/HORTSCI.40.5.1360

Ware, JE 2004, 'SF-36 health survey update' in ME Maruish (ed), *The use of psychological testing for treatment planning and outcomes assessment: Instruments for adults*, Lawrence Erlbaum Associates, Mahwah, NJ, pp. 693–718.

Ware, JE, Kosinski, M and Keller, S 2001, *SF-36 Physical and mental health summary scales. A user's manual, 1994*, Health Assessment Lab, Boston, MA.

Ware, JE and Sherbourne, CD 1992, 'The MOS 36-item Short-Form Health Survey (SF-36): I. Conceptual framework and item selection', *Medical Care*, vol. 30, no. 6, pp. 473–483. http://www.jstor.org/stable/3765916

Waterman, AS, Schwartz, SJ, Zamboanga, BL, Ravert, RD, Williams, MK, Bede Agocha, V, Kim, YS and Brent Donnellan, M 2010, 'The questionnaire for eudaimonic well-being: Psychometric properties, demographic comparisons, and evidence of validity', *The Journal of Positive Psychology*, vol. 5, no. 1, pp. 41–61. https://doi.org/10.1080/17439760903435208

Webber, J, Hinds, J and Camic, PM 2015, 'The well-being of allotment gardeners: A mixed methodological study', *Ecopsychology*, vol. 7, no. 1, pp. 20–28. https://doi.org/10.1089/eco.2014.0058

Wendelboe-Nelson, C, Kelly, S, Kennedy, M and Cherrie, JW 2019, 'A scoping review mapping research on green space and associated mental health benefits', *International Journal of Environmental Research and Public Health*, vol. 16, no. 12, article 2081. https://doi.org/10.3390/ijerph16122081

WHOQOL Group, 1998, 'Development of the World Health Organization WHOQOL-BREF quality of life assessment', *Psychological Medicine*, vol. 28, no. 3, pp. 551–558. https://doi.org/10.1017/S0033291798006667

Wilson, EO 1984, *Biophilia*, Harvard University Press, Cambridge, MA.

Wood, CJ, Pretty, J and Griffin, M 2016, 'A case–control study of the health and well-being benefits of allotment gardening', *Journal of Public Health*, vol. 38, no. 3, e336–e344. https://doi.org/10.1093/pubmed/fdv146

World Health Organization 2013, *Mental health action plan 2013-2020*, World Health Organization, Geneva. https://www.who.int/publications/i/item/9789241506021

World Health Organization 2017, *Depression and other common mental disorders: Global health estimates*, World Health Organization, Geneva. https://apps.who.int/iris/bitstream/handle/10665/254610/W?sequence=1

http://www.who.int/gho/ncd/mortality_morbidity/en/

World Health Organization 2018, 'Mental health: Strengthening our response', *World Health Organization*. https://www.who.int/news-room/fact-sheets/detail/mental-health-strengthening-our-response

Young, C, Hofmann, M, Frey, D, Moretti, M and Bauer, N 2020, 'Psychological restoration in urban gardens related to garden type, biodiversity and garden-related stress', *Landscape and Urban Planning*, vol. 198, article 103777. https://doi.org/10.1016/j.landurbplan.2020.103777

Zigmond, AS and Snaith, RP 1983, 'The hospital anxiety and depression scale', *Acta Psychiatrica Scandinavica*, vol. 67, pp. 361–370. https://doi.org/10.1111/j.1600-0447.1983.tb09716.x

Part II

Companion planting

Cultivating human wellbeing

6 Critically exploring public realm greenspace as a therapeutic landscape and the role of Green Social Prescribing

Jessica Thompson, Michelle Howarth, Michael Hardman, and Penny Cook

> *Everybody needs beauty as well as bread, places to play in and pray in, where Nature may heal and cheer and give strength to body and soul. This natural beauty-hunger is displayed in poor folks' window-gardens made up of a few geranium slips in broken cups, as well as in the costly lily gardens of the rich, the thousands of spacious city parks and botanical gardens, and our magnificent National Parks.*
>
> John Muir (1908)

Introduction

Nineteenth-century pioneer conservationist, John Muir, is perhaps best known for his writing on the virtues of being immersed in the North American wilderness. Muir (1908) was a true advocate of the therapeutic properties of the wild landscape – provided there weren't too many people or tourists! – away from the hustle and bustle of the emerging towns and cities. And yet, in the quote which opens this chapter, Muir moves away from the wilderness and brings to the fore the importance, the innate need, the 'beauty-hunger' felt by rich and poor alike, to have a connection with nature in areas of human habitation. Muir acknowledges, too, the differences between the fortunate and less fortunate in their attempts to forge their own connection with the natural world in their day-to-day lives. Muir was not alone in his thinking. At around the same time, British urban planner Ebenezer Howard founded the garden city movement with his publication, *To-morrow: A peaceful path to real reform* (1902), which aimed to reduce the disconnect experienced by humans and society from nature, particularly amongst the working classes. Howard incorporated the principle of integrating public realm natural environments into British social housing developments for the benefit of those who would become occupants. What both pioneers had in common was insight and passion for the therapeutic properties that an everyday natural environment can offer.

The connection between humans and nature was highlighted in Edward O Wilson's 'biophilia hypothesis' in 1984. Since that time, evidence confirming that natural environments benefit humans has accumulated, and a range of theories further explicate this hypothesis. In 1992, health geographer Wilbert

DOI: 10.4324/9781003355731-9

Gesler introduced the idea of a 'therapeutic landscape' which can promote wellness and healing. Since 2020, the advent of COVID-19 has had a remarkable effect on communities globally and influenced a revived interest in and recognition of the value of the natural world and green spaces for wellbeing (Burnett et al. 2021; Gray and Kellas 2020).

This chapter explores the concept of, and presents the case for, public realm greenspace as a therapeutic landscape. We argue that the act of cultivating this landscape is a means by which to offer opportunity for wellness through access and interaction. Civic environmental participation relies on pathways to nature-based activities designed to cultivate, or nurture, the landscape at a neighbourhood level. Green Social Prescribing (GSP) is a relatively new health initiative to connect those with a health or wellbeing need through a 'prescription' to activities associated with the cultivation of public realm greenspace, particularly in the urban context, with the aim of improving health and reducing the burden on health care systems. This chapter brings together narratives from the environment sector, public health and health geography to understand the relevance of GSP in the green public realm landscape in the wake of COVID, an era now alert to climate and biodiversity emergencies and rife with chronic health and mental health issues. The chapter will discuss why it is that the environmental 'third sector' (i.e., non-governmental and non-profit-making organisations) is driving forward the GSP movement, rather than the health sector. We draw upon case studies based in the northwest of England to reflect on practice issues that may influence the success of GSP in the long term.

Public realm greenspace: Therapeutic or a collection of hot spots and grot spots?

At first glance, the concept of public realm greenspace as a therapeutic landscape might seem a little far-fetched. Public realm is defined as a space that is free and open to anyone (MDAG 2022), while green public realm (defined in this chapter as 'greenspace') consists of the openly accessible green infrastructure found at a neighbourhood level, scattered in and around our towns and cities (Benedict and McMahon 2012). Typically, in the Global North, this is made up of component parts, such as formal parks, urban countryside, and Local Nature Reserves, but also informal recreational grounds, small-scale incidental woodlands, and community orchards. Its collective sense of scale as a 'landscape' is probably rarely considered outside of Local Government Planning departments or strategic bodies centred on green infrastructure and natural capital. To most users, public realm greenspace may be experienced on a practical level, and consciously perceived as singular places found at pavement scale. Perhaps the more enthusiastic user may seek to discover and utilise any linkages and/or green connections between component parts, in pursuit of safer, greener active travel (walking, cycling) routes. Furthermore, particularly within the urban setting, public realm is not always of good quality, but often a product of poor design and lack of management and maintenance, giving a 'grotty' (unpleasant, poor quality) appearance. This begs the question of whether this type of landscape could indeed offer therapeutic properties to the user? Urban greenspace is subject to the complexities of social economic factors and anti-social behaviour that may lead, at best, to a smattering of litter and, at worst, areas of 'fly-tipping' (illegal dumping of large amounts of waste). This is a stark contrast to Muir's 'costly lily gardens of the rich' (1908, p. 5), many of which are now private or charity-run formal gardens. Relatively speaking, these are well resourced, well staffed, and well managed. Such enclosed places have been meticulously planned and maintained to present a landscape where nature is cultivated, sculpted, and designed to be beautiful, calming, restorative, and therapeutic.

Whilst public greenspace may not always score highly aesthetically, its value lies in the opportunities it offers local residents to engage in civic environmental participation that is based on cultivating the landscape at the neighbourhood level: for meaningful activity, for placemaking, and for improved health and wellbeing. We argue that there is value in both the process (nature-based activities) and the product (better quality green provision), influencing the fabric of people's lives and livelihoods.

Cultivating the landscape through civic environmentalism

Civic environmental activity is the process by which people voluntarily take part in in developing, improving, nurturing, and maintaining public realm urban green spaces. The ways in which people take part, the activities they partake in, and the pathways to participation vary. Activities are nature-based and may include, but are not limited to, planting trees, low-level woodland management, habitat improvement, and street greening. What links all these activities is a drive to improve the quality of greenspace and green provision found at the neighbourhood level. This may be defined as *civic environmentalism* – 'voluntary communal actions undertaken to promote ecosystem sustainability' (Townsend 2006, p. 111). These actions are driven not only by reasons related to ecosystem sustainability or re-wilding, but also for wider benefits, including social cohesion, learning new skills, and outdoor exercise (Hansen-Ketchum and Halpenny 2011; Lovell et al. 2015; O'Brien et al. 2011).

The journey to social prescribing: A shift in paradigm

Social prescribing is part of this personalised care approach, which uses a streamlined system designed to refer people with a non-clinical need to an asset in the community for support. The UK National Academy of Social Prescribing (NASP) defines social prescribing as a way of 'supporting people, via social prescribing link workers, to make community connections and discover new opportunities, building on individual strengths and preferences, to improve health and wellbeing' (NASP 2020, p. 7). This approach is predicated on a 'salutogenic' paradigm that asks, 'What makes people healthy?' rather than,

'How do we treat disease?' (Antonovsky 1987). Salutogenic approaches are those based on a person's strengths rather than their deficits (Henry and Howarth 2018). Social prescribing has emerged as one way to promote the benefits of nature to communities.

The notion that nature can promote wellness has influenced contemporary national and international health care policy and practice. This has largely been inspired by an international social movement which aims to limit the dominant medical paradigm of 'fixing' people through clinical treatments. An example of this movement can be observed in the United Kingdom, through the introduction of the National Health Service (NHS) Plan (2019), which emphasises supporting the health *and* wellbeing of people and communities. Predicated on the need to combat the development of long-term conditions, the NHS Plan has used this paradigm shift to manage demands placed on primary care. Significantly, the NHS Plan signalled a sea change in the way health is promoted, advocating a response which is reliant on a 'personalised' approach that considers the wider determinants of wellbeing and the priorities of the person. In parallel, a social movement to promote non-clinical approaches to wellbeing has been acknowledged as an approach that prioritises what matters and is important to the individual, rather than the more traditional medical model, which advocates a priority over what the matter is (NHS 2019). In doing so, the NHS Plan and subsequent proliferation of 'asset-based' approaches (i.e., where individual and community knowledge, skills, and capacity are acknowledged and valued) have ensured that the wider determinants of health and wellbeing are assessed and considered as part of a health promotion strategy.

Green Social Prescribing: A pathway to wellness

One niche aspect of social prescribing which has been gaining popularity as a non-medical intervention is Green Social Prescribing (GSP), which 'links people to nature-based interventions and activities, such as local walking for health schemes, community gardening and food-growing projects' (NHS England 2022). It is understood that nature-based activities that are socially prescribed can support diverse population groups, in a range of contexts (Howarth, Lawler and Da Silva 2021). For example, GSP can help to reduce health inequalities (NHS England 2022), reduce social isolation (Howarth et al. 2016), and improve physical activity (de Boer et al. 2017). Typically, asset-based GSP approaches use a range of interventions to empower individual resilience through community support, which increasingly includes access to, and engagement with, nature. There is mounting evidence for advocating contact with nature as an effective salutogenic social prescription to promote health and wellbeing (Howarth and Lister 2019). Several papers have evaluated the impact of GSP on people with chronic conditions; for example, Howarth et al.' (2021) qualitative study explored the benefits of a Royal Horticultural Society (RHS)-led wellbeing programme. This found that gardening using structured approaches can have a transformational impact on individuals, by providing space to recover, think, and connect. Other forms of GSP are observed through 'care farming' programmes, which use farming practices to help support the mental, physical, emotional social wellbeing of an individual (Bragg and Leck 2017). In another study, de Boer et al. (2017) reported that the physical activity, social engagement, and active connection of nursing home residents on a green care farm were significantly higher than for residents living in traditional nursing homes.

The global evidence for GSP has grown over the past decades, culminating in the publication of systematic reviews (see Kunpeuk et al. 2020; Lu et al. 2020) and scale-up studies to understand practical and sustainable solutions. In Australia, 'wilderness therapy' has been found to have a range of positive benefits (Nevin et al. 2018). The benefits of nature and, in particular, the positive impact on the relationship dynamic between the therapist and the client, are thought to make the therapy multi-dimensional (Horn 2021). The benefits of nature have also been reported for diverse populations. In Denmark, for example, Poulsen and colleagues' (2018) qualitative study reported that nature-based therapy enhances social activity and employment for veterans suffering from post-traumatic stress disorder (PTSD). In Japan, research into 'forest bathing', which helps restore physical and psychological wellbeing through exposing the five senses to nature, has repeatedly demonstrated positive effects on physical and mental health (Wen et al. 2019). Hence, the value of nature is recognised as a natural asset that can be used to support wellbeing for a range of populations. Access and interaction with nature presents an ideal non-medical opportunity to underpin GSP, enabling access to green spaces and activities that can help improve mental health, reduce stress, and reduce social isolation (Cook et al. 2019; Howarth et al. 2016).

Social Prescribing has recently been integrated into the United Nations (UN) Sustainable Development Goals (SDGs). In 2021, the Global Health Social Prescribing Alliance (GHSPA) was formed to help support the implementation of the UN SDG3, which aims to work globally to ensure 'healthy lives and promote well-being for all at all ages' (UN 2022). The Alliance established a global working group to promote social prescribing through collaborative working and change through innovation (GHSPA 2022).

In this Anthropocene era, the drive to maintain ecological stability in the face of climate change is compelling. Organisations globally are faced with the stark reality that our natural environment impacts human health and wellbeing. Indeed, the highest priority for global public health is to collaborate to promote and protect the natural environment and combat climate change (Cook et al. 2019). Salutogenic approaches to social prescribing that promote nature-based activities provide an opportunity to bring people together in nature.

Developing GSP has been a rocky road, however. Critical practice issues, including a complex mix of evidence standards, a health sector rooted in traditional paradigms, and the funding challenges faced by the third sector, have all played a role in the GSP journey. These critical practice issues are addressed in more detail further into the chapter.

The context: Poor health, climate change, biodiversity decline

Several narratives support the rationale for public realm greenspace as a therapeutic landscape and transform this notion beyond the realm of the conceptual into something tangible and important. These narratives are intrinsically linked and highlight a convergence of health, geographical, socio-economic, and environmental issues.

Public health

The role of public health is to prolong healthy life and healthy communities; in doing so, it must address health inequities (Acheson 1988). A health inequality occurs when particular groups, such as those of lower socio-economic status, have significantly worse health outcomes in comparison to other groups in the broader community. For example, the life expectancy of low socio-economic status individuals can be up to ten years lower than for individuals from wealthy backgrounds, even after controlling for risk behaviours, such as diet and physical activity (Elo 2009; Marmot 2013). The public health professional seeks to intervene to improve these outcomes; one resource that can be used to promote health and wellbeing is the natural environment. It has long been understood that increased exposure of a population to greenspace is linked to lower rates of morbidity and mortality (de Vries et al. 2003; Dennis et al. 2020; Mitchell and Popham 2008) in all groups, but especially in the most economically and socially disadvantaged groups (de Vries et al. 2003; Mitchell and Popham 2008). Unfortunately, the most socially disadvantaged are also the least likely to have good quality green infrastructure (Cook et al. 2019; Diaz et al. 2006).

Human health and the health of the natural environment are inextricably linked. The public health role therefore falls into two main areas: (i) to influence policy, since globally the single greatest priority is working across governments to act against climate change and protect natural environments (Watts et al. 2015); and (ii) to intervene to connect people to natural environments. On the first of these priorities, public health leaders and policymakers within spheres of government collaborate to incorporate health considerations into environment and sustainability strategies, and vice versa. For example, the UK's environment strategy incorporates human health and wellbeing goals (Department of Environment, Food and Rural Affairs [DEFRA] 2018). Health strategies should also include protecting the environment and promoting access to nature. Public health professionals can play a role in ensuring the availability of green spaces of appropriate size, quality, and accessibility for the health of the population. On the second of these priorities, professionals with responsibility for planning health locally should put in place interventions to connect people to local natural environments; for example, using a GSP model. These interventions have a double benefit: not only do they support the health of the individuals participating (Bertotti et al. 2018), but they can also contribute to protecting our natural environment.

Those who are responsible for improving the public's health (and who hold the financial resources to do so) often ask for evidence that a green intervention produces better health. For example, local Clinical Commissioning Care groups who may fund some green interventions require outcomes data to ensure that the service provided is effective – particularly in relation to preventing inappropriate general practitioner (GP) consultations and/or hospital attendances. However, such 'outcomes/proof' is currently lacking – in part, because these interventions are complex and difficult to evaluate. The field would benefit from further research: public health researchers should attempt to quantify health benefits and cost savings, in order to be able to present this evidence for investment to those who commission or pay for health services (Cook et al. 2019). Similarly, obtaining funding for such research can be challenging because of the complexity. Therefore, those who fund research and health interventions in the green space realm must recognise and embrace the complexity.

Health geography

Health geography is a large field of study, involving work ranging from service access in medically underserved areas to public health policy and, more recently, GSP (Curtis and Riva 2010). Brown et al. (2018, p. 2) argue that health geography in its simplest form is 'how the interaction of humans, materials and the environment shapes and constrains health, wellbeing, survival and flourishing'. Schwanen and Atkinson (2015, p. 99) argue that geographers are uniquely placed to contribute to key debates on health and wellbeing due to their often-interdisciplinary stance and focus on 'context and space'. The sub-discipline has expanded rapidly of late, with geographers engaging heavily in work around the COVID-19 pandemic: from using spatial analysis to explore trends and spread to leading debates around healthier cityscapes post-COVID (see, e.g., Andrews et al. 2021; Florida et al. 2021).

The concept of GSP has been explored by geographers for some time, with reflections on interventions such as care farming and community gardening appearing frequently in the literature (see, e.g., Alkon and Agyeman 2011; Holland 2004; Milbourne 2011). Pitt's (2014) ethnographic study on the therapeutic experiences of community gardening helps to illustrate the unique contributions of geographers in this growing area of research. Comparative investigation across several community gardens reveals how 'spatial characteristics influence the extent to which people achieve therapeutic experiences' (Pitt 2014, p. 88), with the author ultimately providing recommendations on how to achieve the most value from these sites. More recently, Gorman and Cacciatore's (2020) analysis of care farming models reveals that structured programmes have a positive impact on participant health, with the authors recommending an upscaling of these sites to tackle inequalities. We argue that geographers are uniquely positioned to draw on an array of interdisciplinary methodologies and spatial tools to review the power of movements such as GSP.

Indeed, similar studies surrounding the health impacts of practices such as allotment and community gardens have risen rapidly of late. As Mitchell et al. (2021) argue, a case study approach to exploring GSP has been favoured by geographers, with research often focusing on specific measures within urban contexts. In the city context, geographers are arguably at the forefront of radical new approaches and are well positioned to reflect on their effectiveness using an array of tools. Cinderby and Bagwell's (2017) longitudinal study of an innovation in urban green infrastructure in central London reveals the immense benefits of these spaces for office workers. Likewise, Taylor et al. (2015) highlight how even urban treescapes offer opportunities within the GSP context, by using mundane landscapes to reduce reliance on antidepressants in city dwellers. Although these studies perhaps do not necessarily conform to GSP in its strictest sense (which often involves structured programmes linked to the environment), they do highlight the exploratory nature of how geographers are engaging with debates and possibilities in this field, and centring much of their work around the urban environment, due to population, development, and associated pressures.

Perhaps, unlike their colleagues in other fields, geographers tend to place a strong critical lens on the role of GSP. Although not against the approach, Bell et al. (2018, p. 2) argue that standardised models that provide a 'homogenous dose' of nature are not ideal. Rather, their critique highlights the need for more individual and often personalised plans for GSP to be effective. The authors argue that more researchers need to adopt such a lens for exploring GSP; they urge prescribers to consider individual needs, as opposed to a one-size-fits-all approach. Adding to this, Mitchell's (2021) analysis of care farming reveals the financial vulnerability of these schemes and how innovative funding models are often required to sustain activities. Their findings highlight the grant-reliant nature of many GSP initiatives and the negative impact on participants of funding cuts. As Schwanen and Atkinson (2015) argue, this critical lens is an important contribution by health geographers to this field; it will be increasingly important, if GSP is to be mainstreamed and upscaled.

Environmental factors

Historically, the natural environment has played a role of muse for beauty, poetry, mythology, religion, and indigenous identity. In the United Kingdom, the role of the natural landscape has entered a unique era. In more recent years, from a practical perspective at least, the role of the natural landscape could be crudely divided into recreation and production (agriculture and timber). Today, the natural landscape, particularly in and around towns and cities, must work harder than ever to be both productive and therapeutic: it not only provides a place for nature, but is also a carbon sink, a sponge for excess rainwater, provides air conditioning and a space for leisure and play, and functions as a container for active travel routes, for example. The realisation of the multifunctional benefits of the natural environment brings with it increased opportunities for people to become cultivators of the landscape and to access the health and wellbeing benefits of interaction with the natural world. The act of facilitating participation in nurturing public realm environments specifically is not new to the environmental sector; in fact, it has been around since at least the early 1990s (England's Community Forests [ECF] 2022).

In the United Kingdom, the declaration of climate emergencies by both national and local governments, along with challenge of meeting Net Zero across City Regions (GMCA 2022), emphasises the decarbonisation of both commercial and domestic activity. The challenge of meeting Net Zero targets, coupled with real-life, real-time events attributed to a changing climate (such as extreme weather leading to flooding, drought, and wildfires), means that nature-based solutions are now being sought for mitigation and adaptation opportunities. This has led to significant, never-before-seen national investment in initiatives, such as the £640 million Nature for Climate (DEFRA 2021b), which aims to support the acceleration of tree planting and peatland restoration across the United Kingdom. To maximise targets (which in forestry terms amounts to 30,000 hectares per year before the next Parliament), the UK Government has reached beyond the traditional government agencies (such as the Forestry Commission) by commissioning agents, including England's Community Forests network (ECF 2022), to deliver on the acceleration of woodland creation. The shift from traditional forestry practices to valuing a community forestry approach (centred around involving people and geographically linked to towns and cities) is significant. Here, we see a move from a linear, traditional forestry method to a much more multidimensional, outcomes-focused approach that extends beyond the physical geography of the landscape into the social landscape. This provides new opportunities for people to become custodians of the landscape at a neighbourhood level through the act of cultivating; for example, by planting trees or nurturing greenspaces. These nature-based activities have been shown to have multiple health benefits (Hansen-Ketchum and Halpenny 2011; Lovell et al. 2015) and are ripe for the emerging GSP movement.

The climate emergency is not the only pressing environmental issue. A biodiversity crisis has also been declared (DEFRA 2021b) and provides another strategic angle important to the discussion. Scientists predict a loss of one million plant and animal species, some within decades (Tollefson 2019); Nature Recovery strategies have been piloted with the aim of targeting nature restoration across landscapes. Policy is responding, introducing mechanisms such as Biodiversity Net Gain to address the biodiversity agenda (to a degree) in the planning process, with the drive for nature recovery sparking debate around land use and land designation.

It should not be assumed that the strategic agendas concerning climate change and biodiversity neatly align with landscapes and landscape use. In order to achieve either at scale, there is a tipping point of compromise. This becomes a critical practice issue when deciding, designing, creating, and changing landscapes. Until very recently, scheme objectives were largely centred around the wants of the landowner/land manager, with economic return in rural and semi-rural areas (based on the European Union (EU) Common Agricultural Policy) and regeneration outcomes in urban and peri-urban areas being key drivers. Since the UK left the EU in 2020, a new policy to replace the Common Agricultural Policy has been in development. The anticipated Environmental Land Management (ELM) policy will see a shift to 'payments on public land for public good' and numerous government-commissioned Test & Trials are currently underway. If introduced, ELM will provide a framework, order, and process to landscape and land management, and a shift in paradigm towards land for 'public good' (DEFRA 2021a). It does not yet offer a neat solution to the conflict of land use.

Whilst the issues highlighted here are neither new nor surprising to the environmental sector, the role of the landscape in the context of climate change and nature recovery has stimulated interest from a broader range of sectors, for example, from the public sector and public health through to utility companies and the private sector. Each sector is turning towards nature-based solutions to achieve, in its own way, the triple bottom line pillars of sustainability: economic, social, and environmental factors. This, in turn, casts a spotlight on the potential and capacity of environmental organisations, which are well versed in nature restoration and engaging people in the process of cultivating the natural landscape.

The context highlights pressing, interrelated issues around ill health, a changing climate, and declining biodiversity – each of which identifies the potential for cultivating public greenspaces. The opportunity to improve health and wellbeing through nature-based activities that not only facilitate a positive connection to nearby natural environment, but also *improve* the quality of the natural environment for the wider population, whilst creating a more climate-resilient, species-rich landscape is compelling. There would appear to be a new alignment of common aims and desired outcomes that should neatly fit together. However, although various health and environmental emergencies have been declared, these are not new problems, and the environmental third sector has been running environmental activities to achieve health for decades (Nolan and Vaughan 2001). Given the longstanding association between the benefits of nature and human health – in contemporary evidence-based practice, coupled with the hunches, writings, and actions of the likes of Muir and Howard, it might be assumed that further progress should have been made. An investigation into the critical practice issues may provide some explanation of why nature-based health practices are not an integrated part of public and preventative health. In the following section, we ask why a green health agenda and the concept of green referrals have taken so long to gain momentum.

Public realm greenspace as a therapeutic landscape: Critical practice issues

Many critical practice issues relate to the role and functionality of public realm greenspace and the implications for engaging communities for health and wellbeing outcomes. The perspectives from public health, health geography, and the environmental sector have introduced some of these practical considerations, such as conflicts and changes in land use. Public realm greenspace holds the potential to be a functioning therapeutic landscape, a medium for nature-based activities, and GSP is a potential mechanism by which people can access and interact with therapeutic properties. However, a number of critical practice issues (discussed next) highlight the complexities and practicalities in realising public greenspace potential as a medium for health and wellbeing. Critical practice issues that have also challenged GSP activity and progression.

Poor neighbourhoods, poor health, and poor-quality environments

Geographically, there is a correlation between areas of poor health having low levels of green provision and/or poor-quality green provision (Dennis et al. 2020). This signifies that health inequalities, which are rooted in structural social economic issues, are linked to environmental inequalities. Therefore, health issues and inequalities are intrinsically linked to environmental justice. Hence, it would be assumed that the green health agenda would be championed by both the health sector and the environmental sector in equal merit. However, this agenda has been largely driven by the environmental sector (Bragg and Leck 2017) – although this is changing, with the NHS, Public Health, and government now turning their attention to the possibilities of GSP (Cook et al. 2019). One critical issue is the failure to appreciate public realm greenspace a 'therapeutic' landscape and therefore a potential asset to address health issues and inequalities.

Understanding evidence: Square peg, round hole

A key reason for the health sector's reluctance to embrace green referrals stems from conflict over standards of evidence (Bragg and Leck 2017). Debate over what constitutes scientific evidence is not new and harks back to the 'science wars' of the 1990s (Ashman and Baringer 2001). The health sector still regards a randomised controlled trial as the 'gold standard' for assessing whether nature-based initiatives as an intervention 'works' and, therefore, whether it is worthy of investment from a health perspective. However, the reality is that assessing the impacts of an activity or intervention designed to engage community participants in nature-based activities within green public realm does not fit into an experimental or quasi-experimental design. Such engagement initiatives are essentially social programmes, based on theories of change (Pawson and Tilley 1997); therefore, implementing controls is not appropriate in evaluation design. Social programmes are open, complex systems subject to many variables (Pawson and Tilley 1997). In the case of nature-based engagement programmes, these variables are many and are influenced by external and internal factors, often beyond the control of the delivery organisation. Factors such as funding, staff capacity, motivations for participation, participation retention, and access to land are just some examples of variables that may influence a community engagement programme, and each is subject to many variables of its own. Furthermore, often the desired outcome of engagement programmes is improved wellbeing; again, this is subjective because it is a personal state of mind (Deci and Ryan 2008). However, there is growing recognition that the reliance on traditional standards of evidence needs to change to accommodate the social prescribing movement (Bragg and Leck 2017).

Green Social Prescribing: A pathway to nature-based activities?

In this chapter, public realm greenspace has been conceptualised as a therapeutic landscape, a medium through which nature-based activities can be carried out to help cultivate the landscape. GSP has been discussed as a pathway to reach the destination activity, which in turn leads to a pathway of increased wellness for the participant. We recognise that this is an oversimplified version of this theory of change. The schools of thought presented in this chapter, bringing together perspectives from public health and health geography, and the background to social prescribing, have each articulated that a one-size-fits-all approach, a single delivery model, is neither conducive to the principles of social prescribing nor appealing to large numbers of people.

In the next section, we use case studies to explore how several projects in the north-west of England are taking a multi-disciplinary, multi-partnership approach to addressing some of the practical issues around GSP delivery models. Case Study 1 outlines four green public realm landscape programmes to elucidate different models of engagement and their intrinsic links to funding models. Case Study 2 highlights the complex picture of critical practice issues and GSP by examining a national initiative currently underway in Greater Manchester.

Case Studies

Case Study 1: Different engagement models

Engagement models are driven by a number of factors, including meeting the needs of people with specific health issues, sourcing funding, and framing the activity in the context of wider health issues. The practice issues addressed in these first case studies include: How delivery models are framed, whether as therapy or as prevention? How should specific needs be addressed? What are the best ways to reach out to participants? What are the implications for funding models? Case study 1 highlights four different engagement or delivery models, all based in the north-west of England; all use the green public realm landscape as a medium for delivery.

The Mersey Forest's Nature for Health programme: A 'dose of treatment' model

The first model is an example of how the nature-based activities, framed as 'a dosage', operate as a treatment for specific health needs. The Mersey Forest is one of 13 England's Community Forests, and runs Nature for Health, a programme under the banner of 'The Natural Health Service' (The Mersey Forest 2022). This engagement model takes an approach that offers five distinct 'products': (i) Health Walks, designed to meet individual fitness levels and taking a target-based approach to increase physical activity levels and improve wellbeing; (ii) Horticulture Therapy, which includes growing activities for physical activity in a sociable environment; (iii) Mindfulness Practice in nature to tackle stress and anxiety, with a focus on self-management for longer-term conditions; (iv) Forest School, a Scandinavian pedagogy based outside the classroom for children and younger people to achieve increased activity levels and positive nature connections; and (v) Healthy Conservation, which includes conservation-based activities to improve stamina and fitness, increase confidence, and learn skills.

Activities are run as programmes over a set number of weeks, and participants are expected to finish the course, or 'dose'. This particular model has been designed to connect into the NHS commissioning frameworks but is also funded through grant-giving organisations, such as the National Lottery. The focus is product-based and advertised as evidence-based doses of treatment to address various health issues experienced by participants. Challenges to this type of model are recruiting and retaining participants for a set number of weeks to full completion of a programme, especially when the funding is based on 'payment by results'.

City of Trees' Citizen Forester programme

City of Trees is the Community Forest organisation that works across Greater Manchester. The organisation supports a movement to improve the quality of life for people through nurturing a tree and woodland culture. Under City of Trees' 'Citizen Forester' programme, volunteers can help plant trees and look after existing woodlands and greenspaces across Greater Manchester (City of Trees 2022). In line with City of Trees' aspiration to enable every person to play their part in a movement to improve the landscape across Greater Manchester, Citizen Forester is open to everyone. The audience is broad, ranging from local residents, climate activists, corporate groups, and people seeking vocational and/or therapeutic rehabilitation to those with a defined health need (e.g., people affected by dementia). The programme's ethos is that everyone taking part is a Citizen Forester, regardless of their pathway, motivation, background, or health need. Volunteers take part as and when they wish, based on a calendar of events taking places across Greater Manchester. The programme is not specifically promoted as a 'health service' but uses messaging around joining in the City of Trees movement to create a greener Greater Manchester. From a health perspective, it is in line with Public Health messaging and campaigns around prevention and self-care, using the '5 Ways to Wellbeing' (Aked et al. 2008) as a basis for evaluation. Citizen Forester combines investment in national capital infrastructure programmes for woodland creation, with philanthropic funding, grants, and trusts, as well as donations from the private sector. A challenge to this model is the roving nature of activities. Activities are centred around seasonal tree planting and specific site-by-site maintenance requirements, which means the location of activities shifts as part of a dynamic pipeline of schemes.

Wildlife Trust for Lancashire, Manchester and North Merseyside: My Place

The Wildlife Trust for Lancashire, Manchester and North Merseyside (2022) delivers the My Place programme in partnership with the Lancashire and South Cumbria NHS Foundation Trust. Promoted as ecotherapy, My Place targets people with defined health needs, with a focus on mental health. Participants can be referred or self-refer and sign up to the scheme for a six-week programme. For participants with specific mental health needs, My Place offers face-to-face sessions, practical conservation activities, and training courses, as well as online sessions to provide virtual nature-based activities for those unable to leave their home. With a focus on young people, the programme also delivers a strand called 'MyPlace for Gamers', which encourages Minecraft activity to build virtual green worlds. The programme is funded through a blend of European Social Funds, lottery grants, and government grants for 'Green Recovery'. A challenge for this type of delivery model is the skills capacity of staff; those employed as environmental project officers often lack the skills required of mental health professionals.

Northern Roots

Northern Roots is an ambitious project to create the largest urban farm and eco-park on 160 acres of green space in the Greater Manchester district of Oldham. Its mission is to transform the landscape into a 'destination for learning, growing, and leisure activities' that will benefit both the environment and the quality of life for those that live near or interact with the space (Northern Roots 2022). The land is currently owned by Oldham Metropolitan Borough Council, but a charitable company has been set up to take on the lease and management of the site. At the heart of the Northern Roots project is a business plan to establish a truly sustainable funding model – drawing income from market gardening, mountain biking, a woodland wedding venue, and a Forest School – to support both the physical maintenance of the site and community engagement opportunities. The proposed funding arrangements are complex and, together with Northern Roots' ability to create its own circular economy, are yet to be tested. However, there is enormous drive for this to become a success. If this is achieved, the outcome will be ground-breaking as a model for landscape-scale public realm greenspace as a means to therapeutic activity.

Case Study 2: Critical practice issues

Green Social Prescribing Test and Learn

Case Study 2 highlights critical practice issues in relation to green social prescribing by examining a national initiative taking place within Greater Manchester in 2022. The Green Social Prescribing Test and Learn project aims to tackle poor mental health across Greater Manchester, to ease demand on the health and social care system, reduce health inequalities, and set and share best practice at a local level for health and environmental practitioners. This project tackles critical practice issues head on, by identifying what infrastructure is needed to support a thriving GSP scene across Greater Manchester, recognising unrealised opportunities, and addressing any barriers.

At a national strategic level, the ambition for this nationwide test project stems from the UK Government's commitment to transform mental health services nationally (NHS 2022). We see an intersection of this policy with the Government's 25 Year Environment Plan (DEFRA 2018), which recognises the links between the natural world and human health and commits to taking steps to support all people to have everyday access to the natural environment to realise the health benefits.

In spring 2021, Greater Manchester's Health and Social Care Partnership was successful in its bid to secure £500,000 to deliver a 'Test and Learn' Green Social Prescribing Programme across Greater Manchester, as one of seven test projects taking place nationally between October 2020 and March 2023. The project is delivered as a programme of 'test and learn' strands, four of which are site based. The test sites seek to explore the impact of nature-based activities in public realm greenspace on mental health needs, ranging from low-level emotional mental health needs to diagnosed mental health conditions. These tests sites also aim to gain a greater understanding of the pathways that may lead people to the activities on offer, as well as how the pathways could be scaled up to improve accessibility at the neighbourhood level. The test sites offer nature-based activities that will connect with a variety of audiences with a range of needs. The sites offer activities for self-referrals, as well as referrals from various settings, such as primary and secondary care, third sector services, and local community provision.

A fifth test strand seeks to understand and identify the infrastructure required to support a GSP network, with a focus on the practicalities from the perspectives of clients, health workers, and environmental practitioners. This will identify the various elements required to build the networks and systems operating at different levels, for example, data requirements for software systems used by GPs and Social Prescribing link workers, communications tools and channels to promote activities, and resource sharing and best practice to develop standards. The test site strands are led by a consortium of key partners from the third sector, including City of Trees, Lancashire Wildlife Trust, Petrus, Sow the City, and Salford CVS. The partners share a collective ambition to embed their nature-based activities in the health sector, through the social prescribing pathway. There is an emphasis on employing a partnership approach; a wider partnership is directly involved in the project, made up of a mix of mental health, environmental, and specialist services. The project runs until March 2023 and will inform national policy and practice.

We have seen how delivery models can differ, depending on how they are framed, what they set out to achieve, and who they are looking to engage or target. Furthermore, how delivery models are funded can have a significant influence on how a delivery model is designed and how successful the model is in achieving its outcomes. Funding is a perennial challenge for organisations and the success of GSP (Bragg and Leck 2017). The nature of third sector funding is volatile, typically characterised by a mix of funding streams that support the delivery of activities in the short- to medium-term. Funding often comes from grants, philanthropic sources, and/or the private sector. The health sector's current commissioning process is highly competitive, with relatively small amounts granted over short periods of time. This is not favourable, practical, or reasonable in terms of organisational business planning, and it prevents sustained activity in a certain place over a period of time. This poses a critical problem for GSP. The public realm landscape already hosts a number of nature-based activities designed to cultivate the landscape, but the landscape is dynamic, in that where and for how long those activities are available is subject to frequent change due to funding structures.

Conclusion

This chapter makes the case for public realm greenspace as a therapeutic landscape and explores the role for green social prescribing to generate connections between environmental and health sectors. Public realm greenspace plays an important role in servicing the need for people to have day-to-day access and exposure to the natural world – described by Muir (1908) as 'beauty-hunger'. Community engagement activities, typically delivered by non-profit third sector organisations, aim to facilitate human interaction with nature through activities that help to cultivate the landscape at a neighbourhood level. The importance of cultivating the physical landscape is only increasing, as the climate emergency escalates and biodiversity declines. There is also growing interest from the health sector, which is seeking solutions to chronic health and mental health issues and health inequalities. GSP is a potential pathway to encourage greater participation in nature-based activities for health and wellbeing outcomes, but critical practice issues – particularly funding – will determine the success of GSP in the long term.

This chapter also demonstrates that environmental justice, civic environmental nature-based activities, and the growing GSP movement can align to meet health needs and the appetite for 'beauty-hunger', which is especially salient in the COVID-19 era. It is to be hoped that the GSP movement will provide a tangible way to bring the environmental and health sectors closer together to achieve common aims in relation to health and wellbeing. This alignment has the potential to help redress health inequalities linked to environmental inequalities. It is evident that the inequalities observed by Muir in the early twentieth century – poor folks' geranium slips in broken cups versus the expensive lily gardens of the wealthy – still resonate today.

References

Acheson, D 1988, *Public health in England. The report of the Committee of Inquiry into the future development of the public health function*, HMSO, London.

Aked, J, Marks, N, Cordon, C and Thompson, S 2008, *Five ways to well-being: The evidence*, New Economics Foundation, London.

Alkon, AH and Agyeman, J 2011, *Cultivating food justice: Race, class and sustainability*, MIT Press, Cambridge, MA.

Andrews, GJ, Crooks, VA, Pearce, JR and Messina, JP 2021, *COVID-19 and similar futures: Pandemic geographies*, Springer, London.

Antonovsky, A 1979, *Health, stress, and coping*, Jossey-Bass, San Francisco.

Antonovsky, A 1987, 'The salutogenic perspective: Toward a new view of health and illness', *Advances*, vol. 4, no. 1, pp. 47–55.

Ashman, KM and Baringer, PS 2001, *After the science wars*, Psychology Press, London.

Bell, SL, Leyshon, C, Foley, R and Kearns, RA 2018, 'The "healthy dose" of nature: A cautionary tale', *Geography Compass*, vol. 13, no. 1, pp. 2–14. https://doi.org/10.1111/gec3.12415

Benedict, MA and McMahon, ET 2012, *Green infrastructure: Linking landscapes and communities*, Island Press, Washington, DC.

Bertotti, M, Frostick, C, Hutt, P, Sohanpal, R and Carnes, D 2018, 'A realist evaluation of social prescribing: An exploration into the context and mechanisms underpinning a pathway linking primary care with the voluntary sector', *Primary Health Care Research & Development*, vol. 19, no. 3, pp. 232–245. https://doi.org/10.1017/s1463423617000706

Bragg, R and Leck, C 2017, *Good practice in social prescribing for mental health: The role of nature-based interventions*. Natural England Commissioned Report NECR228, Natural England, York.

Brown, T, Andrews, GJ, Cummins, S, Greenhough, B, Lewis, D and Power, A 2018, *Health geographies: A critical introduction*, Wiley, Oxford.

Burnett, H, Olsen, JR, Nicholls, N and Mitchell, R 2021, 'Change in time spent visiting and experiences of green space following restrictions on movement during the COVID-19 pandemic: A nationally representative cross-sectional study of UK adults', *BMJ Open*, vol. 11, no. 3, e044067. https://doi.org/10.1136/bmjopen-2020-044067

Cinderby, S and Bagwell, S 2017, 'Exploring the co-benefits of urban green infrastructure improvements for business and workers' wellbeing', *Area*, vol. 50, no. 1, pp. 126–135.

City of Trees 2022, *Manchester City of Trees*. https://www.cityoftrees.org.uk/

Cook, PA, Howarth, M and Wheater, CP 2019, 'Biodiversity and health in the face of climate change: Implications for public health' in M Marselle, J Stadler, H Korn, K Irvine and A Bonn (eds), *Biodiversity and health in the face of climate change*, Springer, Cham, pp. 251–281.

Curtis, S and Riva, M 2010, 'Health geographies I: Complexity theory and human health', *Progress in Human Geography*, vol. 34, no. 2, pp. 215–223. https://doi.org/10.1177/0309132509336026

de Boer, B, Hamers, JP, Zwakhalen, SM, Tan, FE, Beerens, HC and Verbeek, H 2017, 'Green care farms as innovative nursing homes, promoting activities and social interaction for people with dementia', *Journal of the American Medical Directors Association*, vol. 18, no. 1, pp. 40–46. https://doi.org/10.1016/j.jamda.2016.10.013

de Vries, S, Verheij, RA, Groenewegen, PP and Spreeuwenberg, P 2003, 'Natural environments – healthy environments? An exploratory analysis of the relationship between greenspace and health', *Environment and Planning A: Economy and Space*, vol. 35, no. 10, pp. 1717–1731. https://doi.org/10.1068/a35111

Deci, EL and Ryan, RM 2008, 'Hedonia, eudaimonia, and well-being: An introduction', *Journal of Happiness Studies*, vol. 9, no. 1, pp. 1–11.

DEFRA 2021a, 'Guidance: Environmental land management schemes: Overview' (15 March 2021). https://www.gov.uk/government/publications/environmental-land-management-schemes-overview/environmental-land-management-scheme-overview

DEFRA 2021b, 'Policy paper: Nature for people, climate and wildlife' (18 May 2021). https://www.gov.uk/government/publications/nature-for-people-climate-and-wildlife/nature-for-people-climate-and-wildlife

Dennis, M, Cook, PA, James, P, Wheater, CP and Lindley, SJ 2020, 'Relationships between health outcomes in older populations and urban green infrastructure size, quality and proximity', *BMC Public Health*, vol. 20, no. 1, pp. 1–15. https://doi.org/10.1186/s12889-020-08762-x

Department of Environment, Food and Rural Affairs [DEFRA] 2018, *A green future: Our 25 year plan to improve the environment*, HM Government, London.

Diaz, S, Fargione, J, Chapin, FS and Tilman, D 2006, 'Biodiversity loss threatens human well-being', *PLoS Biology*, vol. 4, no. 8, pp. 1300–1305. https://doi.org/10.1371/journal.pbio.0040277

Elo, IT 2009, 'Social class differentials in health and mortality: Patterns and explanations in comparative perspective', *Annual Review of Sociology*, vol. 35, pp. 553–572. https://doi.org/10.1146/annurev-soc-070308-115929

England's Community Forests 2022, *England's community forests*. https://englandscommunityforests.org.uk/

Florida, R, Rodríquez-Pose, A and Storper, M 2021, 'Cities in a post-COVID world', *Urban Studies*, online first. https://doi.org/10.1177/00420980211018072

Global Health Social Prescribing Alliance 2022, 'Global social prescribing alliance: Supporting SDG3 "good health and wellbeing"'. https://www.gspalliance.com/

Gorman, R and Cacciatore, J 2020, 'Care-farming as a catalyst for healthy and sustainable lifestyle choices in those affected by traumatic grief', *NJAS: Wageningen Journal of Life Sciences*, vol. 92, no. 1, pp. 1–7.

Gray, S and Kellas, A 2020, 'Covid-19 has highlighted the inadequate, and unequal, access to high quality green spaces', *The BMJ Opinion* (3 July 2020). https://blogs.bmj.com/bmj/2020/07/03/covid-19-has-highlighted-the-inadequate-and-unequal-access-to-high-quality-green-spaces/

Greater Manchester Combined Authority 2022, 'This is Greater Manchester'. https://www.greatermanchester-ca.gov.uk/

Hansen-Ketchum, PA and Halpenny, EA 2011, 'Engaging with nature to promote health: Bridging research silos to examine the evidence', *Health Promotion International*, vol. 26, no. 1, pp. 100–108. https://doi.org/10.1093/heapro/daq053

Harper, NJ, Gabrielsen, LE and Carpenter, C 2018, 'A cross-cultural exploration of 'wild' in wilderness therapy: Canada, Norway and Australia', *Journal of Adventure Education and Outdoor Learning*, vol. 18, no. 2, pp. 148–164. https://doi.org/10.1080/14729679.2017.1384743

Henry, H and Howarth, ML 2018, 'An overview of using an asset-based approach to nursing', *General Practice Nursing*, vol. 4, no. 4, pp. 61–66.

Holland, L 2004, 'Diversity and connections in community gardens: A contribution to local sustainability', *Local Environment: The International Journal of Justice and Sustainability*, vol. 9, no. 3, pp. 285–305.

Howarth, M and Lister, C 2019, 'Social prescribing in cardiology: Rediscovering the nature within us. *British Journal of Cardiac Nursing*, vol. 14, no. 8, pp. 1–9.

Howarth, M, Lawler, C and da Silva, A 2021, 'Creating a transformative space for change: A qualitative evaluation of the RHS Wellbeing Programme for people with long term conditions', *Health & Place*, vol. 71, no. 3 p. 102654. http://doi.org/10.1016/j.healthplace.2021.102654

Howarth, ML, McQuarrie, C, Withnell, N and Smith, E 2016, 'The influence of therapeutic horticulture on social integration', *Journal of Public Mental Health*, vol. 15, no. 3, pp. 136–140. http://doi.org/10.1108/JPMH-12-2015-0050

Kunpeuk, W, Spence, W, Phulkerd, S, Suphanchaimat, R and Pitayarangsarit, S 2020, 'The impact of gardening on nutrition and physical health outcomes: A systematic review and meta-analysis', *Health Promotion International*, vol. 35, no. 2, pp. 397–408. https://doi.org/10.1093/heapro/daz027

Lovell, R, Husk, K, Cooper, C, Stahl-Timmins, W and Garside, R 2015, 'Understanding how environmental enhancement and conservation activities may benefit health and wellbeing: A systematic review', *BMC Public Health*, vol. 15, no. 1, pp. 1–18. https://doi.org/10.1186/s12889-015-2214-3

Lu, L-C, Lan, S-H, Hsieh, Y-P, Yen, Y-Y, Chen, J-C and Lan, S-J 2020, 'Horticultural therapy in patients with dementia: A systematic review and meta-analysis', *American Journal of Alzheimer's Disease & Other Dementias*, vol. 35. https://doi.org/10.1177/1533317519883498

Marmot, M 2013, *Review of social determinants and the health divide in the WHO European Region: Final report*, World Health Organization, Copenhagen.

Mayor's Design Advisory Group 2022, *Public London: Creating the best public realm*. https://www.london.gov.uk/sites/default/files/mdag_agenda_public_london.pdf

Milbourne, P 2011, 'Everyday (in)justices and ordinary environmentalisms: Community gardening in disadvantaged urban neighbourhoods', *Local Environment: The International Journal of Justice and Sustainability*, vol. 17, no. 9, pp. 1–15. https://doi.org/10.1080/13549839.2011.607158

Mitchell, L, Hardman, M, Cook, P and Howarth, ML 2021, 'Enabling urban social farming: The need for radical green infrastructure in the city', *Cogent Social Sciences*, vol. 7, no. 1, pp. 1–15. https://doi.org/10.1080/23311886.2021.1976481

Mitchell, R and Popham, F 2008, 'Effect of exposure to natural environment on health inequalities: An observational population study', *The Lancet*, vol. 372, no. 9650, pp. 1655–1660. https://doi.org/10.1016/S0140-6736(08)61689-X

Muir, J 1908, 'The Hetch-Hetchy Valley', *Sierra Club Bulletin*, vol. 6, no. 4, pp. 211–222. https://vault.sierraclub.org/ca/hetchhetchy/hetch_hetchy_muir_scb_1908.html

National Academy of Social Prescribing 2020, *A social revolution in wellbeing. Strategic plan 2020–23*. https://socialprescribingacademy.org.uk/wp-content/uploads/2020/03/NASP_strategic-plan_web.pdf

NHS 2019, *The NHS long term plan*. https://www.longtermplan.nhs.uk/

NHS England 2022, *Green Social Prescribing*. https://www.england.nhs.uk/personalisedcare/social-prescribing/green-social-prescribing/

Nolan, P and Vaughan, J 2001, 'Forests of the streets – The regeneration role of England's Community Forests', *ECOS*, vol. 22, no. 3–4, pp. 39–46.

Northern Roots 2022, 'Vision'. https://northern-roots.uk/

O'Brien, L, Burls, A, Townsend, M and Ebden, M 2011, 'Volunteering in nature as a way of enabling people to reintegrate into society', *Perspectives in Public Health*, vol. 131, no. 2, pp. 71–81. https://doi.org/10.1177/1757913910384048

Pawson, R and Tilley, N 1997, *Realistic evaluation*, Sage, London.

Pitt, H 2014, 'Therapeutic experiences of community gardens: Putting flow in its place', *Health & Place*, vol. 27, pp. 84–91. https://doi.org/10.1016/j.healthplace.2014.02.006

Poulsen, DV, Stigsdotter, UK and Davidsen, AS 2018, '"That guy, is he really sick at all?" An analysis of how veterans with PTSD experience nature-based therapy', *Healthcare*, vol. 6, no. 2, p. 64. https://doi.org/10.3390/healthcare6020064

Schwanen, T and Atkinson, S 2015, 'Geographies of wellbeing: An introduction', *The Geographical Journal*, vol. 181, no. 2, pp. 98–101.

Taylor, MS, Wheeler, BW, White, MP, Economou, T and Osborne, NJ 2015, 'Research note: Urban street tree density and antidepressant prescription rates: A cross-sectional study in London, UK', *Landscape and Urban Planning*, vol. 136, pp. 174–179. https://doi.org/10.1016/j.landurbplan.2014.12.005

The Mersey Forest 2022, 'Natural Health Service'. https://www.merseyforest.org.uk/our-work/natural-health-service/

Tollefson, J 2019, 'Humans are driving one million species to extinction', *Nature*, vol. 569, no. 7755, pp. 171–172. https://doi.org/10.1038/d41586-019-01448-4

Townsend, M 2006, 'Feel blue? Touch green! Participation in forest/woodland management as a treatment for depression', *Urban Forestry & Urban Greening*, vol. 5, no. 3, pp. 111–120. https://doi.org/10.1016/j.ufug.2006.02.001

United Nations 2022, *Sustainable Development Goals*. https://sdgs.un.org/goals

Watts, N, Adger, WN, Agnolucci, P et al. 2015, 'Health and climate change: Policy responses to protect public health', *The Lancet*, vol. 386, no. 10006, pp. 1861–1914. https://doi.org/10.1016/S0140-6736(15)60854-6

Wen, Y, Yan, Q, Pan, Y et al. 2019, 'Medical empirical research on forest bathing (*Shinrin-yoku*): A systematic review', *Environmental Health and Preventive Medicince*, vol. 24, article 70. https://doi.org/10.1186/s12199-019-0822-8

Wildlife Trust for Lancashire, Manchester and North Merseyside 2022, *My Place: A Natural Way to Wellbeing*. https://www.lancswt.org.uk/our-work/projects/myplace

7 Creating a therapeutic garden for people with Huntington's Disease and other neurological conditions

Josephine Spring

Introduction

In 2007, at the Royal Hospital for Neuro-disability (RHN) in London, a small, enclosed garden was available for a residential unit for clients with Huntington's Disease (HD). At the time, my colleagues and I could find no evidence base for gardening as therapy for HD, except for a single report from Canada of two brothers gardening. This report described how one brother 'enjoys working in the garden and is continually in a happy frame of mind', while the other was 'constantly in good spirits and has worked assiduously in the hospital garden' (Hall 1954, pp. 493–494). This dearth of detail showed that a research study was needed to ensure the RHN's small, enclosed garden was an effective space for therapy and leisure gardening in order to define the benefits and challenges for clients, while also exploring use by relatives and staff (Spring et al. 2014). Ethics approval was obtained from the NHS Charing Cross research ethics committee and funding was secured from the Neuro-disability Research and Hamamelis Trusts.

In our study, to accommodate to communication challenges, we devised a pictorial questionnaire for clients. HD clients' happiness when gardening was measured on a linear analogue scale. We found they had a mean score of 72% (ranging from 52% to 96%). We also discovered that residents had a variety of interests related to the garden. These ranged from growing plants, being outside, watching birds, painting pictures of the garden or a plant, and watching gardening or nature-related programmes on television. Some resident clients enjoyed growing flowers and vegetables, labelling plants, being outside in the sun, smoking, and being in the quiet of the garden. The most interesting finding was HD clients' preference for the red end of the colour spectrum. Exploring this further in the published literature revealed evidence that colour vision is affected in HD, and clients may not see blue well (Paulus et al. 1993; Büttner et al. 1994) due to the thinning of the temporal retinal nerve fibre layer, cell loss, and thinning in the visual cortex (Kersten et al. 2015; Dhalla, Pallikadavath and Hutchinson 2019). In addition, through questionnaires and interviews with staff and relatives, staff reported that the benefits of gardening for clients included the outdoor activity, sense of achievement, constructive endeavour,

DOI: 10.4324/9781003355731-10

and social interaction. More importantly, half the staff stated that gardening with clients posed no problems, and one-third of the staff used the garden for other therapies (Spring et al. 2014). The minor problems reported by staff related to clients becoming hot or cold (thermoregulation is affected by HD) and getting dirty or tired.

Due to an ageing population in the United Kingdom, there is growing interest in gardening as a way to reduce pressure on health services (Thompson 2018). Gardening is viewed in a positive light in general by healthcare professionals; for example, an online survey of occupational therapists in the United States found that gardening with clients was meaningful and purposeful (94%), motivating (80%), and fun (62%) (Wagenfeld and Atchison 2014).

The literature

The evidence for physical, cognitive, social, and psychological benefits of gardening and horticultural therapy for people with neurological conditions (such as stroke and brain injury), mental health issues (schizophrenia, dementia), and HD, in particular, is now well established. For example, horticultural activities with stroke patients have been found to improve upper limb function, grip strength, pinch force, fine motor skills, balance, and activities of daily living (Lee et al. 2018), as well as providing skills training and purposeful occupation (Patil et al. 2019). Horticultural therapy has been found to improve speech intelligibility, problem-solving, memory, and attention for clients with brain injury, as measured on the Functional Independence Measure/Functional Assessment Measure (FIM/FAM) (Hashimoto et al. 2006), whilst gardening groups for post-stroke therapy provided mental rest and connection to past experiences (Patil et al. 2019). Gardening improved the behaviour of clients with schizophrenia, as measured on the Positive and Negative Syndrome Scale, as well as their clinical symptoms, as measured on the Brief Psychiatric Rating Scale (Oh, Park, and Ahn 2018). A pilot randomised trial of horticultural therapy on apathy in nursing home residents with dementia who received three months of horticultural therapy showed reduced apathy, as measured on the Apathy Evaluation Scale informal version (AES-I), and improved cognitive function, as measured on the Mini Mental State Examination (MMSE) (Yang et al. 2022).

A qualitative study of a post-stroke gardening intervention found that clients highlighted a positive experience of nature, in which the garden was seen as a protected space for self-expression, where contact with nature boosted self-efficacy, and plants catalysed the patient/therapist relationship and bridged the hospital with the outside world (Lee et al. 2018). In another study, gardening fostered clients' active participation in their own care, enabling a positive and proactive attitude to disease management (Barello et al. 2016). For clients with acquired brain injury, horticultural therapy was found to improve social integration and productive activity, as measured by the Community Integration Questionnaire (CIQ): nine of the 25 clients in this study returned to work or school afterwards (Hashimoto et al. 2006). A mixed-methods study of horticulture for clients with mental health issues found that it was effective in increasing mental wellbeing and engagement in meaningful activities but did not result in significant changes in affect during therapy, or in an increase in social exchanges among people with mental illness (Siu, Kam and Mok 2020).

The restorative effects of environments can be measured. A review of 15 scales concluded that only three scales, including the Perceived Restorativeness Scale (PRS), have been adequately validated (Han 2018). The PRS is the most sensitive and generalisable instrument for measuring the attention restoration capacity of places, utilising the Kaplan theory of the mentally restorative benefits from attention and fascination in nature (Hartig et al. 1997; Han 2018). The research illustrates characteristics that can aid the therapy function of a garden. Depending on the client group, the need for security, seclusion, and activity will vary. The site itself may limit which characteristics can be incorporated, but an awareness of the suite of characteristics could be useful in planning garden evolution and appropriate activities with pointers to hazards and useful facilities.

A controlled trial of horticultural therapy was conducted with Asian older adults, where the intervention was weekly horticultural therapy for three months, followed by monthly sessions for a further three months (Ng et al. 2018). The results of this study suggest that horticultural therapy could potentially be useful in decreasing inflammation (reduction of IL-6 levels) and protecting neuronal functions (plasma CXCL12, CXCL5 and BDNF levels were maintained). This suggests that horticultural therapy may have some lasting effect at a cellular level that could underpin its overall benefit, although the mechanisms are not yet well understood.

This chapter is written for those who work 'hands-on and get their hands dirty'. The aims of a 'working' garden include hosting therapies, facilitating activity, and celebrating achievement. Gardening involves physical work and poses cognitive challenges. It facilitates activity in a sensory-rich environment in a quiet space. Therapy gardens aim to harness the benefits of gardening, but what makes a garden therapeutic? Are all garden features therapeutic? Can some or all of them be incorporated into working gardens to support therapeutic activities?

In what follows, I discuss how the benefits of gardening for clients with neuro-disabilities match particular therapeutic properties of gardens, based on evidence from scientific publications and clinical experience. I suggest that approaching the garden as an engineer might – looking at whole production systems and learning how several small improvements can make the system work better – could be relevant when working in a garden. This 'systems approach' to gardening can help to design a client-focused gardening system. Cultivation methods and creative activities associated with gardening are described, and reference is made to UK awards schemes as examples of ways to reward achievement and engage clients with the local community. Methods for evaluating effectiveness and auditing outcomes are presented. Finally, impacts of climate change and pandemic restrictions are briefly discussed.

Therapeutic properties of gardens

Gonzalez and Kirkevold (2016) surveyed 121 Norwegian nursing homes with gardens via an email questionnaire addressed to the managers. The results indicated that features such as sensory elements, stable walkways, landmarks, accessibility, visibility from indoors, and seating were therapeutic. Improvements suggested by managers included awareness of edible plants and greenhouses, awareness of toxic plants, maintenance, access to alarm systems, signposts to toilets, and lighting for the evenings.

A more systematic approach was taken by Scandinavian Life Sciences Universities, collaborating with care providers, who looked at which attributes of gardens make the space therapeutic. Stigsdotter and Grahn (2002) identify eight characteristics that make a garden a healing garden. Depending on the client group and available space, the garden should have a series of 'rooms' (sections/spaces) which include some or all of the characteristics listed in Table 7.1.

These findings were put into practice in Sweden, where the group set about creating a healing garden for people who experienced burnout (Stigsdotter and Grahn, 2003). Different garden areas or rooms were designed to provide contrast between the areas for rest and contemplation and those for activities or work. The first area was for rest and contact with the garden; the second was to help clients become more open, with activities and communication (e.g., sowing seeds, which require nurturing and care). This phased approach aimed to give participants scope for therapeutic meaning and action (Stigsdotter and Grahn 2003).

Berentsen et al. (2008) designed therapeutic gardens for dementia clients using a different approach to the garden rooms. Their design promoted activity, social interaction, and reminiscence, with popular plants that are pleasing to the eye. Walking paths in the form of paved, easy-to-follow, circular walks encouraged physical activity. A social centre with outside seating and a shelter, gazebo, or summer house was also included. These features complemented a green, secure room (the main garden area) surrounded by fences or shrubs with historical or cultural links (e.g., apple and fir trees, which are typically grown in this area). This enclosed garden included a reminiscence facility, with popular plants from past decades chosen for their stimulating forms, colours, and

Table 7.1 Therapeutic characteristics of a garden

Characteristic	*Room/space/section*
Serene	Peaceful, silent, and caring
Wild	Facilitates fascination with nature
Rich in species	Offering a variety of species and plants
Space	'Other world' gives a restful feeling
The common	A green open space for vistas and visits
Pleasure garden	Enclosed, safe, and secluded
Festive	Place for festivity and pleasure
Culture or history	Facilitates fascination with passing of time

scents. Suitable plants could be selected from the documented collection in the 'Great Granny's Garden' in the Oslo Botanical gardens. Developed as a living archive and sensory garden, this collection is a useful resource (Borgen and Guldahl 2011).

Simplicity in design

The evidence for the therapeutic features of gardens raises several questions about design. How can planting by therapy participants be facilitated? What do garden users want and experience? Can a design be flexible? Can some characteristics be incorporated to aid therapy?

In 2014, the HD residential unit at the RHN London, introduced at the beginning of this chapter, required refurbishment. Soil settlement had caused the hard paving to become cracked and uneven. The height of existing raised garden beds was too low. This created a falls risk for ambulant patients, given that balance is a problem for clients with HD. Many clients in the mid- to late stages of HD move about using large wheelchairs. Raised beds with vertical sides of inappropriate height led to unergonomic gardening, with wheelchairs facing sideways to the soil. While the unit was being refurbished, clients were relocated. Fortunately, plants in pots, planters, and containers were portable and could be taken to the temporary accommodation, which had a patio area where gardening could continue. Refurbishment created an opportunity to re-design the garden. A flexible design is important, as patient populations may change or units be re-assigned for other uses. Planting in pots allows day clients to take plants home with them and to prepare plants as presents or for sale.

In my study of Scandinavian therapy gardens (supported by the Churchill Trust), I found that just over half of the 14 gardens I visited had simple designs with flowers, lawns, shrubs, and trees. Potted plants were used in almost all of the gardens. A social centre was important, especially for dementia clients (Spring 2016). Raised beds or planters were used in half the gardens. Soil (ground-level) beds were cultivated at over half the sites. Half the gardens were enclosed with sensory plants and had a water feature. Almost all were adapted for wheelchairs and over a third contained art or sculpture.

A post-occupancy evaluation study of gardens for the aged in Canada, including clients with cognitive impairment, found that visitors to therapeutic gardens do not necessarily experience the garden in the way designers had intended (Heath 2004). The gardens were visited weekly during the study (but rarely for gardening). The gardens had been fully planted, with no provision for residents to cultivate. Visitors felt the gardens were helpful but would have appreciated directions for access. Staff (who were not allowed to use the gardens) expressed more reservations; for example, there were reports of residents getting wedged into or falling over rails and of residents falling into water features. Smoking was controversial, while swing seats and rocks were viewed as potential hazards. Shade was the unmet need raised most frequently, followed by a rainproof cover over part of the garden. Residents wanted a simple, flexible, open space with soft features, including a lawn where they could picnic (Heath and Gifford 2001; Heath 2004).

Gardens may need to evolve to accommodate changing populations or uses, so flexibility needs consideration at the design stage.

A 'systems engineering' approach to optimise use

Developed in the manufacturing sector, a 'systems engineering' approach seeks to optimise complex systems from the design stage, for example, by examining the height of a workbench, the weight and accessibility of work tools, and whether workers are involved in improving processes. Around the world, this approach is already applied in healthcare settings (Fanjiang et al. 2005; Kopach-Konrad et al. 2007; Liu 2018). A garden is a complex system, and its therapeutic success requires a systemic approach to activities that will take place there. In this section, I explore two key questions: How can clients with neuro-disabilities be enabled to garden in a systematic way? What are the components of a 'garden system'? I outline how we applied a 'systems engineering' approach to the HD client garden at RHN London by examining the site, facilities, tools, and people (clients and staff) and the cultivation of plants which comprise the 'garden system'.

The site

When considering the site, the first item to look at is the flow through the garden to answer a range of questions: Can the garden be accessed in wheelchairs, as well as by ambulant clients? Are the entrances wide enough, so that doors can be negotiated easily? Are there ramps, if necessary? Are handrails needed? Is the surface suitable for wheelchairs? Is the surface suitable for ambulant clients' exercise? Is there enough space for a group with members taking turns to plant? Is the area suitable for other therapy groups? Can clients have a break in the shade or under cover if it rains? When clients get tired, is it easy for staff to accompany them to their rooms?

Facilities

Facilities for growing include water sources, a storage area for compost, pots, and tools, seating, and suitable areas and/or containers for cultivation. Seating for clients, visitors, and staff, shelter, shade, and a social centre need consideration. Think how the plants will be displayed; the view from the windows is important. Cascade planters allow displays that are tiered and facilitate seasonal planting. For clients with balance problems or who are wheelchair based, containers such as freestanding manger-type beds or corner planters facilitate cultivation as chairs fit underneath (see Figure 7.1). Alternatively, a table, allowing a wheelchair or a stable chair to fit underneath, allows a 'gardening workstation' to be set up.

Figure 7.1 Gardening at a manger bed.

Note that the client rests their arms on the rim of the garden bed which reduces choreic movement, and uses a small watering can.

Tools

Items that facilitate cultivation at the workstation include compost, water, tools, pots or containers, plants, bulbs, or seeds, all within reach of the client. As HD is a movement disorder, clients often loose grip power and some control in their hands. One solution is to use potting tools that are small, light in weight, and easy to grip. A small plastic dibber helps planting out and a small plastic watering can may be lifted easily. Digging compost using a small plastic pot (easier to grasp than a trowel) will improve efficiency. It is important to know which is the client's dominant hand and place compost and water on that side. On the advice of an ergonomist, no adapted tools were used in the RHN garden in London; few were observed in Scandinavia.

People

Clients' sensory as well as motor abilities may be affected by neurological conditions. Some important questions to consider include: Has their colour vision been distorted? Can they hear the birds singing or winds rustling leaves? Can they smell flowers? Can they taste vegetables or fruit? Can they feel different textures?

Occupational therapists recognise two other senses: vestibular (important for balance) (Lopez 2016) and proprioception (feedback from pressure sensors in joints) (Han et al. 2016). For ambulant clients, sweeping with a broom stimulates the vestibular system through rhythmic movements of the head. Pressing down compost will stimulate proprioception sensors in the hands. I observed HD clients enjoying the experience of kneading the compost with their knuckles; it is possible that proprioception was involved.

Bulbs and seeds have pictorial labelling, which is helpful when clients are choosing what to plant. Men were more likely to request vegetables for cultivation. Growing potatoes in large pots was very popular and the clients enjoyed burrowing in the compost to harvest the tubers (see Figure 7.2).

Figure 7.2 Planting potatoes in a pot, seated at a table.

Note the concentration of the client. Assistance was just provided to take the weight of the compost.

Choice was important in a communal setting. Labelling by clients showed ownership. Planting in containers, pots, troughs, or planters allowed for portability. These allowed plants to be displayed in cascade planters, taken home, or given as presents to relatives or friends. Plastic pots are light; if clients throw one, they do not cause much damage. Adapted, ergonomic gardening achieves visible results, in which participants can take pride.

Residents were given choice of plants and activities. The garden provided exposure to nature and ownership in a communal setting. Staff should be encouraged to use the garden for external breaks for their own wellbeing and to conduct other therapies in the garden. Relatives disliked bench seating and wanted individual chairs where they could talk with residents in wheelchairs. Wooden chairs with arms were obtained.

Plants

The range of plants cultivated by residents aimed to provide seasonal variety and accommodate clients' movement and sensory capacities; for example, bulbs and larger seeds were easier to handle, and red flowers and leaves were preferred. See Box 7.1.

Box 7.1 Plants cultivated by residents

Bulbs and tubers

Research showed that bulbs and tubers were easier for clients to plant than seeds, probably as they were easier to grasp. Spring bulbs (including crocus, miniature narcissi, anemones, hyacinth, and tulips) were planted in pots, with larger narcissi in planters. The bulbs are poisonous but, if planting was supervised and the bulbs covered with soil, this was not a problem. Squirrels dug the bulbs up, so planting was topped off by a plant, such as a primula. Summer bulbs were planted, including gladioli, white agapanthus, begonias, and day lilies. Nerines and kaffir lilies flower in autumn. In winter, amaryllis were planted for indoor display.

Plug plants

A range of small ornamental plants was provided for potting on including: cyclamen, heathers, and viola for winter; wallflowers and primula for spring; pansies, salvia, and lychnis for summer; and asters and rudbeckia for the autumn.

Seeds

Due to the movement disorder with HD, large seeds were easier to plant, for example, nasturtium, beans, and peas.

Vegetables

Vegetables were grown from seed, onion sets, and tubers. Red-coloured vegetables were very popular with the HD clients, such as rainbow chard, beetroot, red onions, and red potatoes.

Fruit

Berry fruits were grown. Alpine strawberries were popular; blueberries grown in a planter were requested by a client.

Sensory plants

Sensory plants included Stachys (lamb's ear) for soft, furry leaves, scented stocks, lavender, rosemary, and poppies with seed heads that rattle. Tall sunflowers swayed in the wind. Red flowers and leaves were preferred, particularly azaleas, begonias, pelargoniums, and acers.

Outside the therapy area, a backdrop of shrubs and garden trees was planted for a succession of flowering and to provide shade. Poisonous and thorny plants were avoided.

Container gardening examples

For wheelchair-based gardeners and those with balance or mobility issues, gardening is easier seated at a desktop-height container. The person's feet fit underneath the unit. Trying to garden sideways is definitely not ergonomic. In Scandinavia, I observed a range of containers which differed from those used in the RHN London garden. Designs are discussed in Box 7.2.

Box 7.2 Container gardening examples

Sweden

In the roof garden at Dalheimers Hus in Goteborg, all gardening is in containers. The wooden units were made in a Swedish prison. They have a lip, an upper tier with soil, and a central pillar. A wide range of plants was cultivated, including ornamentals, vegetables, bush, and tree fruit.

Norway

The terrace garden at the Sunaas Rehabilitation Hospital on Oslo fjord was designed by the Agricultural University of Norway situated nearby. Raised beds of various heights made from contoured concrete allowed access by chair-based clients. Plants grown included ornamentals, herbs, and fruits. There were trellises in the raised beds for climbing plants. The adjacent hospital restaurant allows visitors to dine al fresco with clients in the garden.

United Kingdom

At the RHN London, raised wooden, stand-alone units are now used for growing. The RHN London participated with Thrive (a charity which promotes gardening for clients with disabilities) in a design project with a commercial company, which led to the purchase of the corner planters. These are good for stroke patients who can garden with one hand but need encouragement to use the other hand. The other unit is the V-shaped trough from a seed company which is deep enough to grow vegetables. All these units can be emptied and relocated to other parts of the hospital as patient numbers change, allowing flexibility. The tiered ladder allotments take a standard trough or a series of pots and are at heights which allow a wheelchair-based client to water.

All containers need the growing medium to be replaced as the soil becomes exhausted. The smaller containers may be easier, as soil can be tipped out. Fixed beds require digging out every three years, so that the growing medium can be replaced. Different designs of wheelchair-accessible raised beds were available in Europe.

Practical considerations for climate change and COVID-19 contexts

Climate change is leading to heavy rainfall, storms, drought, and extreme temperatures (Cook, Howarth, and Wheater 2019). The COVID-19 pandemic has led to changes in infection control measures required to keep people safe in garden spaces. This section discusses how resilience can be designed into therapy gardens.

Rainfall and drought

Drainage in the garden needs to be adequate, especially on a sloping or low-lying site. Shrubs and trees take up water, while lawns drain. Porous materials should be considered for patios. Large trees near buildings may cause damage in storms and roots can penetrate drains. A pagoda or veranda can protect outside activities in wet weather and allow outdoor meetings. Water can be captured from downspouts and stored in butts. Drip irrigation uses water sparingly. Soft green waste can be composted and used as soil conditioner; bark mulch can be spread over soil beds to retain moisture. Plants adapted to drier conditions with narrow leaves and thick cuticles, such as lavender and rosemary, can be planted.

Extreme temperatures

Verandas also shade windows; small trees or shrubs can provide shade. Umbrellas or blinds that pull out from buildings may be considered. In cold spells, tender plants can be protected with fleece or be brought into a greenhouse or conservatory.

COVID-19

Since the beginning of the COVID-19 pandemic in 2020, infection control measures have been introduced. Spread of the virus is lower outside (Weed and Foad 2020), and many residential units in the United Kingdom have installed visiting pods in grounds or gardens. Future gardens may need to include more covered outside space for visitors to meet relatives and staff. External meeting areas will also need outside access and signage.

Thought needs to be given when planning gardens to allow for resilience in changing climatic conditions and sudden changes (like an emerging pathogen) making new demands on external spaces.

Motivation of clients

Some clients indicate that they do not wish to garden (Patil et al. 2019), at least when they are initially approached. One who stated, 'I kill all plants' watched us gardening through the window in London, saw potatoes being harvested and later became a keen gardener. HD clients tend to be middle-aged; many have cultivated their own garden and wish to continue gardening. Some find being part of a group attractive; others never want to do hands-on gardening but may have an interest in associated activities. They may express an interest directly, or it may be discovered through speaking with them or their families about their hobbies. In this section, I ask what can encourage clients to participate in gardening? Some clients start gardening because they ask to garden; other clients are unsure. Do garden-related activities meet this need?

A paper garden for use inside on wet days was devised using a green, felt-covered board and pictures of flowers cut out of a seed catalogue backed with Velcro to hold them in place. It was not welcomed by residents. The gardeners said they wanted real soil, water, and living plants. Was there a need to nurture a living organism? Planting in pots addressed this need. However, collage using items cut from gardening magazines was popular with a client who enjoyed art (Spring et al. 2011).

Other activities associated with plant production and gardening

Growing plants is productive and provides opportunities for other creative and therapeutic activities. Not all clients wished to garden but many participated in activities related to the garden. Clients may be drawn into a gardening group for reasons other than wishing to cultivate. For example, their interest in craft and creative activities related to the garden can be explored.

In Scandinavia, the most common activities related to gardening were art, cookery, computing, ceramics, and music. However, a wider range of activities was observed, including constructive crafts involving concrete, wood, and metal that appealed to men and crafts using textiles. There was a strong emphasis on cookery using garden produce. Similar activities engaged clients in London but also included crafts, photography, bird watching, history, and music.

Art

At RHN London, paints were available to decorate pots. Even if clients were only able to hold a paintbrush, they could decorate a pot in an abstract style with the therapist moving the pot. Other clients with more control made beautiful designs: one had been a designer and the work was impressive. Sponges and stencils were used by the occupational therapy art technician (OTAT) to help clients decorate pots and make cards, with a good outcome despite movement disorders. Plants in pots were borrowed for another client who wished to draw plants. Christmas cards were made using materials from the garden. For the redesigned garden, the OTAT made a ceramic totem pole with the HD clients. Featuring ladybirds, butterflies, and flowers, it was unveiled by the Chief Executive of the Churchill Trust. In Scandinavia, stones were painted with ladybirds, seed packets and labels were designed, and concrete sculptures were made.

Photography

Photographs of the garden were taken by the horticulturist each month and a blank calendar was purchased in London. Clients made their own calendar, choosing pictures from a limited selection and pasting them up with a glue stick. Labels were added and the calendar was hung up. The garden was photographed after snowfall and special hangings were made. A calendar and photography were also observed in Sweden.

Bird watching and nature study

In the United Kingdom, each January, the Royal Society for the Protection of Birds (RSPB) co-ordinates a national bird watch. A client used the simple, brightly coloured, pictorial bird recognition chart to count the birds in the garden over the course of one hour. The results were sent to RSPB by computer. Engagement with nature was encouraged in Scandinavia: tree bark was touched, bird calls were identified, and a lens used to examine organisms. Dalheimers Hus had a nature room for study indoors.

Music and singing

Music therapy was also enjoyed by HD clients in London and Tangkær in Denmark. Tangkær had a group called Dizzy which went to festivals and music was made around the fire pit in the garden in summer. In London, a music therapist held a small concert in the garden. From this programme, clients adopted the Hippopotamus song refrain ('Mud, mud, glorious mud') and this was subsequently sung at most gardening sessions.

Gardens to mark religious festivals and life events

The chaplain in London encouraged gardening. Residents wished to mark Easter with a container-based Easter garden. A trug was filled with compost to create a hill containing a tomb. This was topped with three crosses and the garden was planted with herbs, moss, and gravel for paths. The gardeners wished to share it in the chapel. In Scandinavia, Easter decorations were made, as were Christmas wreaths and baskets. At RHN London, decorated pots of flowers were taken to the chapel at Christmas; the Imam agreed that a pot of cactus and shrubby plants representing the wilderness was appropriate for Eid al Adha.

In London, when gardening HD clients died, pots they had decorated or planted up were offered to relatives. At the Annual Memorial Service, lavender cuttings which had been potted up were offered to relatives to plant at their homes.

Gardening indoors

In Northern Europe, winter has higher rainfall and lower temperatures. In Scandinavia, gardening moved into conservatories, nature rooms, or greenhouses. These were heated and many contained grape vines. Some locations also had craft workshops. In London, gardening initially moved into a day room; redevelopment led to a bespoke 'garden room' where crafts and cultivation could be practised. An indoor area where gardening and crafts can be conducted during inclement weather helps the garden group continue throughout the year and may aid participation.

Gardening in pots

At RHN London, gardening in pots was conducted in the day room. This posed challenges, particularly as compost caused the floor to discolour and wheelchair-based clients had to garden on their tray tables. During the refurbishment, a small garden room immediately adjacent to the garden was fitted with work benches suitable for wheelchair use. Non-slip flooring (used for ship's decking) was installed, as were shelves for potted plant displays and boards to display artwork, collage, a calendar, and articles about the group.

Crafts

At RHN London, lavender was collected and dried. Bags were cut using a variety of fabrics and shapes. They were decorated with appliquéd shapes stuck with glue and filled with lavender. Each person chose a different theme, which stimulated discussion.

A wider range of crafts was tackled in Scandinavia. At one site where bees were kept, candles and hive frames were made. Carpentry was conducted, with one site having both carpentry and mechanical workshops, and willow weaving was popular. Clients produced textiles by weaving, and made cushions, wall hangings and wool crafts. Jewellery was made at one site.

Computing

At RHN London, the garden reached more residents via digital photographs viewed on a computer screen, either in the computer room or through the 'computer on wheels' service. The pictures invited clients to use a switch to change from one picture to another. Another HD client used the horticulturist's monthly garden tour to improve their typing skills, using a rollerball for the mouse and an adapted keyboard (Spring et al. 2011). An HD client made a book about the Dalheimers Hus roof garden in Sweden using an iPad.

Cooking

In Scandinavia, there was emphasis on cooking, with therapy kitchens located close to garden areas. Cooking was led either by an occupational therapist or by a professional chef, was aimed at healthy eating, and incorporated garden produce. At Lian, Trondheim, a hemiplegic client cooked waffles, served with cream and jam, in the garden for the gardening group. At RHN London, vegetables were prepared under the supervision of the occupational therapist, but the texture had to be modified due to swallowing problems caused by HD.

History

At RHN London, an HD client had a strong interest in history. The UK horticulturist lent him and a visitor a booklet on the Roman villa garden at Fishbourne. On their own initiative, they produced a wall hanging about the Roman garden, edged with a border of Roman-style hedging which was eye-catching. History and archaeology were also group activities at the historic Gunnebo Slott in Sweden.

Other therapies in the garden

A number of therapies can be conducted in a garden, and it may be a useful venue for building relationships with clients and families.

In Scandinavia, garden groups were attended usually by occupational therapists and often by other staff, including physiotherapists, psychologists, music therapists, and nurses. A Norwegian occupational therapist stated, 'You get highly visible results. It [gardening] provides an opportunity for staff and therapists to relate to clients. It provides good opportunities for cognitive and physical therapy'.

At RHN London, therapists and staff used the garden for sessions of psychology, occupational therapy, speech and language therapy, and music therapy. The paved surface was unsuitable for physiotherapy. Nursing staff used the garden as a less formal setting to speak with families.

Garden planning needs to be flexible and engage with staff to allow a range of therapies to be conducted outdoors.

Redesign for activity, therapy, and achievement

If an opportunity arises to redesign a garden, a review of relevant literature, research, or audit on the site and the users' views need consideration and communication to designers. Redesign may provide opportunities to allow the garden to provide more therapies and meet user needs.

The Huntington's Disease therapy garden at RHN in London was redesigned using information from research conducted on site, from Scandinavia, and from a literature review. The following characteristics were specified: secure, a soft surface which could support heavy wheelchairs, green, a social centre with individual chairs with arms, a table that wheelchairs would fit underneath, space for a range of group activities, raised accessible units for cultivation and plant display, quiet, and shade. An architect was briefed, the distortion of colour vision for HD clients was explained, and he was shown a picture of a garden painted by an HD client of red azaleas with yellow stamens on a green background. This led him to suggest a rubberised surface made from recycled tyres, coloured green with a central red and yellow compass rose motif, and coloured edges to emphasise entrances. A lawn would not support the heavy wheelchairs used by most clients.

The green rubber surface was laid, allowing for physiotherapy in the garden. Shrubbery was separated from the garden with a fence instead of low, raised beds, giving more space. Freestanding wooden V-shaped corner and cascade planters were provided for plant display and the cultivation of plants by clients. The social centre was a round table which accommodated the wheelchairs and included stable wooden armchairs and an umbrella for shade. The ceramic totem pole was installed behind the low fence. The restored garden was simple, flexible, and spacious.

Marking achievement with awards

Activity in the garden which produced plants led to achievement; achievement can engender pride and increase motivation. In London, a visitor from the

Royal Horticultural Society (RHS) suggested the gardening group enter the 'It's Your Neighbourhood' competition, part of Britain in Bloom. This led to helpful feedback.

Britain in Bloom

Started in 1963, Britain in Bloom has become the biggest horticultural competition in Europe. It has three pillars: horticultural achievement; environmental responsibility; and community participation. Its objectives are to combine horticultural skill, enthusiasm, and fun to improve and regenerate local environments with trees, shrubs, flowers, and edible crops. It also aims to conserve nature, recycle materials, remove litter and graffiti, and discourage vandalism. Groups work to regenerate neglected public open spaces and heritage sites. The scheme is operated on a regional basis.

It's Your Neighbourhood

It's Your Neighbourhood is a friendly competition where the three pillars are assessed rather than judged. It is designed to highlight achievement and encourage development. There is support and mentoring for groups, support packs are available, and the competition has five levels: establishing; improving; advancing; thriving; and outstanding.

A 'neighbourhood' is flexible but is characterised by the group being hands-on, involved in community gardening activities, and covers groups from parks, flats, schools, and many local organisations. The group must engage with the community and benefit it. The group must be sustainable over time, owned by the community, responsible for work, and involve volunteers. The scheme includes groups in care homes, hospitals, and on allotments or other gardens where therapeutic gardening is practised.

Some entries are from sites supporting clients experiencing physical or mental health conditions, learning difficulties, addiction, homelessness, or social exclusion. In the United Kingdom, doctors are now encouraged to make 'social prescriptions', which include gardening. Examples include the Plymouth Horticultural Therapy Trust, sited on an allotment, which grows fruit, vegetables, and flowers and teaches art and crafts. The site continued during the COVID-19 lockdowns and was featured on national radio and in RHS publications. Shekinah Grow takes homeless clients and teaches growing, crafts, and life skills. Some town Bloom groups have accepted clients with social prescriptions. All It's Your Neighbourhood entries receive a certificate.

Inclusive Gardening – discretionary award

As more groups are set up for therapy or accept social prescription clients, an advisory document for assessors was written outlining various conditions, communication strategies, and safeguarding for vulnerable clients. Some regions have discretionary awards for particular aspects of gardening. Visiting assessors write why they think an entry has excelled in this category and a panel reviews entries on a regional basis. The Inclusive Gardening award commenced in 2019 to confirm achievement by gardeners in therapy groups. Shekinah Grow received the first award and Plymouth Horticultural Therapy Trust with Yeovilable2achieve and Ivybridge Bloomers received awards in 2021. Clients may attend a local presentation ceremony to be awarded their framed certificate, which is a good chance for social interaction with other gardeners. If their condition does not allow for attendance, a presentation may be made to clients in their own garden by a local judge.

Positive feedback and suggestions

The assessors' reports have led to helpful suggestions for the groups; for example, having a garden visitors' book. In London, this was signed by civic dignitaries and academic visitors, who wrote admiring and encouraging comments. Clients were proud of the garden and enjoyed showing visitors their plants. Hints on growing, sustainability, and community involvement were also given by the assessor.

Recognition of good cultivation through an established scheme with trained assessors may bring useful information and contacts with the local horticultural community and re-enforce clients' achievements.

Evaluation of effectiveness

Therapeutic gardens may be assessed using standard methodologies to see if they meet the design objectives, elucidate unintended consequences, and formulate further improvements.

Great effort goes into designing a therapeutic garden: consulting users, doing research, planning, fundraising, and designing for the site and users. An audit establishes whether changes (the establishment of a new garden or changes to an existing garden) have met intended objectives. It can unearth unintended consequences and suggest improvements. Results could be fed back to funders to re-assure them that their money has been spent appropriately. It is best to use an established methodology. A health organisation may have an audit co-ordinator who could advise.

Post-occupation evaluation

An established methodology is post-occupation evaluation (Li, Froese, and Brager 2018) developed by architects to assess whether new buildings met the needs of users. This methodology uses a questionnaire. Heath and Gifford (2001) adapted this method for therapeutic gardens; Spring (2014) modified it for research and further adapted it for the audit of the garden at the RHN London in 2015, supported with funding from the Stanley Smith (UK)

Horticultural Trust. Weather conditions, date, and time when the questionnaire was completed were recorded. Had the respondent looked out on or had been outside in the garden? Information was sought on the purpose of garden visits, length of visit, who visited the garden with the respondent and their opinions of the features and plantings in the garden were recorded using five-point Likert-type scales. Questions were posed with yes or no answers or Likert scales were used. Open questions were included on impressions, problems, or improvements. Staff, such as cleaners and healthcare assistants, whose work or experiences might have been affected by changes in the garden were included. Views were sought from volunteers and visitors, as well as clinical staff and, most importantly, the clients. The audit provided evidence that the redesigned garden met outputs of the provision of a safe outdoor environment, a space for reflection and relaxation, facilitated resident gardening, encouraged social contact, and provided an outdoor area for therapy. Respondents requested more art and more colour in the garden.

Healthcare Garden Evaluation Toolkit

In 2017, the Healthcare Garden Evaluation Toolkit (HGET) was developed, providing a standardised method for evaluation of gardens in acute care settings (Sachs 2017). The study developed the Garden Assessment Tool for Evaluators, Staff and Patient/Visitor Surveys, Behaviour Mapping protocol and Stakeholder Interviews. In the United Kingdom and Scandinavia, much therapeutic gardening occurs outside care facilities, on allotments or in parks. Evaluation is still needed, but methodologies require adaptation for differing settings.

Evaluation tools, including post-occupation evaluation and the HGET, may be deployed in clinical settings. These methodologies may require adaptation for community settings and audit advice can be sought.

Conclusion

Evidence from research, including controlled trials, is accruing on the benefits and limitations of horticultural therapy for neuro-disability. In order to harness knowledge for therapeutic purposes, a wide range of professional, practical, and craft skills are needed to design, operate, and evaluate therapy gardens. Flexibility is important, as changes in climate, client populations, and treatments emerge.

Acknowledgements

Funders of research conducted at the Royal Hospital for Neuro-disability London: Neuro-disability Research Trust, Churchill Trust, Hamamelis Trust, Childwick Trust, Stanley Smith (UK) Horticultural Trust, Worshipful Company of Gardeners, and Suttons Seeds.

Thanks to Dr Duport (RHN London) as principal advisor and mentor, Geoff Hyde, Graham Price, Terry Porter, Britain in Bloom, Tessa Clarke, Cottenham Park Allotments, and Peter Spring, Librarian, for their critique of the draft text.

References

Barello, S, Graffigna, G, Menichetti, J, Sozzi, M, Savarese, M, Albino Claudio, B and Corbo, M 2016, 'The value of a therapeutic gardening intervention for post-stroke patients' engagement during rehabilitation: An exploratory qualitative study', *Journal of Participatory Medicine*, vol. 8, e9.

Berentsen, VD, Grefsrød, EE and Eek, A 2008, *Gardens for people with dementia: Design and use*, Forlaget Aldring og helse, Tønsberg.

Borgen, L and Guldahl, AS 2011, 'Great-granny's garden: A living archive and a sensory garden', *Biodiversity and Conservation*, vol. 20, no. 2, pp. 441–449. https://doi.org/10.1007/s10531-010-9931-9

Büttner, T, Schulz, S, Kuhn, W, Blumenschein, A and Przuntek, H 1994, 'Impaired colour discrimination in Huntington's disease', *European Journal of Neurology*, vol. 1, no. 2, pp. 153–157.

Cook, PA, Howarth, M and Wheater, CP 2019, 'Biodiversity and health in the face of climate change: Implications for public health' in M Marselle, J Stadler, H Korn, K Irvine and A Bonn (eds), *Biodiversity and health in the face of climate change*, Springer, Cham, pp. 251–281.

Dhalla, A, Pallikadavath, S, Hutchinson, CV 2019, 'Visual dysfunction in Huntington's disease: A systematic review', *Journal of Huntington's Disease*, vol. 8, no. 2, pp. 233–242. https://doi.org/10.3233/JHD-180340

Fanjiang, G, Grossman, JH, Compton, WD and Reid, PP (eds) 2005, *Building a better delivery system: A new engineering/health care partnership*, National Academy of Engineering and Institute of Medicine, National Academy Press, Washington, DC.

Gonzalez, MT and Kirkevold, M 2016, 'Design characteristics of sensory gardens in Norwegian nursing homes: A cross-sectional e-mail survey', *Journal of Housing for the Elderly*, vol. 30, no. 2, pp. 141–155. https://doi.org/10.1080/02763893.2016.1162252

Hall, M 1954, 'Huntington's chorea: Four cases in one family', *Canadian Medical Association Journal*, vol. 71, no. 5, pp. 493–494.

Han, J, Waddington, G, Adams, R, Anson, J and Liu, Y 2016, 'Assessing proprioception: A critical review of methods', *Journal of Sport and Health Science*, vol. 5, no. 1, pp. 80–90. https://doi.org/10.1016/j.jshs.2014.10.004

Han, KT 2018, 'A review of self-report scales on restoration and/or restorativeness in the natural environment', *Journal of Leisure Research*, vol. 49, no. 3–5, pp. 151–176. https://doi.org/10.1080/00222216.2018.1505159

Hartig, T, Korpela, K, Evans, GW and Gärling, T 1997, 'A measure of restorative quality in environments', *Scandinavian Housing and Planning Research*, vol. 14, no. 4, pp. 175–194. https://doi.org/10.1080/02815739708730435

Hashimoto, K, Okamoto, T, Watanabe, S and Ohashi, M 2006, 'Effectiveness of a comprehensive day treatment program for rehabilitation of patients with acquired brain injury in Japan', *Journal of Rehabilitation Medicine*, vol. 38, no. 1, pp. 20–25. https://doi.org/10.1080/16501970510038473

Heath, Y 2004, 'Evaluating the effect of therapeutic gardens', *American Journal of Alzheimer's Disease and Other Dementias*, vol. 19, no. 4, pp. 239–242. https://doi.org/10.1177/153331750401900410

Heath, Y and Gifford, R 2001, 'Post-occupancy evaluation of therapeutic gardens in a multi-level care facility for the aged', *Activities, Adaptation & Aging*, vol. 25, no. 2, pp. 21–43. https://doi.org/10.1300/J016v25n02_02

Kersten, HM, Danesh-Meyer, HV, Kilfoyle, DH and Roxburgh, RH 2015, 'Optical coherence tomography findings in Huntington's disease: A potential biomarker of disease progression', *Journal of Neurology*, vol. 262, no. 11, pp. 2457–2465. https://doi.org/10.1002/brb3.2592

Kopach-Konrad, R, Lawley, M, Criswell, M, Hasan, I, Chakraborty, S, Pekny, J and Doebbeling, BN 2007, 'Applying systems engineering principles in improving health care delivery', *Journal of General Internal Medicine*, vol. 22, no. 3, pp. 431–437. https://doi.org/10.1007/s11606-007-0292-3

Lee, AY, Park, SA, Park, HG and Son, KC 2018, 'Determining the effects of a horticultural therapy program for improving the upper limb function and balance ability of stroke patients', *HortScience*, vol. 53, no. 1, pp. 110–119. https://doi.org/10.21273/HORTSCI12639-17

Li, P, Froese, TM and Brager, G 2018, 'Post-occupancy evaluation: State-of-the-art analysis and state-of-the-practice review', *Building and Environment*, vol. 133, pp. 187–202. https://doi.org/10.1016/j.buildenv.2018.02.024

Liu, D 2018, *Systems engineering: Design principles and models*, CRC Press, Boca Raton.

Lopez, C 2016, 'The vestibular system: Balancing more than just the body', *Current Opinion in Neurology*, vol. 29, no. 1, pp. 74–83.

Ng, KST, Sia, A, Ng, MK, Tan, CT, Chan, HY, Tan, CH, Rawtaer, I, Feng, L, Mahendran, R, Larbi, A and Kua, EH 2018, 'Effects of horticultural therapy on Asian older adults: A randomized controlled trial', *International Journal of Environmental Research and Public Health*, vol. 15, no. 8, p. 1705. https://doi.org/10.3390/ijerph15081705

Oh, YA, Park, SA and Ahn, BE 2018, 'Assessment of the psychopathological effects of a horticultural therapy program in patients with schizophrenia', *Complementary Therapies in Medicine*, vol. 36, pp. 54–58. https://doi.org/10.1016/j.ctim.2017.11.019

Patil, G, Asbjørnslett, M, Aurlien, K and Levin, N 2019, 'Gardening as a meaningful occupation in initial stroke rehabilitation: An occupational therapist perspective', *The Open Journal of Occupational Therapy*, vol. 7, no. 3, pp. 1–15. https://scholarworks.wmich.edu/ojot/vol7/iss3/8/

Paulus, W, Schwarz, G, Werner, A, Lange, H, Bayer, A, Hofschuster, M, Müller, N and Zrenner, E 1993, 'Impairment of retinal increment thresholds in Huntington's disease', *Annals of Neurology: Official Journal of the American Neurological Association and the Child Neurology Society*, vol. 34, no. 4, pp. 574–578. https://doi.org/10.1002/ana.410340411

Sachs, NA 2017, *The healthcare garden evaluation toolkit: A standardized method for evaluation, research, and design of gardens in healthcare facilities* (Doctoral dissertation, University of Maryland). https://doi.org/10.13140/RG.2.2.19314.43209

Siu, AM, Kam, M and Mok, I 2020, 'Horticultural therapy program for people with mental illness: A mixed-method evaluation', *International Journal of Environmental Research and Public Health*, vol. 17, no. 3, p. 711. https://doi.org/10.3390/ijerph17030711

Spring, JA 2016, 'Design of evidence-based gardens and garden therapy for neurodisability in Scandinavia: Data from 14 sites', *Neurodegenerative Disease Management*, vol. 6, no. 2, pp. 87–98. https://doi.org/10.2217/nmt.16.2

Spring, JA, Baker, M, Dauya, L, Ewemade, I, Marsh, N, Patel, P, Scott, A, Stoy, N, Turner, H, Viera, M and Will, D 2011, 'Gardening with Huntington's disease clients–creating a programme of winter activities', *Disability and Rehabilitation*, vol. 33, no. 2, pp. 159–164. https://doi.org/10.3109/09638288.2010.487924

Spring, JA, Viera, M, Bowen, C and Marsh, N 2014, 'Is gardening a stimulating activity for people with advanced Huntington's disease?' *Dementia*, vol. 13, no. 6, pp. 819–833. https://doi.org/10.1177/1471301213486661

Stigsdotter, U and Grahn, P 2002, 'What makes a garden a healing garden?', *Journal of Therapeutic Horticulture*, vol. 13, no. 2, pp. 60–69.

Stigsdotter, U and Grahn, P 2003, 'Experiencing a garden: A healing garden for people suffering from burnout diseases', *Journal of Therapeutic Horticulture*, vol. 14, no. 5, pp. 38–48.

Thompson, R 2018, 'Gardening for health: A regular dose of gardening', *Clinical Medicine*, vol. 18, no. 3, pp. 201–205. https://doi.org/10.7861/clinmedicine.18-3-201

Wagenfeld, A and Atchison, B 2014, '"Putting the occupation back in occupational therapy": A survey of occupational therapy practitioners' use of gardening as an intervention', *Open Journal of Occupational Therapy*, vol. 2, no. 4, p. 4. https://doi.org/10.15453/2168-6408.1128

Weed, M and Foad, A 2020, 'Rapid scoping review of evidence of outdoor transmission of COVID-19', *MedRxiv*. https://doi.org/10.1101/2020.09.04.20188417

Yang, Y, Kwan, RY, Zhai, HM, Xiong, Y, Zhao, T, Fang, KL and Zhang, HQ 2022, 'Effect of horticultural therapy on apathy in nursing home residents with dementia: A pilot randomized controlled trial', *Aging & Mental Health*, vol. 26, no. 4, pp. 745–753. https://doi.org/10.1080/13607863.2021.1907304

8 Green places in red spaces

Broadening understanding of therapeutic gardening within rural Australia

Amy Baker, Alejandra Aguilar, and Ben Sellar

Introduction

Colour features prominently in representations of gardens and gardening. Bell et al. (2018) highlight the dominance of 'green' space in therapeutic landscapes literature, which has more recently included or been hybridised with 'blue' space. They argue that green and blue spaces, where land and water meet, are frequently cited as the spaces for leisure and recreation activities that support a healthy lifestyle, and thus dominate health education, promotion, and tourism discourse. Very little research examines colours outside this spectrum. Lengen (2015, p. 170) examined the associations psychiatric patients made with 'grey and white spaces such as rocks and snowfields', while Brooke and Williams (2021) reflect on the whites of Iceland. Wang and colleagues (2018) examine the yellows of the desert in Xinjiang, in which sand therapy is practised. Ultimately, the palette in therapeutic landscapes literature is very narrow, with no reds, minimal yellows, and predominantly the blues, greens, whites, and greys of European landscapes. Rural landscapes in Australia demand specific attention to the issue of colour in therapeutic gardening and cultivation.

Bell and colleagues (2018, p. 124) do critique colour-centric approaches as a whole, for failing to acknowledge the multi-sensory nature of landscapes, and propose instead the concept of palettic 'sensescapes' as more inclusive of diverse embodied experiences. The issue remains: Which colours are even included in the narrow visual field to begin with? While rural landscapes close to water or in tropical regions will find representation in this blue/green dominance, red and yellow landscapes, and the people and practices inhabiting them, find far less representation. This absence risks rural landscapes being understood as having limited therapeutic value – or, worse, being understood as places of hostility, where life struggles to survive (Maclean 2009).

Colour is just one critical component of therapeutic landscapes research that is challenged by rural landscapes. In this chapter, we suggest that much therapeutic gardening research has not only focused on blue/green zones, but has also been overly concerned with the benefits of domestic or community gardens and gardening in metropolitan areas, and how these benefits translate into rural areas. We argue that this focus has narrowed the perspective of

DOI: 10.4324/9781003355731-11

gardening and cultivation to the exclusion of important places and practices particular to rural contexts.

In this chapter, we survey the literature for evidence about the benefits and challenges of therapeutic gardening specifically in rural contexts. We examine how rurality forces a broadening of the definition of therapeutic gardening that can accommodate not only gardening, but also cultivation, which can occur across vast scales, and which entangles complex commercial, land management, cultural, and conservation practices. We argue that, while therapeutic gardening has been examined to a limited extent in rural contexts, future research needs to be shaped by rurality itself to acknowledge the ecology of practices particular to rural landscapes.

At this point, it is important to declare our viewpoints as authors. We are based in South Australia, which, like most states or territories in Australia, comprises predominantly rural landscapes. Our backgrounds in occupational therapy and occupational science offer certain perspectives which inevitably differ from other contributors to this edited collection. From our perspectives, we take the view that human action is a product of transactions between the person, the environment, and the occupational or activity demands at the time (Law et al. 1996). The core components of person, environment and occupation are central to theoretical models in occupational therapy and occupational science, such as the PEO (Person-Environment-Occupation) model (Law et al. 1996) and Canadian Model of Occupational Performance and Engagement (Townsend and Polatajko 2007). These models regard human occupation as a product of a complex network of relationships, elements, and entities that must be analysed together (Dewey and Bentley 1975), such that human action and agency are distributed amongst a set of more or less functionally coordinated relationships between contributing actants (Garrison 2001, p. 278). As such, human occupation and its therapeutic benefits cannot be understood outside the environments and demands that shape what people do, how they do it, and what people understand themselves to be doing. This perspective has led us to investigate topics such as what community gardening means to people in metropolitan and rural areas, and how cultural factors affect children's playful engagement in 'natural' spaces.

The rural landscape

The term 'rurality' has been defined in many ways, with definitions and measures of rurality changing over time (Nelson et al. 2021). Some approaches to defining rurality are based on what is observable and measurable, whereas other definitions revolve around people's perceptions of rurality and sociocultural characteristics (Nelson et al. 2021). The varied functions and meanings that have been attributed to rural spaces mean that the concept of 'the rural is a messy and slippery idea that eludes easy definition' (Woods 2011a, p. 1). Measurements, such as distance from urban centres or population numbers, are often widely used to define the concept of 'rural' (Nelson et al. 2021).

Although widely adopted by governments, a definition which understands rurality as a dichotomy between urban and non-urban has been critiqued as an artificial construct (Woods 2011c) that does not adequately represent 'the inherent differences and range of differences between these spaces' (Nelson et al. 2021, p. 355). Furthermore, as Cloke (2006, p. 18) has observed, there is also a 'blurring of conventional boundaries between country and city' which works in both directions. Some characteristics of urban space can be found in most 'rural' places; likewise, elements of rural space can often be found in urban areas (Woods 2011b). Therefore, there can be many different ways to describe and understand rurality (Woods 2011c). Amidst this complexity, for the purposes of this chapter, we take rurality to include any location outside a major city, encompassing inner regional areas to very remote locations.

Rural Australia is diverse in its landscape and people, being home to First Nations Australians, new migrants and refugees, regional migrants, international travellers, farming families, small business owners, workers (including healthcare workers, miners, engineers, prospectors, tourism operators), contractors, and retirees (Alston 2009; Duffy-Jones and Connell 2014; Radford 2017). In 2020, there was growth in regional migration, with 43,000 Australians moving from capital cities to regional areas, the largest net inflow to the regions since 2001 (Australian Bureau of Statistics [ABS] 2021).

Around 7 million people – approximately 28% of the population – now live in rural Australia (ABS cited in Australian Institute of Health and Welfare 2020). Population growth has been experienced in several rural areas, including in Victoria, where there is more suitable soil for agriculture and crop production (Duffy-Jones and Connell 2014). As of August 2019, 318,600 people were employed in agriculture, forestry, and fisheries in Australia, accounting for 2.5% of the national workforce (National Farmers Federation 2019). However, Australian farmers experience increasing climate variability and uncertainty, with extreme weather events, such as bushfires (wildfires), flooding, and hurricanes, with the resulting losses affecting their mental health and contributing to higher rates of suicide in rural areas (Perceval et al. 2019).

The mining industry is growing in Australia, particularly in parts of Queensland and Western Australia (Duffy-Jones and Connell 2014). Mining has brought several thousand international migrant workers to rural Australia (Duffy-Jones and Connell 2014). Thus, there has been an increase in temporary migration, with migrants filling job vacancies in regional areas, including in mining and health (Alston 2009; Maidment and Bay 2020). Interestingly, small rural towns have become sought-after retirement locations, with older Australians seeking a 'tree change' rather than the traditional 'sea change' (Duffy-Jones and Connell 2014).

Migrants and refugees seeking safety and a better future have also become part of rural Australia's population. Historically, migrants and refugees were settled in Australian cities, but the Australian Government has more recently prioritised rural settlement (Ziersch et al. 2020). Migrants arrive from a range of countries and cultures, and enter through various visa categories, including skilled, family, working holiday maker, refugee, and student (Alston 2009; Maidment and Bay 2020). It is not uncommon for people from the same region to be settled within the same Australian rural area (Wilding and Nunn 2018). For example, families from Southeast Asia and Africa have been resettled together in rural South Australian towns (Ziersch et al. 2020).

As a result of colonisation, First Nations Australians – the traditional owners of the continent now called Australia – were forcibly removed from their land and moved to missions and towns, a practice that has significant traumatic effects across communities and generations today (Human Rights and Equal Opportunity Commission 1997). Recognition of land rights and access to income support has provided a means for some to maintain or re-establish attachment to traditional country (Duffy-Jones and Connell 2014). For many First Nations populations, the 'shift to the bush' has meant living in some of the remotest places in Australia (Duffy-Jones and Connell 2014). According to the Australian Institute of Health and Welfare (2021), it was predicted that in 2021, 38% of First Nations Australians lived in major cities, whilst 62% lived in regional, remote, and very remote areas combined.

It should also be acknowledged that not all rural areas are growing in population. Smaller towns are declining, losing residents and industries (Duffy-Jones and Connell 2014). Young people are leaving in large numbers, attracted to jobs in the city and the opportunity for different experiences (Maidment and Bay 2020). The rural population is also ageing, with many farmers reaching retirement age (Duffy-Jones and Connell 2014). Changes to the rural population over the past decades are not unique to Australia. Across the globe, nations are experiencing the same rapid changes in geographical distribution of their populations, particularly in the Western World (Alston 2009).

As different populations enter the landscape, the ways in which rural land is used and shaped also change. The incoming population has diversified farming, from the traditional family-run farms to small horticultural ventures, boutique wineries, and niche food industries, through to dairy, the production of grain and other crops, and large cattle stations (Alston 2009). Thus, farming and its associated land clearing and use have shaped the landscape, with livestock grazing occurring in most areas of Australia, while cropping and horticulture tends to occur closer to the coast (Department of Agriculture, Water and the Environment 2021). New technologies, such as those used in the extraction of coal seam gas on farming land, have also changed how agricultural land is used and structured (Perceval et al. 2019).

Additionally, the mining industry has not only changed the land physically, but has also been responsible for building infrastructure and services for thousands of workers. Furthermore, the ageing and 'tree change' populations have led to the construction of accessible spaces for healthcare and socialisation (Alston 2009; Maidment and Bay 2020). Migrants and refugees bring with them connections to their previous home and several communities have used the land to continue the connection, by growing familiar crops from home. For example, Indian migrants have grown bananas in Coffs Harbour, while

Vietnamese refugees have engaged in market gardening near Perth (Alston 2009). The growing rural population has created the need for more housing, and recreational and entertainment spaces, and there is still great need for health (including mental health) and social services (Perceval et al. 2019). Thus, rural Australia is a place of change in both its people and landscape. The people influence the landscape, but the land, in turn, can shape who lives in rural Australia. If the events of recent decades continue, then rural Australia will continue to experience changes in demography, economy, culture, and practice.

It is interesting to note that, during the COVID-19 pandemic of 2020, rural Australia experienced the highest growth in regional migration since 2001 (ABS 2021). This could represent a desire to move away from crowded cities, where COVID-19 is prevalent and residents face the ongoing threat of snap lockdowns to control the spread of the virus (McManus 2022). It could also reflect a greater opportunity to move rurally, afforded by technologies and policies that encourage working from home (Terzon 2021). It is not yet known if those who moved to rural Australia during the pandemic will stay. Some predict that, once the virus is under control, the lack of infrastructure and services in rural Australia will result in people returning to urban areas (Terzon 2021).

Therapeutic gardening in rural Australia

In this section, we explore the characteristics and reported benefits and challenges of therapeutic gardening practices in rural Australian locations. It should be noted that our survey of the literature is limited to rural Australian therapeutic gardening initiatives where the authors have specifically reported on issues related to rurality. Additional examples which report on therapeutic gardening practices in rural Australia but where rurality did not feature in the study findings or discussion of findings were noted, including gardening in the context of a therapeutic day rehabilitation programme for people recovering from substance misuse (Missen et al. 2021) and communal gardening within an aged care residential setting (Fielder and Marsh 2021).

Participants from a rural community garden in the state of Victoria, Australia, noted that community gardening enabled them to feel a sense of belonging and connection to the local community, as well as a sense of ownership in the local area (Sanchez and Liamputtong 2017). This was described as especially important in the rural context; in particular, for people who were new to the region. As one participant explained, '[Living in a] rural community can often be isolating and I think it [the garden] has helped me integrate here ... the garden has been a big part of me settling in here' (Sanchez and Liamputtong 2017, p. 277). This rural community garden also welcomed non-gardeners from the local community, thus creating 'a garden for all to enjoy' and a 'gateway to the community' (Liamputtong and Sanchez 2018, p. 133). Participants also noted how this community garden offered opportunities to learn and share through working together, with the rural environment otherwise regarded as holding limited pathways for learning such skills and knowledge

(Liamputtong and Sanchez 2018). Finally, the community garden enabled participants to consume a wider range of vegetables and fruit, including varieties not available in supermarkets in rural areas (Sanchez and Liamputtong 2017). As such, this rural community garden offered some important, meaningful, and practical responses to a range of challenges that can be faced by people who live rurally, including the challenges of isolation, integration, skills and knowledge development, and the availability of fresh, diverse produce.

In contrast to the social experience of community gardens, rural backyard gardens have been noted to act as informal spaces of care for women with depression (Fullagar and O'Brien 2018). For one woman, the relationship with nature produced an energy – of pleasure, satisfaction, and joy – that sustained her recovery, created through the embodied practice of gardening (Fullagar and O'Brien 2018). This participant drew on the metaphor of 'nurturing herself through nurturing her garden', explaining that 'gardening kept her grounded', whilst also noticing that her garden kept her 'living in this particular place' (Fullagar and O'Brien 2018, p. 16). Yet for women who felt emotionally depleted or unmotivated to go outside, gardening was seen to assume the 'status of a burden of care' (Fullagar and O'Brien 2018, p. 16). Furthermore, a lack of available space for gardening also posed a challenge for some women. They 'spoke of a longing for a garden but were unable to access one with their present living arrangements' due to affordable housing which had no backyard – pointing to a need for more public gardening opportunities (Fullagar and O'Brien 2018, p. 16). Despite these challenges, it was noted that, for many women who live rurally and who experience depression, the 'organisation of formal spaces of expert care often impeded their recovery', with recovery instead being 'experienced through diverse care spaces, such as moving into the garden' (Fullagar and O'Brien 2018, p. 18). Thus, in the context of rural spaces where health and care services may be lacking or less than ideal, gardens and gardening can offer an opportunity for improving wellbeing and connection to place, but only if and when suitable spaces and resources are available.

At a broader level, involvement in gardening practices as part of a rural Men's Shed in regional Victoria – which incorporated a community garden and eco-house and included activities such as organic gardening, composting, water cultivation and worm farming – was seen as leading to greater awareness and promotion of environmental sustainability. It was noted that having green infrastructure onsite 'simultaneously promoted both human health and environmental sustainability' (Ayres et al. 2018, p. 66). This was described as particularly pertinent for regional settings, where the impact of extreme weather events, natural disasters, and climate change are 'felt more strongly' (Ayres et al. 2018, p. 66). This example highlights the potential of health promotion programmes, underpinned by a broader socio-ecological approach, to help respond to wider challenges facing both humans and non-human beings.

This brief overview indicates that therapeutic gardening is happening in rural areas, where unique challenges and opportunities are presented. Yet, it also raises important questions about the limits of the term 'therapeutic gardening' and the range of spaces and activities it might include in rural settings. Domestic and community gardens, as well as Men's Sheds, can all be found in metropolitan areas. Does the absence of more rural-specific places and practices associated with commercial farms, conservation areas, and traditional Aboriginal lands (for example) indicate that practices and benefits that are more prominent in rural settings might be overlooked?

Broadening the definition of therapeutic gardening to respond to rural settings

In this section, we examine what 'therapeutic gardening' can mean in terms of its nature and core features as a concept. In doing so, we show that it can be conceptualised sufficiently broadly to engage with a range of practices and places that are currently unexamined in the field.

Perhaps it is best to begin with a term that is embedded in the phrase 'therapeutic gardening' – the garden itself. The environment or setting of the garden or place of therapeutic gardening is, of course, vital to the experience of gardening. From what we know of people's experiences within garden spaces, this seems to be, at least partly, attributable to whether a gardening experience may be deemed therapeutic (e.g., Marsh et al. 2017; Ong et al. 2019). Yet, what are the limits to what may be considered a 'garden', and can therapeutic gardening only take place within the confines of what we refer to as a 'garden'?

Countless types of garden spring to mind – community gardens, home gardens, healing gardens, therapeutic gardens, kitchen gardens, school gardens, guerrilla gardens, botanical gardens, market gardens – to name just a few spaces that include the word garden. Other examples include arboretums, parks, orchards, or undercover spaces where gardens may exist, such as conservatories and indoor gardens. Any or all of these could qualify as a place of therapeutic gardening. In characterising a 'garden', Cooper (2006, p. 19) offers this insight: 'Gardens are not "virtual" or "ideal" places, somehow or other connected with physical "chunks" of land: they are simply places, albeit ones that invite a rich and varied range of description, experience, and comportment'. In considering the many and varied experiences and behaviours a garden can afford us, Cooper (2006) identifies the need to consider more than garden spaces or places to understand therapeutic gardening: 'For there can be little prospect, surely, of explaining the significance of gardens that ignores what human beings do in them' (p. 85). On this view, gardens are places that afford gardening practices, while being constituted and re-constituted as gardens through those very practices.

Unlike a 'therapeutic landscape' – on which there has been much attention since Gesler's seminal work (1992) – when we speak of therapeutic gardening, we refer to the varied *practices* of gardening. As such, the 'gardening' in therapeutic gardening refers not to the garden as an empirical space, but to the relationship(s) people form with and within it. The occupations people undertake in a garden space and the way people occupy that space can transform it into a place of gardening. Such occupations can be expected to vary greatly from person to person and from place to place. Cooper (2006, p. 68) provides some examples, ranging from what might be thought of as typical garden activities – digging, pruning, and propagating seeds – to gardening activities which may extend beyond the garden itself, such as ordering bulbs from catalogues or sketching out a new fishpond. From Cooper's (2006, p. 68) view at least – a perspective with which we agree – gardening is 'just about any activity geared to the design, cultivation, and care of the garden'.

So far, we have considered two essential components of therapeutic gardening: the environment or setting within which gardening takes place and the occupation or activity of gardening itself. It can be presumed that what connects these two components is the person, or people, who – like the occupation of gardening and the environment within which gardening takes place – may be complex and varied.

Another important consideration in defining 'therapeutic gardening' is the phrase 'therapeutic'. For this, we draw from Milligan's (2007, p. 255) suggestion as it appears within therapeutic landscape research: 'It can be argued that attaching the prefix "therapeutic" to a geographical interpretation of landscape, whether natural or created, brings with it an assumption that certain places have inherent qualities that promote physical or mental well-being'. In bringing together the essence of 'therapeutic' in Milligan's (2007) words and 'gardening' from Cooper's (2006) perspective – and by acknowledging that gardens or places of gardening may be highly varied – for the purposes of this chapter, we propose therapeutic gardening to describe 'any activities geared to the design, cultivation, and care of any place described or experienced as a garden, which promote wellbeing'.

When applied to rural settings, this definition of therapeutic gardening could include cultivation and agriculture at very large scales, undertaken commercially but with health-promoting effects, and large-scale Landcare and resource management practices with combinations of ecological, spiritual, and cultural significance. Can gardening be 'therapeutic' if it is primarily for work purposes – for example, for market gardeners? What if the place of cultivation is hundreds or thousands of hectares in size, such as many farms in rural Australia? The definition of therapeutic gardening proposed above suggests these examples could be considered therapeutic, if such practices promote wellbeing. Another question stems from the occupations or activities seen as inherent to therapeutic gardening. Can we be engaging in therapeutic gardening while not 'doing' anything at all, such as 'taking in' and enjoying the atmosphere, or 'being' in a garden, whether out of choice or necessity? Whilst resting or reflecting in a place of gardening may promote wellbeing, does this qualify as an 'activity'? We would argue that resting or reflecting could certainly lead to changes in the design, cultivation, or care of the garden, even if action is not the focus in that moment.

Another question centres on human beings' relationship to the places and beings within the place of therapeutic gardening. Many descriptions of gardening or therapeutic gardening practices make an anthropocentric assumption that humans are at the centre of this experience. Yet, what of the other beings with whom humans share these places of gardening, and who are, in fact, essential to places of gardening – such as insects, plants, and soil-dwelling microbes? How does their presence influence therapeutic gardening and what implications might this hold for them? For example, without an awareness of, and care for, soil organisms and soil as a whole, there would eventually be no place in which to garden therapeutically. The emergence of regenerative farming or regenerative agricultural practices offers one example of how farming can occur on larger scales, but still emphasise care for the place of cultivation. Within regenerative agriculture, farmers adopt a relational approach based on the interdependence between human and non-human organisms, in which farmers constantly co-evolve with farm ecosystems (Gordon et al. 2022). Mang and Reed (2012, p. 31) note that regenerative farming begins through the recognition that 'each place is a dynamic entity with its own unique history and future – growing and evolving, forming and decomposing, continuously influenced by the larger system in which it is embedded'.

Thus, there is great diversity in the description and nomenclature of approaches considered to constitute 'therapeutic gardening' or 'therapeutic cultivation' more broadly. Many initiatives and practices explored in this chapter are not classified as therapeutic gardening per se, yet they appear to align with what we (and others argue) this term can encompass. Examples include 'community gardening', 'care farming' (e.g., Norwood et al. 2019), and 'community agricultural initiatives' (e.g., Besterman-Dahan, Chavez and Njoh 2018). Whilst we welcome and celebrate the diversity and richness of approaches, such breadth of names generates challenges for pooling and describing existing research as the basis for progressing work in this area.

We argue that therapeutic gardening includes a wide range of active and passive practices of cultivation, taking place across urban–rural gradients, from which participants draw health-promoting benefits.

Implications

We claim that broadening the scope of therapeutic gardening in the face of rural places and practices is important. We began the chapter by describing how colour already demands this reconsideration. Below, we discuss three further opportunities presented to the field by this critical engagement:

1 Examining commercial and agricultural practices;
2 Engaging with Aboriginal and cross-cultural land management and care practices; and
3 Critically examining core conceptual foundations about humans' entanglement with landscapes.

This discussion might be critiqued for diluting the concept and misrepresenting the places and practices discussed below; however, we argue that such a discussion might be more likely to strengthen and expand the conceptual boundaries currently restricting therapeutic gardening.

Garden or land: Therapeutic cultivation at larger scales

Rural Australia increases the scale at which therapeutic gardening or cultivation can be practised, raising issues for which therapeutic spaces, practices, and experiences it might include. 'Urban Agriculture', with which community gardens are associated, is claimed to promote social resilience, employment and skills, food security, and reduced climate impacts of food supply chains (Azunre et al. 2019; Kirby et al. 2021). Emerging from a problematisation of unsustainable urbanisation in developing and developed nations, the Urban Agriculture movement has sought to bridge the divide between the urban and rural regions. This logic testifies to the potential therapeutic benefits of agriculture itself, once the urban context has been removed. Over the last few decades, 'care farms', which involve the therapeutic use of farming practices (Social Farms and Gardens 2018), have been established in Europe and North America (Hassink and van Dijk 2006; Hine, Peacock and Pretty 2008; Leck, Evans, and Upton 2014) – although there is limited literature from Australia on care farming. This research typically focuses on the therapeutic benefits of farms and farming for non-farmers, with specific disability, health, and mental health needs.

Baldwin, Smith, and Jacobson (2017) researched farmers themselves and their connection to place. Farmers spoke in terms consistent with healthy spaces, places, and practices, by describing powerful affective relationships to the land that were akin to religious experiences, functional practices of land management and improvement, and ongoing learning to understand and respond to the land's needs. From their view, farming as a practice and the farm as a place, are more than commercial or vocational. The farm is loved, cared for, and a host to practices reflective of this love. From this perspective, farming practices could conceivably be included in the scope of therapeutic gardening and cultivation.

Discursively, the shift from garden to farm or land potentially risks locating 'farming', 'agriculture', and 'land management' with vocational or obligatory pursuits, rather than the self-directed and recreative associations often conjured by gardening. Nonetheless, such practices fit within the broader notion of therapeutic cultivation.

Farming or firing: Therapeutic cultivation across cultures

Non-Western land management practices require consideration here. Kearns and Milligan (2020) identify a lack of application of therapeutic landscape ideas in non-Western settings, including Aboriginal community and country. In the Australian context, Aboriginal land and sea management, often referred to as Natural Resource Management (NRM) or 'caring for country' (Altman and Whitehead 2003; Hill et al. 2013), should also be considered as a form of therapeutic gardening or cultivation. Yibarbuk et al. (2002, p. 341) argue, for example, that fire management practices in tropical northern Australia are not only effective at maintaining ecological integrity, but also 'play important social and spiritual roles ... and hence in sustaining a culture characterized by intimate association with land'.

Australian and international literature suggests that indigenous fire management programmes deliver individual emotional and spiritual health benefits (Burgess et al. 2005; Campbell et al. 2011; Norgaard 2014). Burgess et al. (2005) show that participation of Aboriginal people in environmental management (including fire management) enhanced physical activity, reduced psychological distress, and increased connection to the land. Campbell et al. (2011) showed that programmes deliver 'significant and substantial savings in primary health care expenditure for the management of chronic disease'. Garnett and Sithole (2007) demonstrated that increased engagement in Indigenous Natural and Cultural Resource Management was correlated with reduced Body Mass Index (BMI), non-insulin-dependent diabetes, and coronary heart disease. Campbell et al. (2011) validated these results, showing that Aboriginal people's involvement in land management can result in very concrete health benefits.

Altman and Whitehead's (2003) description of 'Caring for Country' amongst Aboriginal people in the Northern Territory suggests that both customary and commercialised land management practices amongst Aboriginal communities must be better recognised as legitimate NRM. They argue that governmental land management policies have typically not recognised traditional land management practices – or, worse, represented them as a threat to ecological integrity and sustainability. Unsurprisingly, this has negative impacts on the collective wellbeing of Aboriginal people. Importantly, care must be taken to not conflate Aboriginal practices of caring for country with other practices of gardening and cultivation described in this chapter. Significant cultural, linguistic, epistemological, and ontological differences in the relationship to country, community, and self must be respected (Verran 2008). But, as several authors have called for greater recognition of the complex contributions such practices make to ecology, spirituality and health, further investigation is warranted.

Cultivated, natural, wild, or adapted

Finally, rurality raises the question of how cultivation relates to purification in historical and contemporary gardening practices. Many European traditions of gardening work by purifying spaces of unwanted flora and fauna. To a certain extent, modern agriculture operates on a similar logic by purifying landscapes of pests and plants that compromise the yield and quality of harvest. Often such practices function on an implicit binary between 'nature' or 'wilderness', which is represented as empty or uncultivated. As Burgess et al. (2005, p. 119) point out, in Australia the notion of 'wilderness' is deeply rooted in the 'fallacious notions of "Terra Nullius" [empty land] … often applied to Aboriginal estates'. Moreover, it fails to acknowledge the critical role of Aboriginal NRM in maintaining ecological integrity, contributing to the economy, and promoting individual and community health described above.

This notion of wilderness points to a second notion of purification in rural cultivation, that is, purifying spaces of introduced species, or returning them to the 'wild'. As Plumwood (2005) suggests, such approaches invest in the same myth of outback Australia as an uninhabited wilderness, by removing introduced or exotic crops in a practice of 'reversal' to return the garden to 'nature'. At times, this involves a specific focus on plants indigenous to the specific region.

Both efforts at purification invest in a rigidly binarised worldview, which is divided into human and non-human, or social and natural realms. Plumwood (2005) argues instead for the notion of an 'adapted garden' which is deeply entwined in, and responsive to, the ecological demands of its place. These include the economic and social demands of contemporary human life, the needs of other species for habitat and food, as well as changing environmental demands, such as climate change, drought, and fire. She describes how in her own garden she neither tries to keep wombats out nor seeks to create a 'wilderness' for them. Instead, she pursues a mutually beneficial gardening mode in which 'wombat lawnmowers' produce a 'marsupial lawn' that is 'beautiful to behold'. As such, she acknowledges the prior presence of native flora and fauna within an inescapably 'humanised' space.

Plumwood's (2005) critique highlights a critical issue for therapeutic gardening and cultivation in rural areas. The term 'therapeutic' implies that the spaces, places, and practices of gardening and cultivation are 'for someone'. Understanding how the therapeutic effects and benefits of gardening and cultivation are distributed across the landscape, across communities, across cultures, and across species, requires further research equipped with a complex concept of cultivation that includes more diverse perspectives and representations of the land, the landscape, and the practices of care performed upon it.

A call for future research

In discussing the rural environment, we acknowledge that characterising the term 'rural' is not an easy task. Even within Australia, there are significant and meaningful differences to consider when researching rurality and practices such as gardening and cultivation, including the nature of the land itself; natural forces at play, such as weather; the history and current circumstances of local people; and the wider ecosystems and economies of which they are part, amongst numerous other factors. Whilst some authors provide a clear description to illustrate their study setting, we encourage future research to include clear contextual information about the nature of the rural setting in which research has taken place.

In this chapter, we have argued that therapeutic gardening research in rural areas typically includes private or communal gardening for aesthetic or nutritional value, with incidental or intentional therapeutic benefits. However, the rural context raises questions about whether this literature exhausts all of the healthy spaces, places, and practices (Bell et al. 2018) that could be included within such a definition. Specifically, we argue for research that examines and responds to:

1 the under-representation of engagement with 'sensescapes' that reflect rural landscapes and their non-urban colour palettes;
2 the place of farms, farming, and land management as therapeutic places and practices;
3 the relationship between Aboriginal land management and cultural practice and therapeutic gardening and cultivation; and
4 the influence of assumptions about nature and culture on gardening research and practice.

Whilst there have been previous calls for further research into initiatives such as community gardening in rural settings (Ong et al. 2019), our exploration of the literature helps to confirm that very little is currently reported about the nature, scope, benefits, and challenges of therapeutic gardening practices happening across rural Australia. Given that almost 30% of the Australian population lives rurally, we expect there is a lot more to be discovered and reported in this field. Further research, particularly qualitative research, is needed to better understand the contexts in which therapeutic gardening occurs, the processes and approaches underlying therapeutic gardening as a practice, and the experiences and meanings of therapeutic gardening for people living in rural contexts.

Conclusion

In this chapter, we have explored a range of examples of therapeutic gardening and cultivation occurring in rural Australia. Our observations of the existing literature, alongside exploration of the term 'therapeutic gardening', raise important questions and assumptions about the nature, scope, and possibilities of therapeutic gardening within rural settings. The diversity of rural 'sensescapes', including characteristics such as colour and size, demands further exploration in the context of therapeutic gardening. We argue that rurality itself requires us to broaden the conceptual boundaries of therapeutic gardening to consider the diverse cultivation practices occurring across rural landscapes. Future research is needed to explore the places and practices particular to therapeutic gardening in rural contexts, with careful consideration towards geographical, cultural, and historical dimensions.

References

Alston, M 2009, *Innovative human services practice: Australia's changing landscape*, Palgrave Macmillan, Melbourne.

Altman, JC and Whitehead, PJ 2003, *Caring for country and sustainable Indigenous development: Opportunities, constraints and innovation*, Centre for Aboriginal Economic Policy Research, Australian National University, Canberra.

Australian Bureau of Statistics 2021, 'Net migration to the regions highest on record' (Media release, 4 May 2021). https://www.abs.gov.au/media-centre/media-releases/net-migration-regions-highest-record

Australian Institute of Health and Welfare 2020, *Rural and remote health*, Australian Government, Canberra.

Australian Institute of Health and Welfare 2021, *Profile of Indigenous Australians*, Australian Government, Canberra.

Ayres, L, Patrick, R and Capetola, T 2018, 'Health and environmental impacts of a regional Australian Men's Shed program', *Australian Journal of Rural Health*, vol. 26, no. 1, pp. 65–67. https://doi.org/10.1111/ajr.12373

Azunre, GA, Amponsah, O, Peprah, C, Takyi, SA and Braimah, I 2019, 'A review of the role of urban agriculture in the sustainable city discourse', *Cities*, vol. 93, pp. 104–119. https://doi.org/10.1016/j.cities.2019.04.006

Baldwin, C, Smith, T and Jacobson, C 2017, 'Love of the land: Social-ecological connectivity of rural landholders', *Journal of Rural Studies*, vol. 51, pp. 37–52. https://doi.org/10.1016/j.jrurstud.2017.01.012

Bell, SL, Foley, R, Houghton, F, Maddrell, A and Williams, AM 2018, 'From therapeutic landscapes to healthy spaces, places and practices: A scoping review', *Social Science and Medicine*, vol. 196, pp. 123–130. https://doi.org/10.1016/j.socscimed.2017.11.035

Besterman-Dahan, K, Chavez, M and Njoh, E 2018, 'Rooted in the community: Assessing the reintegration effects of agriculture on rural veterans', *Archives of Physical Medicine and Rehabilitation*, vol. 99, no. (Suppl 2), pp. S72–S78. https://doi.org/10.1016/j.apmr.2017.06.035

Brooke, K and Williams, A 2021, 'Iceland as a therapeutic landscape: White wilderness spaces for well-being', *GeoJournal*, vol. 86, pp. 1275–1285. https://doi.org/10.1007/s10708-019-10128-9

Burgess, CP, Johnston, FH, Bowman, DMJS and Whitehead, PJ 2005, 'Healthy country: Healthy people? Exploring the health benefits of Indigenous natural resource management', *Australian and New Zealand Journal of Public Health*, vol. 29, no. 2, p. 117–122. https://doi.org/10.1111/j.1467-842X.2005.tb00060.x

Campbell, D, Purgess, CP, Garnett, ST and Wakerman, J 2011, 'Potential primary health care savings for chronic disease care associated with Australian Aboriginal involvement in land management', *Health Policy*, vol. 99, pp. 83–89. https://doi.org/10.1016/j.healthpol.2010.07.009

Cloke, P 2006, 'Conceptualizing rurality' in P Cloke, T Marsden and P Mooney (eds), *Handbook of rural studies*, SAGE Publications, London, pp. 18–28.

Cooper, DE 2006, *A philosophy of gardens*, Oxford University Press, Oxford.

Department of Agriculture, Water and the Environment 2021, *Snapshot of Australian Agriculture 2021*, Australian Government. https://www.awe.gov.au/abares/products/insights/snapshot-of-australian-agriculture-2021#the-farm-population-is-diverse-and-constantly-changing

Dewey, J and Bentley, AF 1975, *Knowing and the known*, Greenwood Press, Westport.

Duffy-Jones, R and Connell, J (eds) 2014, *Rural change in Australia: Population, economy, environment*, Routledge, London.

Fielder, H and Marsh, P 2021, '"I used to be a gardener": Connecting aged care residents to gardening and to each other through communal garden sites', *Australasian Journal on Ageing*, vol. 40, pp. e29–e36. https://doi.org/10.1111/ajag.12841

Fullagar, S and O'Brien, W 2018, 'Rethinking women's experiences of depression and recovery as emplacement: Spatiality, care and gender relations in rural Australia', *Journal of Rural Studies*, vol. 58, pp. 12–19. https://doi.org/10.1016/j.jrurstud.2017.12.024

Garnett, ST and Sithole, B 2007, *Sustainable northern landscapes and the nexus with Indigenous health: Healthy country, healthy people*, Land and Water Australia, Canberra.

Garrison, J 2001, 'An introduction to Dewey's theory of functional "trans-action": An alternative paradigm for activity,' *Theory, Mind, Culture, and Activity*, vol. 8, no. 4, pp. 275–296. https://doi.org/10.1207/S15327884MCA0804_02

Gesler, WM 1992, 'Therapeutic landscapes: Medical issues in light of the new cultural geography', *Social Science and Medicine*, vol. 34, no. 7, pp. 735–746. https://doi.org/10.1016/0277-9536(92)90360-3

Gordon, E, Davila, F and Riedy, C 2022, 'Transforming landscapes and mindscapes through regenerative agriculture', *Agriculture and Human Values*, vol. 39, pp. 809–826. https://doi.org/10.1007/s10460-021-10276-0

Hassink, J and van Dijk, M 2006, *Farming for health: Green-care farming across Europe and the United States of America*, Springer, Dordrecht.

Hill, R, Pert, PI, Davies, J, Walsh, FJ and Falco-Mammone, F 2013, *Indigenous land management in Australia: Extent, scope, diversity, barriers and success factors*, CSIRO Ecosystem Sciences, Cairns. https://doi.org/10.4225/08/584ee74971137

Hine, R, Peacock, J and Pretty, J 2008, 'Care farming in the UK: Contexts, benefits and links with therapeutic communities', *Therapeutic Communities*, vol. 29, no. 3, pp. 245–260.

Human Rights and Equal Opportunity Commission 1997, *Bringing them home: National inquiry into the separation of Aboriginal and Torres Strait Islander children from their families*, Human Rights and Equal Opportunity Commission, Sydney.

Kearns, R and Milligan, C 2020, 'Placing therapeutic landscape as theoretical development in Health & Place', *Health & Place*, vol. 61, article 102224. https://doi.org/10.1016/j.healthplace.2019.102224

Kirby, CK, Specht, K, Fox-Kämper, R, Hawes, JK, Cohen, N, Caputo, S, Ilieva, RT, Lelièvre, A, Poniży, L, Schoen, V and Blythe, C 2021, 'Differences in motivations and social impacts across urban agriculture types: Case studies in Europe and the US', *Landscape and Urban Planning*, vol. 212, article 104110. https://doi.org/10.1016/j.landurbplan.2021.104110

Law, M, Cooper, B, Strong, S, Stewart, D, Rigby, P and Letts, L 1996, 'The person-environment-occupation model: A transactive approach to occupational performance', *Canadian Journal of Occupational Therapy*, vol. 63, no. 1, pp. 9–23. https://doi.org/10.1177/000841749606300103

Leck, C, Evans, N and Upton, D 2014, 'Agriculture – Who cares? An investigation of "care farming" in the UK', *Journal of Rural Studies*, vol. 34, no. 1, pp. 313–325. https://doi.org/10.1016/j.jrurstud.2014.01.012

Lengen, C 2015, 'The effects of colours, shapes and boundaries of landscapes on perception, emotion and mentalising processes promoting health and well-being', *Health & Place*, vol. 35, pp. 166–177. https://doi.org/10.1016/j.healthplace.2015.05.016

Liamputtong, P and Sanchez, EL 2018, 'Cultivating community: Perceptions of community garden and reasons for participating in a rural Victorian town', *Activities, Adaptation and Aging*, vol. 42, no. 2, pp. 124–142. https://doi.org/10.1080/01924788.2017.1398038

Maclean, K 2009, 'Re-conceptualising desert landscapes: Unpacking historical narratives and contemporary realities for sustainable livelihood development in central Australia', *GeoJournal*, vol. 74, article 451. https://doi.org/10.1007/s10708-008-9234-9

Maidment, J and Bay, U 2020, *Social work in rural Australia: Enabling practice*, Routledge, London.

Mang, P and Reed, B 2012, 'Designing from place: A regenerative framework and methodology', *Building Research and Information*, vol. 40, no. 1, pp. 23–38. https://doi.org/10.1080/09613218.2012.621341

Marsh, P, Gartrell, G, Egg, G, Nolan, A and Cross, M 2017, 'End-of-life care in a community garden: Findings from a participatory action research project in regional Australia', *Health & Place*, vol. 45, pp. 110–116. https://doi.org/10.1016/j.healthplace.2017.03.006

McManus, P 2022, 'Counterurbanisation, demographic change and discourses of rural revival in Australia during Covid-19', *Australian Geographer*. https://doi.org/10.1080/00049182.2022.2042037

Milligan, C 2007, 'Restoration or risk? Exploring the place of the common place' in A Williams (ed), *Therapeutic landscapes*, Routledge, London, pp. 255–272.

Missen, K, Alindogan, MA, Forrest, S and Waller, S 2021, 'Evaluating the effects of a therapeutic day rehabilitation program and inclusion of gardening in an Australian Rural Community Health Services', *Australian Journal of Primary Health*, vol. 27, no. 6, pp. 496–502. https://doi.org/10.1071/PY20294

National Farmers Federation 2019, 'Farm Facts'. https://nff.org.au/media-centre/farm-facts/#:~:text=As%20of%20August%202019%2C%20318%2C600,2.5%25%20of%20the%20national%20workforce.andtext=The%20gross%20value%20of%20Australian,2018%2D19%20was%20%2462.208%20billion

Nelson, KS, Nguyen, TD, Brownstein, NA, Garcia, D, Walker, HC, Watson, JT and Xin, A 2021, 'Definitions, measures, and uses of reality: A systematic review of the empirical and quantitative literature', *Journal of Rural Studies*, vol. 82, pp. 351–365. https://doi.org/10.1016/j.jrurstud.2021.01.035

Norgaard, KM 2014, 'The politics of fire and the social impacts of fire exclusion on the Klamath', *Humboldt Journal of Social Relations*, vol. 36, pp. 77–101.

Norwood, MF, Lakhani, A, Maujean, A, Downes, M, Fullagar, S, McIntyre, M, Byrne, J, Stewart, A, Barber, BL and Kendall, E 2019, 'Assessing emotional and social health using photographs: An innovative research method for rural studies and its applicability in a care-farming program for youth', *Evaluation and Program Planning*, vol. 17, article 101707. https://doi.org/10.1016/j.evalprogplan.2019.101707

Ong, M, Baker, A, Aguilar, A and Stanley, M 2019, 'The meanings attributed to community gardening: A qualitative study', *Health & Place*, vol. 59, pp. 1–9. https://doi.org/10.1016/j.healthplace.2019.102190

Perceval, M, Kolves, K, Ross, V, Reddy, P and De Leo, D 2019, 'Environmental factors and suicide in Australian farmers: A qualitative study', *Archives of Environmental and Occupational Health*, vol. 74, no. 5, pp. 279–286. https://doi.org/10.1080/19338244.2018.1453774

Plumwood, V 2005, 'Decolonising Australian gardens: Gardening and the ethics of place', *Ecological Humanities*, vol. 36. http://australianhumanitiesreview.org/2005/07/01/decolonising-australian-gardens-gardening-and-the-ethics-of-place/

Radford, D 2017, 'Space, place and identity: Intercultural encounters, affect and belonging in rural Australian spaces', *Journal of Intercultural Studies*, vol. 38, no. 5, pp. 495–513. https://doi.org/10.1080/07256868.2017.1363166

Sanchez, EL and Liamputtong, P 2017, 'Community gardening and health-related benefits for a rural Victorian town', *Leisure Studies*, vol. 36, no. 2, pp. 269–281. https://doi.org/10.1080/02614367.2016.1250805

Social Farms & Gardens 2018, 'What is care farming?'. https://www.farmgarden.org.uk/knowledge-base/article/what-care-farming

Terzon, E 2021, 'Covid-19 has made a tree change more alluring – but that may not last', (*ABC News*, 25 June 2021). https://www.abc.net.au/news/2021-06-25/covid-regional-australia-population-housing-services/100235562

Townsend, E and Polatajko, H 2007, *Enabling occupation II: Advancing an occupational therapy vision for health, well-being, and justice through occupation*, Canadian Association of Occupational Therapists, Ottawa.

Verran, H 2008, 'Science and the Dreaming', *Issues Magazine*, vol. 82, pp. 23–26.

Wang, K, Cui, Q and Xu, H 2018, 'Desert as therapeutic space: Cultural interpretation of embodied experience in sand therapy in Xinjiang, China', *Health & Place*, vol. 53, pp. 173–181. https://doi.org/10.1016/j.healthplace.2018.08.005

Wilding, R and Nunn, C 2018, 'Non-metropolitan productions of multiculturalism: Refugee settlement in rural Australia', *Ethnic and Racial Studies*, vol. 41, no. 14, pp. 2542–2560. https://doi.org/10.1080/01419870.2017.1394479

Woods, M 2011a, 'Approaching the rural' in M Woods (ed), *Rural*, Routledge, New York, pp. 1–15.

Woods, M 2011b, 'Imagining the rural' in M Woods (ed), *Rural*, Routledge, New York, pp. 16–49.

Woods, M 2011c, 'Re-making the rural' in M Woods (ed), *Rural*, Routledge, New York, pp. 264–294.

Yibarbuk, D, Whitehead, PJ, Russell-Smith, J, Jackson, D, Godjuwa, C, Fisher, A, Cooke, P, Choquenot, D and Bowman, DMJS 2002, 'Fire ecology and Aboriginal land management in central Arnhem Land, northern Australia: A tradition of ecosystem management', *Journal of Biogeography*, vol. 28, pp. 325–343. https://doi.org/10.1046/j.1365-2699.2001.00555.x

Ziersch, A, Miller, E, Baak, M and Mwanri, L 2020, 'Integration and social determinants of health and wellbeing for people from refugee backgrounds resettled in a rural town in South Australia: A qualitative study', *BMC Public Health*, vol. 20, p. 1700. https://doi.org/10.1186/s12889-020-09724-z

9 Health and wellbeing benefits of therapeutic gardens and gardening activities for older people living in residential aged care settings

Theresa L Scott

If you want to be happy, plant a garden

(Chinese Proverb)

Introduction

For many, entry to a residential aged care home is preceded by coping with the loss of home and close relationships. The incidence of depression is high (Teresi et al. 2001). An environment that provides aesthetic pleasure and opportunities for meaningful social interaction with other residents and staff can reduce anxiety and enhance wellbeing (Devlin and Arneill 2003; Dijkstra et al. 2006; Scott et al. 2014b). Conversely, highly institutionalised settings, in which physical health and safety take precedence over social and emotional health, can aggravate social isolation, and encourage adverse reactions (Brawley 2007; Calkins 2001; Ulrich 1995). Perceiving the environment as more aesthetically pleasing is positively related to improved mood and wellbeing for residents of aged care facilities (Beukeboom et al. 2012; Dahlkvist et al. 2016; Dijkstra et al. 2006). Gardens are not only aesthetically pleasing, but also have symbolic meaning for older people, with links to 'homemaking' (Bhatti and Church 2001, 2004). Gardening is a meaningful activity that promotes wellbeing in residential care settings. Person-centred gardening-based programmes provide activity that encourages socialisation through a shared appreciation of the aesthetics of nature (Fielder and Marsh 2021; Haslam et al. 2010; Scott et al. 2014a; Tse 2010). Gardens and their natural elements stimulate the senses and encourage social interaction through mutual appreciation of the associated sights, sounds, and smells (Fielder and Marsh 2021; Scott and Pachana 2015; Scott et al. 2014b), and through shared recollections of favourite trees and plants, or past gardens and experiences in nature. Outdoor gardens can provide access to sunshine, fresh air, and exercise, all of which help to regulate circadian rhythms, which support appetite and sleeping patterns (Aschoff 1965; Van Someren 2000). Outdoor gardens offer a place to socialise (by providing access to social partners), sensory stimulation, and a safe place to ambulate for people with dementia.

DOI: 10.4324/9781003355731-12

A brief history

The benefits of the aesthetics of gardens have been acknowledged for many centuries. Gardens and plants have been used as remedies for restoration as far back as the earliest documented use, when court physicians prescribed 'a walk in the palace garden' for princes suffering from mental fatigue and illness (Lewis 1976). During the Late Middle Ages (circa 1500), monks used their gardens to grow medicinal plants and provide places for the sick to reside and recuperate. Monastic gardens included flowers, herbs, orchards, vegetable beds, and fishponds, which not only provided monks with their livelihood, but also enabled them to care for the sick and poor as part of their charitable duty. The decline of monasticism saw the decline of restorative gardens, until the late-eighteenth and nineteenth centuries, when hospitals and asylums included gardens for their patients as a soothing distraction from their illnesses (Turner 2005). For example, male patients were initially engaged in the gardens and kitchens of asylums for economic reasons, to reduce the costs of caring for patients (Wilfert-Portal 2013). Nevertheless, the happiness and calming effect that their gardening duties bestowed led to recognition of the garden and grounds as a therapeutic tool; accordingly, male patients were later directed to be engaged as much as possible in the garden to 'promote happiness' (Wilfert-Portal 2013). However, the benefits of gardens and greenspaces go beyond aesthetics. A growing body of theoretical and empirical evidence shows the profound effect of gardens and gardening on our psychological and physiological wellbeing (Dijkstra et al. 2006; Fielder and Marsh, 2021; Kingsley et al. 2009; Scott et al. 2014a; 2022). One reason for these wide-ranging benefits for residents of aged care facilities is, according to the Biophilia hypothesis, an innate appreciation of nature (Wilson 1984). In the next section, I explain this and other theories underpinning therapeutic horticulture.

Theoretical explanations

Biophilia hypothesis

The Biophilia hypothesis proposes that humans' inborn affinity with the natural world arises from our evolutionary history, when our ancestors roamed the savannahs (Wilson 1984). During this period of evolution, green environments would have provided shelter from predators and a safe haven in which to rest and recover. Therefore, as the hypothesis suggests, all humans have evolved to prefer and respond positively to natural environments and their elements, such as trees, plants, water, and wildlife. Over millions of years, repeated experiences in natural environments thus encoded humans with a behavioural response – attraction to natural environments (Wilson 1993). The aesthetic experience of nature is fundamental to the theory, which proposes that our emotional health and wellbeing depend on having access to nature, or at least to views of nature.

Attention restoration theory

Attention restoration theory (ART) provides a psychophysiological explanation for the link between natural environments and human wellbeing. According to ART, natural environments can positively influence stress recovery and reduce anxiety because natural environments and their elements require effortless attention and act as a distraction from the things that 'ail' us. Further, they offer a break from 'directed attention' – a key concept of the theory which describes the prolonged periods of focused attention that we engage in our daily lives, which are followed by mental fatigue (Kaplan and Kaplan 1989). Spending time in nature is said to allow people a break from periods of prolonged mental effort and concentration and the capacity to recover focused attention (Kaplan and Kaplan 1989). Wandering through a garden, tending to indoor plants, or simply admiring a flower, provides an opportunity to 'recharge', according to this theory. Numerous studies have built a body of evidence supporting ART, using multiple modalities to measure the human stress response (such as blood pressure, pulse rate, salivary cortisol level, electrodermal activity), as well as self-reported mood states, to demonstrate that experiencing or even viewing nature, gardens and plants, helps regulate emotion, lower stress levels, and improve mood (Detweiler et al. 2012; Hassan et al. 2018; Ulrich 1984; Ulrich et al. 1991).

Given the accumulated evidence, it is discouraging that not all aged care homes in Australia have outdoor gardens; even when they do, use of this space varies. Highly institutionalised settings devoted above all to safety over aesthetic, social, and emotional needs, can exacerbate social isolation and depression. Gardens, outdoor spaces, and common areas lead to more time spent in them and less time spent in bedrooms, and thus promote socialisation (Devlin and Arneill 2003). Moreover, gardens provide a more home-like environment and encourage socialisation between residents, and between residents and staff, through a shared appreciation of the aesthetics of nature. However, while gardens are considered desirable, they are seen by many aged care providers as unachievable (Ulrich 2002); therefore, it may be said that gardens are on everybody's agenda, and nobody's budget.

The International Psychogeriatric Association's (IPA) Task Force on Mental Health Services in Long Term Care Facilities advocates explicit focus on quality of care that promotes quality of life for residents as a primary objective (Gibson et al. 2010). Moreover, the World Health Organization's Ottawa Charter (World Health Organization 2015) mandated psychologically healing environments as necessary for patient wellbeing. Psychologically healing environments are characterised by their health-promoting elements. The effect of the physical environment results from people's direct appraisal of it; for example, responding positively to attractive environments because they represent nurturing (Carver 1990). Conversely, physical features and aspects of the indoor environment of residential aged care facilities – such as 'dehumanizing' long sterile corridors and sparsely furnished rooms (Zeiss 2005) – have negative symbolic meaning and act to define a sick role and increase dependence (Veitch and Arkkelin 1995), rather than create a home-like atmosphere. In residential care settings, creating psychologically healing environments involves less 'hospital-like' environments and, instead, includes gardens, natural light, armchairs near windows, views of trees, nature sounds, and indoor plants.

Residential aged care: The Australian context

Residential aged care in Australia is a complex system that includes a range of policies, programmes, and services that has developed over time in a piecemeal way (Royal Commission into Aged Care Quality and Safety 2020). Residential aged care is a broad term used here to describe the range of accommodation and support systems provided to older adults who may have diminished capacity to care for themselves and whose needs can no longer be met in the community (Australian Institute of Health and Welfare [AIHW] 2012). These facilities are owned and/or managed by the not-for-profit, government, and private sectors. Many long-established facilities are modelled on the corridor design seen in hospitals, first introduced in the early 1960s to maximise staff efficiency and patient monitoring.

Box 9.1 Levels of residential care in Australia

Respite care

Residential respite care provides short-term planned and emergency places to individuals who require temporary care but intend to return to their home or usual place of residence. Respite care may be provided to people who are transitioning from hospital to home, or whose carers need a break from their caring duties. The care provided may be either low- or high-care (AIHW 2012).

Low-level residential care

Permanent care is offered on two levels (low and high) and, ideally, depends on the assessment of the individual's physical, medical, psychological, cultural, and social needs (AIHW 2012). Residents receiving low care are usually considered more mobile but still require assistance with personal care. Accommodation is usually in single occupancy rooms with private or shared bathroom facilities.

High-level nursing care

In addition to assistance with personal care and activities of daily living, high-level care units should provide 24-hour nursing care and operate within a medical model. Chronic conditions (such as dementia, diabetes, and frailty) are a major cause of entry to a high-level residential care/nursing home (AIHW 2012). In Australia, the rates of these chronic conditions and multi-morbidities are increasing and will have enormous fiscal and social implications for our aged care sector and a major impact on the quality of care on offer (Productivity Commission 2011).

Independence and autonomy in residential aged care settings

Older people prefer to age in place, living in the community and in their own homes (Gillsjö et al. 2011). At a minimum, older people have emphasised the importance of exercising choice and control over their care (Cleland et al. 2021). Reluctance to move into a residential care facility is usually due to the desire to retain independence and autonomy (Cleland et al. 2021), and to negative perceptions of formal residential aged care settings (Gillsjö et al. 2011). Aged care facilities in Australia differ in terms of the type of resident being cared for and, therefore, the intensity and type of services provided to the older adult resident. Care may include assistance with activities of daily living (such as meals, laundry, and social activities), personal care (such as bathing, dressing, eating, and toileting), and health care (such as medication upkeep, physiotherapy, dentistry, dietetics, and accommodation) on a permanent or temporary respite basis. The focus on efficient and functional delivery of care has been perceived as requiring rigid routine and rules, and although it may be well intended, these care routines and professional practices result in loss of privacy, autonomy, and dignity, and negatively impact an individual's quality of life (van Dijck-Heinen et al. 2014).

Autonomy in residential aged care is poorly understood by staff and management. Residents' abilities and opportunities to govern themselves are linked with increased quality of life, their human value and dignity (Moilanen et al. 2021). Perceived autonomy in residential aged care refers to an individual's right to choice and to dignity (Moilanen et al. 2021; Welford et al. 2012). Individual capacity, dependence, and level of care notwithstanding, residents must be afforded some level of autonomy and opportunities to govern themselves. While physical and mental capacity is associated with autonomy, and there may be legitimate reason to intervene in another's actions (for example, in the case of potential for harm to self or other (Welford et al. 2012)), the capacity to make decisions should be decision specific. Allowing individual choice in daily routines, such as what to wear, what to eat and when, how to decorate one's room, are just some examples of where the opportunities to govern oneself can be preserved (Moilanen et al. 2021). Even simple actions can support perceptions of autonomy, such as allowing residents independent choice and control concerning the care of a potted plant, as one seminal study showed (Langer and Rodin 1976).

Early empirical evidence

Caring for a potted plant

Influential research from the 1970s showed that active participation in caring for a potted plant led to significant improvements in residents' engagement and self-reported happiness (Langer and Rodin 1976). Combining the effect of access to nature, opportunities to stay engaged, and perceived autonomy and control, the researchers provided two groups of residents of a large Connecticut nursing home with a potted plant. Prior to this study, the residents of the facility were living in an environment where choice and opportunities to govern their own lives had largely been removed. One group (the 'responsibility group') was instructed to take full responsibility for the plant, while the other group (the 'comparison group') had no such control or care of the plant. The outcomes of the experiment were quite remarkable, including significant improvements in wellbeing and enhanced sociability in the 'responsibility group' compared with the 'comparison group', who were rated as more debilitated across the same period of time.

View from a window

One other of the earliest and most persuasive studies of the physiological and psychological healing benefits derived from the mere observation of nature was an empirical study set in a Pennsylvania hospital (Ulrich 1984). This retrospective study, examining the recovery of post-operative gall-bladder patients, took place during a time when such a procedure was more invasive than it is today, and patients' recovery stays were necessarily much longer. The hospital records of 23 pairs of patients were examined ($n = 46$). These 23 pairs of patient records were retrospectively matched on all health and demographic variables, from gender, age, socioeconomic status, and health behaviours (e.g., diet, history of alcohol and tobacco use) to attending doctors and nurses, and the colour of the rooms. The only difference between the matched patients was the window view from their hospital bed, which was either: (i) a canopy of trees; or (ii) the wall of a brick building. Results showed that, compared with patients with the brick wall view, patients with the nature view had significantly shorter hospital stays, fewer post-operative complications, and negative evaluative comments from nurses, and they took fewer strong analgesics to manage pain. The study challenged the traditional view that the physical environment had little influence on health outcomes and provided the impetus for the use of posters of plants, real plants, or aquariums in hospital and medical waiting rooms.

These studies were some of first to provide reliable empirical evidence for the healing benefits of contact with nature and were the catalyst for much further research and the later acceptance of the stress-reducing qualities of nature to improve clinical outcomes. Such findings are germane to contemporary residential aged care settings. As these studies show, a simple intervention (such as caring for a potted plant or having a window view of gardens) may compensate for the loss of direct contact with nature, providing a soothing distraction for people who are infirm, bed-bound or have limited access to outdoors for other reasons.

A hierarchy of benefits

Accumulated evidence reveals a hierarchy of benefits in relation to resident wellbeing, from immersed experiences in an outdoor garden, to tending to a potted plan in one's own room, or simply viewing gardens and trees through a window. Given that the number of older adults accommodated in residential aged care is expected to increase exponentially in the next few decades due to population ageing, implementing and funding activities and experiences that prevent boredom and loneliness and foster improved mood and health for residents must therefore warrant attention from stakeholders.

Depression and dementia are common and serious conditions that negatively impact the quality of life of older people living in residential aged care homes. One suggestion for the high incidence of depression and anxiety among the population of older adults in long-term care, compared with community-dwelling older adults, is that the environment of long-term care facilities is too often 'depressogenic' (Zeiss 2005). That is, they are too noisy, odorous, lacking in warm interpersonal contact (Zeiss 2005), and often feature long and austere, hospital-like corridors. An alternative approach is to adapt an environment more reminiscent of *home* (Dahlkvist et al. 2016; Eijkelenboom et al. 2017; Rappe and Kivelä 2005; van Dijck-Heinen et al. 2014). An outdoor garden, a wandering or multisensory garden, or natural elements indoors may benefit the overall incidence and prevalence rates of depression. People are drawn to natural environments, according to the Biophilia hypothesis (Wilson 1984), and gardens may be considered as agents of social inclusion because of their ability to bring people together to share a common experience and appreciation of the garden's aesthetics. If an outdoor garden is not possible, including nature-based interventions and making them more hands-on, in terms of planting or propagation of plants, may accrue social benefits; for example, group-based activities that provide residents with access to social partners, enhance opportunities for residents to interact with each other (Knight et al. 2010) and thereby improve overall mood and morale. Further, shared social activities have been documented to increase overall wellbeing for residents, in comparison to the wellbeing obtained from individual activities (Haslam et al. 2010); that is, by encouraging repeat interactions and socialisation between residents and residents, and residents and staff. These types of intervention may therefore be an important adjunct to other psychological therapies, such as cognitive behaviour therapy, for people with clinical depression.

An environment that provides aesthetic pleasure can enhance wellbeing and adjustment to institutional living for people living with dementia. For example, memory may be stimulated by being exposed to the sights and smells of nature, such as trees and plants, flowers, birds, water, and insects. Further, gardens may be places of restoration for people living with dementia or physical limitations. That is, when wide, looping paths are built into the design, gardens provide a safe area for wheelchair users and people with dementia to meander in an aesthetically pleasing space.

Sensory experiences

Therapeutic gardens describe a variety of spaces that are designed to be plant-dominated and accessible to accommodate a range of abilities; for example, people who are frail or wheelchair users. The basic features include: wide, level or gently graded and meandering paths that are glare-free and have handrails, where appropriate; raised planter beds or containers; a wide variety of non-toxic plants and shrubs selected for their variety of colours, textures, and fragrances; and trees that provide shade. In addition, comfortable seating that is placed in view of the garden and water features are important to enjoyment of the space (Brawley 2007). The aim is to enhance pleasure by engaging the senses (sight, sound, touch, and sometimes taste) and to stimulate memory (Scott and Pachana 2015). Such sensory experiences are important to institutionalised older adults whose indoor living environment may be quite austere and static. Previous research suggests that the opportunity to go outside and visit gardens has been linked to increased subjective wellbeing and to feelings of 'being away' and of 'fascination' (Dahlkvist et al. 2016). Conversely, not being able to go outside was related to feelings of depression and isolation (Dahlkvist et al. 2016). Therapeutic gardens provide a place to socialise by providing access to social partners (Brown et al. 2004) and a re-creation of memories of home (Brawley 2007) and, therefore, a place to reminisce. In addition, the aesthetic pleasure derived from having a garden to retreat to may help residents cope with the difficult emotional transition to institutional living (Collins and O'Callaghan 2007).

Group-based gardening

Group-based gardening activities provide a means to support residents' psychosocial needs. Active social engagement is not just a matter of providing more social activities (Andersson et al. 2007; Drageset et al. 2010). The quality of the activity is more important than the quantity. Accumulated research shows that residents report being most satisfied by meaningful activities and when opportunities exist for reciprocal nurturing, such as nurturing plants and biodiversity in a garden (Chaudhury et al. 2018; Langer and Rodin 1976; Nolan et al. 2001; Tse 2010). Increasing older adults' engagement in shared activities has important implications for their quality of life. Increasing social cohesion is also particularly important for isolated older adults living in residential care facilities who spend a lot of their time doing nothing (Burgio et al. 1994), or in their own rooms alone (Hague and Heggen 2007).

Gardens provide access to fresh air and sunshine and to the physical benefits of a wandering garden or actively gardening. A well-designed garden accommodates all levels of ability to enable everyone to take part in active gardening through raised garden beds and adaptive tools (Raske 2010). For example, activities may include cultivating flowers, herbs, and vegetables (Collins and O'Callaghan 2007; Raske 2010), watering, weeding, and even tending to chickens. These activities allow the residents to experience mastery, by seeing their cultivated flowers on dining tables, eating the cherry tomatoes and eggs at mealtimes, and thereby enhanced wellbeing. Such mastery – the belief that one has some control over one's outcomes – is an otherwise uncommon experience in institutional living, where people have diminished control over their lives.

Gardens provide a place of restoration. While activity is valued by residents (Collins and O'Callaghan 2007), mere observation, sitting and admiring a facility garden or visiting another garden provides important therapeutic benefits (Rappe et al. 2006). A Swedish study of nursing home residents with a high frequency of hospital visits found that participants benefited greatly from visits to an outdoor garden, as measured by their increased levels of concentration (Ottosson and Grahn 2005). Notably, for residents who had previously reported 'low tolerance' of other residents in group activities, time spent in the garden led to substantial decreases in their stress levels, as measured by changes in heart rate and blood pressure. Gardens therefore may be an important non-pharmacological alternative for managing anxiety in people with dementia residing in nursing homes.

Use of gardens in residential care homes has great social significance. The wellbeing benefits of being outdoors in a garden may be augmented by the garden's attractiveness to visitors. That is, a garden may provide a pleasant meeting place for residents and their family and friends, a place to relax, to linger. Morning teas served in a garden are one way to encourage residents to spend time outdoors and to promote social interactions. An outdoor green environment can enable shared activities and focus, as well as a way for visitors to capture the attention of a loved one with communication difficulties (Bengtsson and Carlsson 2013). These benefits may not be contingent upon installing an elaborate garden. Rather, according to one Swedish study, providing regular opportunities to spend time outdoors, to experience greenery and fresh air are valued by residents (Bengtsson and Carlsson 2013). In this study, residents appreciated views of greenery, either in the nursing home garden, or other green spaces further away from the facility.

Space and financial constraints are major barriers to including gardens in aged care settings; where they exist, such gardens risk becoming unused and neglected spaces if they are created without considering how they might best be used by residents (Gigliotti and Jarrott 2005; Heath and Gifford 2001). Design considerations are critical to reduce the risk of falls and enable free access for all residents, whether physically limited or very frail. Input from residents, family members, and staff into garden design and use is important. An awareness of the principles of dignity of risk, person-centred care, and an appreciation of the benefits against genuine risks will enhance the use of space. Staff education and training should therefore be implemented to support them to manage the facilitators and barriers to garden use. Barriers may include inclement weather, unsuitable clothing for weather conditions, heavy or locked doors, and inaccessible or insufficient entry points to the garden. Staff motivation and commitment to a maintenance programme is also vital, so that the garden thrives. Staff awareness of the garden's potential impact upon residents' wellbeing is essential to its permanence.

Indoor environmental transformations

While there are well documented benefits of including large outdoor gardens, if space and financial constraints prohibit the inclusion of an outdoor garden, bringing nature indoors is one way to obtain wellbeing benefits for residents in aged care settings. Introducing elements of nature (such as plants, shrubs, flowers, and birdsong) to the indoor environment is an achievable way for facilities to augment the physical environment and enhance wellbeing for residents. For example, one early and novel study conducted by Cohen-Mansfield and Werner (1998) examined the effects of an enhanced environment (including garden-| related elements) on the behaviours of nursing home residents. The enhancement, placed in a main corridor of a nursing home, was meant to simulate a garden environment – it included wall murals of garden scenes, aroma diffuser machines, audio effects of nature sounds, bench seating, and artificial plants. The intervention had a positive effect on mood and behaviour, reducing residents' desire to exit the home and a significant decrease in agitation. That is, the residents were observed to stop and sit for longer periods of time at the simulated garden setting. As the authors noted, however, the study was somewhat limited by the design, including the use of small single subject pre- and post-measures and a lack of a control comparison group (Cohen-Mansfield and Werner 1998).

In 2014, I was part of a team that conducted a conceptual replication of the environmental intervention which demonstrated that small improvements in the quality of the physical environment could lead to comparatively large effects on residents' wellbeing (Scott et al. 2014b). Our quasi-experimental study design sought to examine whether exposure to plants, through an indoor simulated garden installation, affected the wellbeing of residents ($n = 33$) and staff ($n = 24$). This replication included some variation to address previously cited methodological issues, such as the lack of a control group comparison, incorporating multiple methods of data collection, increasing sample size, and including living plants. We also investigated the effect of the garden-based intervention on social exchange. Introducing potted plants and nature elements, large murals of tree canopies, audio of birdsong hidden beneath the plants, aroma diffusers, and garden furniture in an arrangement to recreate an outdoor garden indoors, were the stimulus for interaction and engagement for the older adult residents (Scott et al. 2014b). The garden simulation was compared with a reminiscence-based installation and a control (no installation) condition. Both the garden- and reminiscence-based installations led to significantly more social engagement for residents, compared with a control condition, possibly due to the novelty of both. However, in addition, staff and residents reported that they appreciated the aesthetic appeal of natural plants and the audio of birdsong. Residents reported that they felt this simple environmental transformation related to nature and gardens made the facility feel more 'homely'. According to staff reports, some of the residents even shared recollections about their former gardens while the garden-based installation was in place. While administrators might consider therapeutic gardens to be desirable but not achievable (Ulrich 2002), introducing a 'sense of nature' through simple environmental transformation or planting shrubs and trees outside windows is not only effective, but also achievable.

Indoor gardening

Another strategy is to bring elements of nature indoors through horticulture-based activities. Acknowledging the important connection between having meaningful occupation and mental health Adolf Meyer, first Chief of Psychiatry at Johns Hopkins Hospital, observed in 1921 that, 'Man learns to organize time and he does it in terms of *doing* things' (Meyer and Features Submission 1983, p. 86). The residents of aged care facilities have considerable leisure hours available, and residents viewed typical leisure activities, such as carpet bowls and bingo, as having little value to them (Hill and Relf 1982). By contrast, horticulture-based activities may be goal directed and modified to meet the physical and cognitive needs of a range of residents of nursing homes. Horticulture-based activities, such as hands-on propagation of plants, plant cuttings and seedlings, nature crafts such as pressing foliage, visits to public gardens and parks are more engaging than traditional activities, such as games (Gigliotti et al. 2004; Wagenfeld and Atchison 2014).

For aged care home residents living with dementia, the mean percentage of 'time spent doing nothing' was significantly lower during horticultural activities, in comparison to traditional activities (such as games). Horticulture-based activities – planting outdoors, cooking and craft with plants indoors – also led to increased participant engagement and positive affect (Gigliotti et al. 2004). The focus on purposeful and tailored activities was enhanced by the horticulture therapy facilitator's engagement of participants in social interaction, eliciting reminiscence around participants' past activities and involvement in gardening, farming, and cooking. Compared with usual activities, gardening and horticulture-based activities promoted a sense of accomplishment and enhanced self-esteem. These types of person-centred non-pharmacological treatments may therefore be a strong complement to offset some of the psychosocial effects of living with a dementia by harnessing participants' strengths into meaningful productivities.

Horticulture activities can be tailored to meet the needs of frail aged care home residents. Indoor gardening may offset concerns for the safety of residents using outdoor gardens unsupervised, or where staffing prohibits residents with mobility limitations from accessing outdoor gardens. Increases in happiness resulted from participation in horticultural activities indoors for frail and pre-frail residents who took part in an eight-week indoor horticultural activity programme. Tailored activities included fertilising, re-potting plants, watering, trimming, propagating, plant species education, and seeding in small groups (Lai et al. 2018). In another study, an eight-week indoor gardening activity involving potting up seedlings, photography, discussions, and diarising gardening activities, led to significant improvements in residents' life satisfaction and social network, and significant decreases in perceptions of loneliness for the gardening group, compared to a control group (Tse 2010).

Chen and Ji (2015) examined the effect of indoor horticultural activities on psychosocial health in 10 residents of a nursing home in Taiwan. The 90-minute weekly session activities included planting green bean seeds, leaf and stem cutting, discussing proper plant care techniques, and making tea (Asiatic wormwood). Pre- and post-intervention measures of depressive symptoms and loneliness were recorded, and participants were interviewed about their experiences of the programme. After the intervention, depression and loneliness scores decreased. Participants also reported increased social connectedness, hope, and sense of achievement.

In relation to time spent in gardening and related activities, there is a dose-response outcome according to prior research. One study found an effect of 'duration' of an indoor gardening programme, such that a horticulture activity programme conducted once a week for five weeks was more effective than the programme conducted twice a week for two weeks (Brown et al. 2004). The study demonstrated positive benefits of implementing horticulture activities in aged care facilities to socialisation, and given the differences found between the two- and five-week programmes, a need for longer term implementation to realise such benefits.

The effects of horticultural activities for older adults residing in a nursing home in China were explored using robust physiological measurement. The study involved 40 female nursing home residents (mean age = 79.5 years), who were experiencing psychological distress and depression (Hassan et al. 2018). The study used a cross-over design, where participants took part in both experimental and control conditions. The experimental condition activities involved transplanting Chrysanthemum seedlings, while the control condition involved no activity with plants. Physiological assessments involved blood pressure and electroencephalography, used to measure discomfort and stress; psychological measures included the State-Trait Anxiety Inventory and Semantic Differential method (subjective ratings). Blood pressure decreased significantly, and changes in brainwaves were observed, such that Alpha wave activity (generally produced during relaxation) was higher when completing the plant task, as compared to the control task. Participants reported feeling more 'comfortable and relaxed' after the planting task than after the control task. Further, anxiety levels were significantly lower after the planting task, as compared to the control task.

One simple horticultural therapy intervention was effective at managing anxiety and agitation in residents living with dementia at bath time. This novel intervention involved recreating natural environments during residents' bath time, normally a stressful occasion for some residents with dementia (Whall et al. 1997). The multisensory intervention included audio, such as a bubbling stream, birdsong, and other small animal sounds; visuals of large and bright nature pictures; and offering residents food, such as pudding. This intervention resulted in a significant decrease in agitation for the treatment group compared to a control group – who received the usual care at bath time.

Conclusions

While these studies show that just being around plants has wellbeing benefits for residents of aged care homes, by providing opportunities for social exchange around a shared appreciation of the aesthetics of nature, there is evidence to suggest that active participation in caring for plants provides additional benefits. The seminal studies of Ulrich (1984) and Langer and Rodin (1976) demonstrate the restorative quality of plants and identity benefits can also be enjoyed, simply by viewing nature through a window or caring for a potted plant.

Entry to a long-term residential aged care setting is characterised by loneliness, social isolation, and a loss of identity and independence; the incidence of depression is high. Highly institutionalised settings, in which physical health, paternalistic and excessively cautious care take precedence over residents' self-determination, social, and emotional health can aggravate social isolation. Access to nature outdoors is a basic human right, not a privilege. Gardens and gardening stimulate the senses and encourage social interaction through mutual admiration of the associated sights and smells, and through shared recollections of favourite plants or past gardens. Outdoor gardens provide access to sunshine, fresh air, and exercise, all of which help to regulate circadian rhythms, sleep-wake cycles, and appetite. Gardens provide a place to socialise, by providing access to social partners, and to reminisce about home and past gardens.

The extent of the evidence base has not translated to policy and practice. There is an urgent need for improvements in the quality of care provided to residents. To begin, a few are suggested here: improving staff attitudes, knowledge, and awareness of the aesthetic needs of residents; prioritising access to the range of benefits that gardens and gardening provide; and, embedding gardening practices into routine care. The absence of gardens, or restrictions placed on access to them, where residents are afforded access only if accompanied by a family or staff member, removes individuals' most basic human right to self-determination (albeit with adjustments for an individual's characteristics). Gardening is a meaningful activity that can promote social and emotional wellbeing in aged care settings, which may offset the isolation that is so often experienced by residents, and where residents may languish indoors without contact with nature. If institutions are serious about their stated commitment to quality of life of residents, they should prioritise outdoor gardens. The cost versus investment is obvious in terms of the benefits to resident wellbeing and flow-on effects on frontline care staff.

References

Andersson, I, Pettersson, E and Sidenvall, B 2007, 'Daily life after moving into a care home – Experiences from older people, relatives and contact persons', *Journal of Clinical Nursing*, vol. 16, no. 9, pp. 1712–1718. https://doi.org/10.1111/j.1365-2702.2007.01703.x

Aschoff, J 1965, 'Circadian rhythms in man: A self-sustained oscillator with an inherent frequency underlies human 24-hour periodicity', *Science*, vol. 148, no. 3676, pp. 1427–1432.

Australian Institute of Health and Welfare 2012, *Residential aged care in Australia 2010-11: A statistical overview*, AIHW, Canberra.

Bengtsson, A and Carlsson, G 2013, 'Outdoor environments at three nursing homes – Qualitative interviews with residents and next of kin', *Urban Forestry & Urban Green*, vol. 12, no. 3, pp. 393–400. https://doi.org/10.1016/j.ufug.2013.03.008

Beukeboom, CJ, Langeveld, D and Tanja-Dijkstra, K 2012, 'Stress-reducing effects of real and artificial nature in a hospital waiting room', *Journal of Alternative and Complementary Medicine*, vol. 18, no. 4, pp. 329–333. https://doi.org/10.1089/acm.2011.0488

Bhatti, M and Church, A 2001, 'Cultivating natures: Homes and gardens in late modernity', *Sociology*, vol. 35, no. 2, pp. 365–383. https://doi.org/10.1177/S0038038501000177

Bhatti, M and Church, A 2004, 'Home, the culture of nature and meanings of gardens in late modernity', *Housing Studies*, vol. 19, no. 1, pp. 37–51. https://doi.org/10.1080/0267303042000152168

Brawley, EC 2007, 'Designing successful gardens and outdoor spaces for individuals with Alzheimer's disease', *Journal of Housing for the Elderly*, vol. 21, no. 3–4, pp. 265–283. https://doi.org/10.1300/J081v21n03_14

Brown, VM, Allen, AC, Dwozan, M, Mercer, I and Warren, K 2004, 'Indoor gardening and older adults: Effects on socialization, activities of daily living, and loneliness', *Journal of Gerontological Nursing*, vol. 30, no. 10, pp. 34–42. https://doi.org/10.3928/0098-9134-20041001-10

Burgio, LD, Scilley, K, Hardin, JM, Janosky, J, Bonino, P, Slater, SC and Engberg, R 1994, 'Studying disruptive vocalization and contextual factors in the nursing home using computer-assisted real-time observation', *Journal of Gerontology*, vol. 49, no. 5, pp. 230–239. https://doi.org/10.1093/geronj/49.5.p230

Carver, AM 1990, 'Hospital design and working conditions' in R Moran, R Anderson and P Paoli (eds), *Building for people in hospitals*, European Foundation for the Improvement of Living and Working Conditions, Dublin, pp. 85–92.

Chaudhury, H, Cooke, HA, Cowie, H and Razaghi, L 2018, 'The influence of the physical environment on residents with dementia in long-term care settings: A review of the empirical literature', *The Gerontologist*, vol. 58, no. 5, e325–e337. https://doi.org/10.1093/geront/gnw259

Chen, YM and Ji, JY 2015, 'Effects of horticultural therapy on psychosocial health in older nursing home residents: A preliminary study', *Journal of Nursing Research*, vol. 23, no. 3, pp. 167–171. https://doi.org/10.1097/jnr.0000000000000063

Cleland, J, Hutchinson, C, Khadka, J, Milte, R and Ratcliffe, J 2021, 'What defines quality of care for older people in aged care? A comprehensive literature review', *Geriatrics & Gerontology International*, vol. 21, no. 9, pp. 765–778. https://doi.org/10.1111/ggi.14231

Cohen-Mansfield, J and Werner, P 1998, 'The effects of an enhanced environment on nursing home residents who pace', *Gerontologist*, vol. 38, no. 2, pp. 199–208. https://doi.org/10.1093/geront/38.2.199

Collins, C and O'Callaghan, A 2007, 'Healing gardens for assisted living: An interdisciplinary approach to health education', *Journal of Extension*, vol. 45, no. 6, article 6IAW7.

Dahlkvist, E, Hartig, T, Nilsson, A, Högberg, H, Skovdahl, K and Engström, M 2016, 'Garden greenery and the health of older people in residential care facilities: A multi-level cross-sectional study', *Journal of Advanced Nursing*, vol. 72, no. 9, pp. 2065–2076. https://doi.org/10.1111/jan.12968

Detweiler, MB, Sharma, T, Detweiler, JG, Murphy, PF, Lane, S, Carman, J, Chudhary, AS, Halling, MH and Kim, KY 2012, 'What is the evidence to support the use of therapeutic gardens for the elderly?', *Psychiatry Investigation*, vol. 9, no. 2, pp. 100–110. https://doi.org/10.4306/pi.2012.9.2.100

Devlin, AS and Arneill, A 2003, 'Health care environments and patient outcomes: A review of the literature', *Environment and Behavior*, vol. 35, no. 5, pp. 665–694. https://doi.org/10.1177/0013916503255102

Dijkstra, K, Pieterse, M, and Pruyn, A 2006, 'Physical environmental stimuli that turn healthcare facilities into healing environments through psychologically mediated effects: Systematic review', *Journal of Advanced Nursing*, vol. 56, no. 2, pp. 166–181. https://doi.org/10.1111/j.1365-2648.2006.03990.x

Drageset, J, Kirkevold, M and Espehaug, B 2010, 'Loneliness and social support among nursing home residents without cognitive impairment: A questionnaire survey', *International Journal of Nursing Studies*, vol. 48, no. 5, pp. 611–619. https://doi.org/10.1016/j.ijnurstu.2010.09.008

Eijkelenboom, A, Verbeek, H, Felix, E and Van Hoof, J 2017, 'Architectural factors influencing the sense of home in nursing homes: An operationalization for practice', *Frontiers of Architectural Research*, vol. 6, no. 2, pp. 111–122. https://doi.org/10.1016/j.foar.2017.02.004

Fielder, H and Marsh, P 2021, '"I used to be a gardener": Connecting aged care residents to gardening and to each other through communal garden sites', *Australasian Journal on Ageing*, vol. 40, no. 1, e29–e36. https://doi.org/10.1111/ajag.12841

Gibson, MC, Carter, MW, Helmes, E and Edberg, AK 2010, 'Principles of good care for long-term care facilities', *International Psychogeriatrics*, vol. 22, no. 7, pp. 1072–1083. https://doi.org/10.1017/S1041610210000852

Gigliotti, CM and Jarrott, SE 2005, 'Effects of horticulture therapy on engagement and affect', *Canadian Journal on Aging/La Revue canadienne du vieillissement*, vol. 24, no. 4, pp. 367–377.

Gigliotti, CM, Jarrott, SE and Yorgason, J 2004, 'Harvesting health: Effects of three types of horticultural therapy activities for persons with dementia', *Dementia*, vol. 3, no. 2, pp. 161–180. https://doi.org/10.1177/1471301204042335

Gillsjö, C, Schwartz-Barcott, D and Von Post, I 2011, 'Home: The place the older adult can not imagine living without', *BMC Geriatrics*, vol. 11, no. 1, article 10. https://doi.org/10.1186/1471-2318-11-10

Hague, S and Heggen, K 2007, 'The nursing home as a home: A field study of residents' daily life in the common living rooms', *Journal of Clinical Nursing*, vol. 17, no. 4, pp. 460–467. https://doi.org/10.1111/j.1365-2702.2007.02031.x

Haslam, C, Haslam, A, Jetten, J, Bevins, A, Ravenscroft, S and Tonks, J 2010, 'The social treatment: The benefits of group interventions in residential care settings', *Psychology and Aging*, vol. 25, no. 1, pp. 157–167. https://doi.org/10.1037/a0018256

Hassan, A, Qibing, C and Tao, J 2018, 'Physiological and psychological effects of gardening activity in older adults', *Geriatrics & Gerontology International*, vol. 18, no. 8, pp. 1147–1152. https://doi.org/10.1111/ggi.13327

Heath, Y and Gifford, R 2001, 'Post-occupancy evaluation of therapeutic gardens in a multi-level care facility for the aged', *Activities, Adaptation and Aging*, vol. 25, no. 2, pp. 21–43. https://doi.org/10.1300/J016v25n02_02

Hill, CO and Relf, PD 1982, 'Gardening as an outdoor activity in geriatric institutions', *Activities, Adaptation & Aging*, vol. 3, no. 1, pp. 47–54. https://doi.org/10.1300/J016v03n01_07

Kaplan, R and Kaplan, S 1989, *The experience of nature: A psychological perspective*, Cambridge University Press, Cambridge.

Kingsley, JY, Townsend, M and Henderson-Wilson, C 2009, 'Cultivating health and wellbeing: Members' perceptions of the health benefits of a Port Melbourne community garden', *Leisure Studies*, vol. 28, no. 2, pp. 207–219. https://doi.org/10.1080/02614360902769894

Knight, C, Haslam, A and Haslam, C 2010, 'In home or at home? Evidence that collective decision making enhances older adults' social identification, well-being, and use of communal space when moving into a new care facility', *Ageing and Society*, vol. 30, no. 8, pp. 1393–1418. https://doi.org/10.1017/S0144686X10000656

Lai, CK, Kwan, RY, Lo, SK, Fung, CY, Lau, JK and Mimi, MY 2018, 'Effects of horticulture on frail and prefrail nursing home residents: A randomized controlled trial', *Journal of the American Medical Directors Association*, vol. 19, no. 8, pp. 696–702. https://doi.org/10.1016/j.jamda.2018.04.002

Langer, EJ and Rodin, J 1976, 'The effects of choice and enhanced personal responsibility for the aged: A field experiment in an institutional setting', *Journal of Personality and Social psychology*, vol. 34, no. 2, pp. 191–198.

Lewis, CA 1976, 'The evolution of horticulture therapy in the US', Paper presented at the *Fourth Annual Meeting of the National Council for Therapy and Rehabilitation through Horticulture*, 6 September 1976, Philadelphia.

Meyer, A and Features Submission, HC 1983, 'The philosophy of occupational therapy', *Occupational Therapy in Mental Health*, vol. 2, no. 3, pp. 79–86. https://doi.org/10.1300/J004v02n03_05

Moilanen, T, Kangasniemi, M, Papinaho, O, Mynttinen, M, Siipi, H, Suominen, S and Suhonen, R, 2021, 'Older people's perceived autonomy in residential care: An integrative review', *Nursing Ethics*, vol. 28, no. 3, pp. 414–434. https://doi.org/10.1177/0969733020948115

Nolan, M, Davies, N and Grant, G 2001, *Working with older people and their families*, McGraw-Hill Education, Maidenhead.

Ottosson, J and Grahn, P 2005, 'Measures of restoration in geriatric care residences: The influence of nature on elderly people's power of concentration, blood pressure and pulse rate', *Journal of Housing for the Elderly*, vol. 19, no. 3–4, pp. 227–256. https://doi.org/10.1300/J081v19n03_12

Productivity Commission 2011, *Caring for older Australians*, Commonwealth of Australia, Melbourne.

Rappe, E and Kivelä, SL 2005, 'Effects of garden visits on long-term care residents as related to depression', *HortTechnology*, vol. 15, no. 2, pp. 298–303. https://doi.org/10.21273/HORTTECH.15.2.0298

Rappe, E, Kivelä, SL and Rita, H 2006, 'Visiting outdoor green environments positively impacts self-rated health among older people in long-term care', *HortTechnology*, vol. 16, no. 1, 55–59. https://doi.org/10.21273/HORTTECH.16.1.0055

Raske, M 2010, 'Nursing home quality of life: Study of an enabling garden', *Journal of Gerontological Social Work*, vol. 53, no. 4, pp. 336–351. https://doi.org/10.1080/01634371003741482

Royal Commission into Aged Care Quality and Safety 2020, *Aged care and COVID-19: A special report*, Commonwealth of Australia, Canberra. https://agedcare.royalcommission.gov.au/sites/default/files/2020-12/aged-care-and-covid-19-a-special-report.pdf

Scott, TL, Jao, Y-L, Tulloch, K, Yates, E, Kenward, O and Pachana, NA 2022, 'Well-being benefits of horticulture-based activities for community dwelling people with dementia: A systematic review', *International Journal of Environmental Research and Public Health*, vol. 19, no. 17, pp. 1–20. https://doi.org/10.3390/ijerph191710523

Scott, TL, Masser, BM and Pachana, NA 2014a, 'Exploring the health and wellbeing benefits of gardening for older adults', *Ageing & Society*, vol. 35, no. 10, pp. 1–25. https://doi.org/10.1017/S0144686X14000865

Scott, TL, Masser, BM and Pachana, NA 2014b, 'Multi-sensory installations in residential aged-care facilities: Increasing novelty and encouraging social engagement through modest environmental changes', *Journal of Gerontological Nursing*, vol. 40, no. 9, pp. 20–31. https://doi.org/10.3928/00989134-20140731-01

Scott, TL and Pachana, NA 2015, 'Therapeutic gardens and expressive therapies' in Helen Lavretsky, Martha Sajatovic and Charles Reynolds (eds), *Complementary, alternative, and integrative interventions for mental health and aging: Research and practice*, Oxford University Press, New York, pp. 529–546. https://doi.org/10.1093/med/9780199380862.003.0033

Teresi, JA, Abrams, R, Holmes, D, Ramirez, M and Eimicke, J 2001, 'Prevalence of depression and depression recognition in nursing homes', *Social Psychiatry and Psychiatric Epidemiology*, vol. 36, no. 12, pp. 613–620. https://doi.org/10.1007/s127-001-8202-7

Tse, MM 2010, 'Therapeutic effects of an indoor gardening programme for older people living in nursing homes', *Journal of Clinical Nursing*, vol. 19, no. 7–8, pp. 949–958. https://doi.org/10.1111/j.1365-2702.2009.02803.x

Turner, T 2005, *Garden history, philosophy and design, 2000 BC-2000 AD*, Spon Press, New York.

Ulrich, RS 1984, 'View through a window may influence recovery from surgery', *Science*, vol. 224, no. 4647, pp. 420–421. https://doi.org/10.1126/science.6143402

Ulrich, RS 1995, 'Effects of healthcare interior design on wellness: Theory and recent scientific research' in SO Marberry (ed), *Innovations in healthcare design*, Van Nostrand Reinhold, New York, pp. 88–104.

Ulrich, RS 2002, 'Health benefits of gardens in hospitals', Paper presented at *Plants for People Conference*, International Exhibition Floriade, 2002.

Ulrich, RS, Simons, RF, Losito, BD and Fiorito, E 1991, 'Stress recovery during exposure to natural and urban environments', *Journal of Environmental Psychology*, vol. 11, no. 3, pp. 201–230. https://doi.org/10.1016/S0272-4944(05)80184-7

van Dijck-Heinen, CJML, Wouters, EJM, Janssen, BM and van Hoof, J 2014, 'The environmental design of residential care facilities: A sense of home through the eyes of nursing home residents', *International Journal for Innovative Research in Science & Technology*, vol. 1, no. 4, pp. 57–69.

Van Someren, EJ 2000, 'Circadian rhythms and sleep in human aging', *Chronobiology International*, vol. 17, no. 3, pp. 233–243. https://doi.org/10.1081/cbi-100101046

Veitch, R and Arkkelin, D 1995, *Environmental psychology: An interdisciplinary perspective*, Prentice Hall, Englewood Cliffs.

Wagenfeld, A and Atchison, B 2014, '"Putting the occupation back in occupational therapy:" A survey of occupational therapy practitioners' use of gardening as an intervention', *Open Journal of Occupational Therapy*, vol. 2, article 4. https://doi.org/10.15453/2168-6408.1128

Welford, C, Murphy, K, Rodgers, V and Frauenlob, T 2012, 'Autonomy for older people in residential care: A selective literature review', *International Journal of Older People Nursing*, vol. 7, no. 1, pp. 65–69. https://doi.org/10.1111/j.1748-3743.2012.00311.x

Whall, A, Black, M, Groh, C, Yankou, D, Kupferschmid, B and Foster, N 1997, 'The effect of natural environments upon agitation and aggression in late stage dementia patients', *American Journal of Alzheimer's Disease and Other Dementias*, vol. 5, no. 12, pp. 216–220. https://doi.org/10.1177/153331759701200506

Wilfert-Portal, B 2013, 'L'histoire culturelle de l'Europe d'un point de vue transnational', *Revue Sciences/Lettres*, vol. 1. https://doi.org/10.4000/rsl.279

Wilson, EO 1984, *Biophilia*, Harvard University Press, Cambridge, MA.

Wilson, EO 1993. 'Biophilia and the conservation ethic' in SR Kellert and EO Wilson (eds), *The biophilia hypothesis*, Island Press, Washington, DC, pp. 250–258.

World Health Organization 1986, Ottawa charter for health promotion, 1986 (No. WHO/EURO: 1986-4044-43803-61677). World Health Organization. Regional Office for Europe.

World Health Organization 2015, *World report on ageing and health*, WHO, Geneva. https://www.who.int/publications/i/item/9789241565042

Zeiss, AM 2005, 'Depression in long-term care: Contrasting a disease model with attention to environmental impact', *Clinical Psychology: Science and Practice*, vol. 12, no. 3, pp. 300–302. https://doi.org/10.1093/clipsy.bpi038

Part III

Dig deep

Expanding and enriching the cultivated therapeutic landscape

10 Environmental place-making by the 'out of place'

Migrants building connections to new landscapes through structured conservation activities

Pauline Marsh, Suzanne Mallick, Dave Kendal, and Renae Riviere

Introduction

Migration comes with a suite of new experiences, not all of which are positive. New migrants can be marginalised by citizens of the host country and feel 'out of place' as they face language barriers, economic and social hardships, as well as racism and prejudices (Egoz and De Nardi 2017). Building connections to the natural landscapes of the host country is one means by which some migration challenges could be mitigated. Spending time in the natural landscapes of the host country can facilitate place attachment – a sense of emotional belonging to the physical locale – which can lessen the sense of alienation and disconnection (Byrne and Goodall 2013). Place attachment can be developed by visiting nature, and even more so by active pursuits such as 'designing, improving and managing of green spaces' (Sen and Nagendra 2020). This 'environmental placemaking' by migrant communities can occur at multiple points along cultivated-wilderness gradients. Importantly, green spaces on common land, such as community gardens, public parks, and reserves, also provide opportunities for recent migrants to build social connections with local people and form positive relationships outside like-cultural groups (Seeland et al. 2009; Shinew et al. 2004). These cross-cultural interactions can be maximised through structured, purposeful events, designed to facilitate conversation and connection (Egoz and De Nardi 2017; O'Brien et al. 2011). This suggests that active, facilitated, cross-cultural participation in the management of local natural landscapes could improve migration experiences. Through this lens, conservation sites can function as therapeutic landscapes.

A therapeutic landscape is any space that engenders positive interrelationships between location, health, and human subjective experiences (Gesler 2017; Williams 2007). Through increased connection with nature, these positive interrelationships can result in greater wellbeing, purpose, and happiness (Pritchard et al. 2020) – or feeling less 'out of place'. While an individual's contact with the natural environment is an important part of the therapeutic landscape's suite of causal factors, so too are the social and cultural determinants of health and/or restoration that contextualise that contact (Gesler 2017;

DOI: 10.4324/9781003355731-14

Korpela et al. 2008; Mossabir et al. 2021). Research now demonstrates that environmental, spiritual, and social elements can singularly or collectively enable health and wellbeing benefits.

Similarly, there has been much interest in the health and wellbeing benefits of participating in conservation activities that protect and preserve environmental resources. Conservation activities include weeding out undesirable (usually introduced) plant species, propagating and planting native plants, improving habitat for wildlife (such as installing nest boxes), and monitoring the composition and quality of local habitats. Conservation Volunteers Australia (CVA), for example, aims to connect people to local conservation sites 'to re-balance nature for a stronger more resilient future' (Conservation Volunteers Australia 2022). Empirical studies demonstrate that conservation volunteering can lead to improved physical and mental health (Pillemer et al. 2010; O'Brien et al. 2011) and new social connections (Gooch 2005; O'Brien et al. 2011), as well as developing place identity and building individual and community social identity (Gooch 2005). Disadvantaged people or individuals with mental health issues can improve social relations, develop employability skills, and gain experience in working with others who are different through participation in nature-based programmes (O'Brien et al. 2011).

Environmental debates about place attachment often focus on the conservation of 'wild landscapes' and the role of indigenous knowledges and connections to place (Sen and Nagendra 2020), or, as some suggest, the experience of colonial, Western environmentalists (Van Holstein and Head 2018). This is the case in Australia, a country that has enjoyed robust environmental debate for many years and continues to do so in the context of climate change, urban development, and large-scale industrial operations, such as natural resource mining, as well as a biodiverse and unique flora and fauna.

In this chapter, we take the approach that conservation activities are akin to those traditionally described as cultivation. Cultivation, with its roots in medieval Latin terms such as *cultivus* (tilled), *cultus* (care, labour), and *colere* (to respect, to tend), has a raft of meanings. In contemporary language, it is most often used to indicate a garden or farming space that has evidence of human intervention and is designed to yield particular produce or create a certain aesthetic. In contrast to the 'wilds' of conservation sites, 'cultivation' brings to mind manicured spaces. However, conceptual divisions between conservation and cultivation sites collapse when we accept that both involve caring for the environment). There is also growing recognition that wild areas have been tended by indigenous peoples for millennia, and that all ecosystems need active management in the face of global environmental change (Cronon 1996). Through this lens, we clearly see that conservation sites are also cultivated therapeutic landscapes.

Studies of the impacts of participation in conservation activities on culturally diverse people such as migrants are scarce. This partly reflects the fact that environmental organisations and participants in environmental programmes have not attracted a diversity of membership (Van Holstein and Head 2018), despite Australia being a culturally diverse nation. According to the 2021 population census, one-third of the population was born overseas, and the top five languages spoken at home (other than English) were Mandarin, Arabic, Vietnamese, Cantonese, and Punjabi. Islam and Hinduism were amongst the top five religious affiliations (Australian Bureau of Statistics 2022). This is not only the case in Australia. In a detailed study of environmental organisations in the United States, Taylor (2014) described the low level of racial diversity as 'troubling' and found little evidence of cross-cultural collaborations in the sector. No surprise then that 'nuanced' exploration of migrants' engagements with natural environments are only beginning (Klocker and Head 2013, p. 41), yet this is vital to inform understandings of the relationships between conservation, place, and society (Sen and Nagendra 2020).

In this chapter, we report on a mixed-methods study that directly expands our understandings and addresses two key themes that underpin this chapter: (i) the dearth of research into migrant participation in conservation work and (ii) the conceptual and practical notion of conservation work as cultivation. We explore the impacts of engaging in a conservation programme specifically designed for culturally and linguistically diverse (CALD) adult migrants living in southern Tasmania, who were enrolled in a government-funded Adult Migrant English Program (AMEP) at TasTAFE, a state-based vocational education provider.

Participants in our study included adult migrants, teachers, and volunteers who were part of the Cross Cultural Conservation (CCC) Program, delivered by Conservation Volunteers Australia (CVA) in partnership with TasTAFE. CVA is a national, not-for-profit organisation which oversees local projects connecting people with conservation activities. The CCC Program collaboration between TasTAFE and CVA ran from 2017 to early 2021 and was structured to take students out of the TAFE building and provide hands-on experience in a native plant nursery and local conservation areas. In all, it comprised 15 nursery and community project activities, including work in community gardens, and land care and bush care projects. The activities were compatible with the objectives of the TasTAFE AMEP course, to ensure students gained general skills and experience that may help them obtain employment (if desired), expand their social participation, and improve their English language skills.

Box 10.1 Cross Cultural Conservation (CCC) Program

The Cross Cultural Conservation (CCC) Program aims to deliver:

- improved conversational workplace English confidence;
- skills acquisition which may lead to improved employment prospects;
- improved social connections;
- increased understanding of the barriers faced by migrants, refugees, and asylum seekers by the non-migrant community; and
- environmental benefits associated with the propagation and subsequent planting of native species.

Methodology

This study aimed to explore the experiences and understand the impacts of participation in conservation activities in the CCC Program, as well as evaluate the success and impacts of the process and outcomes. The AMEP students had developed some skills in spoken and written English. Throughout 2017–2019, the Program team included local (non-migrant) volunteers who participated in the CCC Program alongside the migrant TAFE participants. Volunteers were important to achieve the goal of providing a safe space for social connections, English language exchange, and increased local community understanding of the challenges migrants, refugees, and people seeking asylum may experience. However, during 2020–2021, in response to COVID-19 requirements, local volunteers were no longer included in the Program.

Methods

The two researchers (Marsh and Mallick) brought health geography and social sciences expertise to this convergent mixed-methods study. Co-designed with TasTAFE AMEP teachers and CVA staff, the design consisted of three methods, conducted simultaneously: (i) participant observation at four conservation sessions; (ii) postcard feedback from 71 AMEP student participants; and (iii) follow-up in-depth interviews with 12 student participants and four key stakeholders.

During the research period (2019–2020), participants undertook a range of activities in various locales in and around the city of Hobart over four sessions (see Table 10.1). Ethics approval was obtained from the University of Tasmania Human Research Ethics Committee.

Participant observation

The two researchers (Marsh and Mallick) attended and participated in four conservation activities, together with the Program team members, participants, and volunteers (in 2019). Observation was overt and participants were informed that part of the researchers' involvement was for them to see the context and how the Program works. The participant observation allowed these

Table 10.1 Scope of activities and locales

Conservation activity	*Locale*
Planting native shrubs and trees and mulching	Suburban public green space
Tending existing native plants in a wetland area, planting and mulching additional native plants	Dairy farm, rural area
Potting on and labelling native plants	Commercial nursery
Collecting micro plastics in the sand and rubbish from the beach	Inner city shoreline

researchers to meet some of the students, volunteers, and staff and foster connections and exchange; experience the type and nature of activities being undertaken; and observe the degree and type of interactions between all those participating in the activities.

This method enabled deeper understandings and awareness of local community knowledges, attitudes, and ideas. Working alongside the participants helped to reduce any hierarchical divide between researchers and participants, a key outcome and aim of this method. No notes were taken during the session, but field notes were documented following each session. The researchers noted the conditions on the day, activities undertaken, and volunteers present, as well as general weather and other observations.

The CCC Feedback Postcard

Co-designed with TasTAFE teaching staff, the CCC Feedback Postcard tool included six simple questions about participants' experiences of the conservation activity. The Postcard was distributed to participants at the end of each session and completed by hand (see Figure 10.1). On average, 10–15 students attended each conservation activity; the majority of participants filled in the Postcard survey. In the week before one of the conservation activities, the researchers attended a TasTAFE class to explain the research, discuss the Postcard activity, provide a written information sheet detailing the aims and design of the research project, and respond to any questions students had at that time about participating in the research.

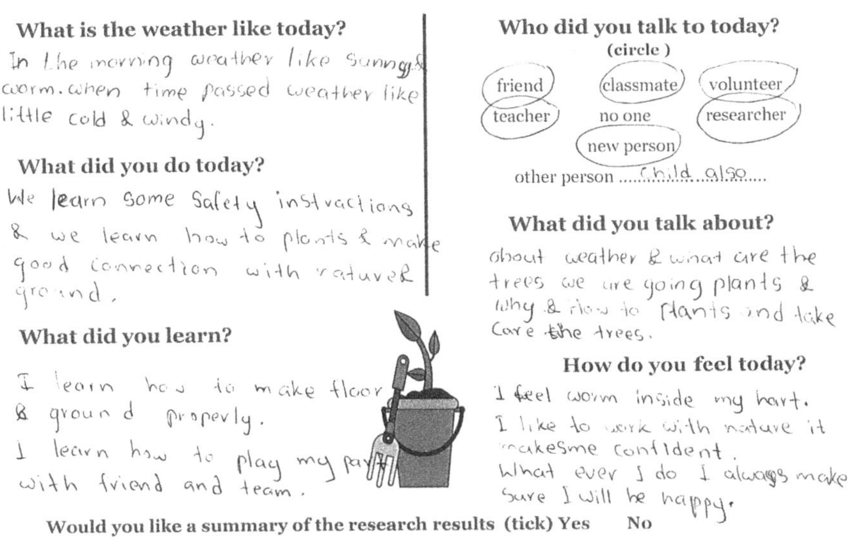

What is the weather like today?
In the morning weather like sunny & worm. when time passed weather like little cold & windy.

What did you do today?
We learn some safety instructions & we learn how to plants & make good connection with nature & ground.

What did you learn?
I learn how to make floor & ground properly.
I learn how to play my part with friend and team.

Who did you talk to today? (circle)
friend — classmate — volunteer — teacher — no one — researcher — new person
other person ...Child also...

What did you talk about?
about weather & what are the trees we are going plants & why & how to plants and take care the trees.

How do you feel today?
I feel worm inside my hart. I like to work with nature it makesme confident. What ever I do I always make sure I will be happy.

Would you like a summary of the research results (tick) Yes No

Figure 10.1 Example of a completed postcard.

Any person attending the conservation sessions had the option of completing a Postcard; it did not matter whether they had attended previous conservation sessions or had previously completed a Postcard. The Postcards were thus designed to collect meaningful data, regardless of the pattern of attendance. For three of the four research days, the researchers were onsite while participants completed their Postcards. Because the fourth research day was extremely hot, participants completed the Postcards in the classroom during the following lesson, without the researchers present. No personal information was recorded on the Postcards. Where necessary, participants received help to fill in the Postcards. The researchers directed participants to (or reminded them about) the information sheet, and advised them that they had the option to withdraw at any time. Consent was indicated by completing a Postcard.

The Postcards allowed for simple and quick data collection with minimal participant burden. They were accessible to all participants, regardless of ability, captured real-time (on-site) experiences, and were useable over an extended timeframe, and there was no restriction on the number of times participants completed the Postcard tool.

In-depth interviews

A total of 16 in-depth interviews were conducted with 12 student participants and four stakeholders (see Figure 10.2). Interviews were first conducted with students (n = 12) who had undertaken at least one of the conservation activities. Participation in the in-depth interviews was by self-selection, with TasTAFE teachers inviting students to meet with the researchers during class time. Pre-COVID-19, interviews were held on-site at the TasTAFE building; during COVID-19 restrictions, they were conducted by telephone. Interviews were conversational and prompts from the interviewer encouraged participants to share their personal reflections on the CCC Program. Interviews lasted between 20 and 75 minutes. Male and female students participated; they came from diverse backgrounds and had arrived in Australia under various migration schemes between one month and five years before participating in the CCC Program (Figure 10.2).

Interviews with stakeholders included one TasTAFE teacher, one CVA volunteer, a CVA staff member, and the CCC Program Manager (n = 4). These were conducted towards the end of the Program to gain contextual information. Stakeholders were asked about the logistics and practical aspects of delivering and participating in the Program, as well as their understanding of its benefits and outcomes.

All interviews were audio-recorded and transcribed. Interviewees were de-identified, given a number, and referred to only as ‘Student Participant’ or ‘Stakeholder’.

Student participants				
No.	Sex	Home country	Arrival type (indicated during interview)	Time since arrival in Tasmania
1	F	Sri Lanka	Partner of skilled migrant	2 years
2	F	China	Partner of international student	1 years
3	F	Sudan	Refugee	4 years
4	F	Iran	Refugee	3.5 years
5	F	China	Partner of international student	5 years
6	F	Thailand	Spouse visa	2 years
7	F	China	Partner of international Student	4 years
8	F	France	Partner of skilled migrant	1month
9	M	Burma	Refugee	7 years
10	M	China	Partner of international Student	1 years
11	M	Ethiopia	Refugee	2 years
12	M	China	Partner of international student	3 years
Stakeholders				
No.	Sex	Position	Stakeholder organisation	
1	M	Teacher	TasTAFE	
2	F	Manager	CVA	
3	F	Coordinator (staff)	CVA	
4	M	Volunteer	CVA	

Figure 10.2 Interview participant profile.

Analysis

Each method was analysed separately using appropriate methods. All Postcard content (n = 71) was entered into an Excel spreadsheet, categorised under question headings. One researcher (Mallick) did a simple content analysis involving ascertaining the words and phrases that arose most often in the Postcard responses. These words and phrases formed the basis for understanding the thematic patterns of the Postcard data. This method gave us a familiarity with the text, such that we could determine the key patterns and identify any outliers or unusual phrases. All interviews (n = 16) were recorded and transcribed in full, and interview data was analysed using reflexive thematic analysis techniques (Braun and Clarke 2019) involving close reading, coding, discussion, review and iterative organisation, and synthesis of codes into themes and sub-themes. We then considered these themes as they aligned with the aims of the CCC Program. Participant observation field notes informed the reflexive analysis of the interview data. In the following section, we present and discuss the findings, structured by the CCC Program aims.

Findings and discussion

Analysis of the interview and postcard data generated six thematic categories: (1) Improved confidence with conversational English; (2) Skills acquisition which may lead to employment; (3) Improved social connections; (4) Increased community understanding of the barriers faced by migrants, refugees, and asylum seekers; (5) Environmental benefits associated with the propagation and subsequent planting of native species; and (6) Miscellaneous benefits – happiness, optimism, and gratitude. In the following section, we discuss each theme in more detail.

Improved confidence with conversational English

Postcard data indicated that almost all participants reported talking in English with a classmate, volunteer, or teacher; some also reported talking in English with a friend, a new person, or a researcher. No participants selected the option of talking to 'no one' (see Figure 10.3). Participants talked with each other about the tasks, food, plants, music, their study, how to speak English fluently, and how to find a job as a volunteer. They also talked about their home country, life in Tasmania, their family, and the future.

Interview data highlighted that some participants had limited opportunities to speak English outside the classroom, particularly if their partner or friends spoke the same native language. Others indicated that they were not enthusiastic about joining a same-language community group or only socialising with people with a similar cultural and linguistic background, as they felt it may further limit their English-speaking opportunities.

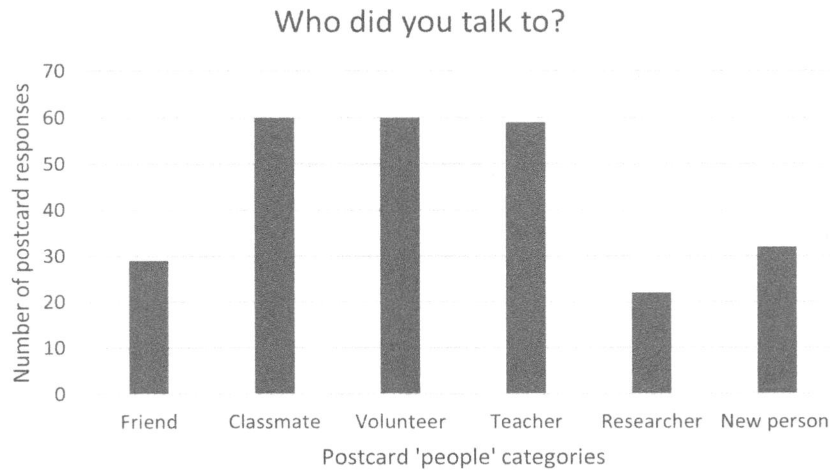

Figure 10.3 Responses to the 'Who did you talk to today?' prompt on the postcards.

> Yes, we can talk with different people, not just my classmates. 'Cause when we study English, we only talk it with my classmates and teacher. And when we go outside, we can talk with more local people and different, different English accent. Always talking here, the teacher they know that maybe our English isn't good, and they will always talking slowly, but outside it's faster.
>
> (Student#7)

Listening and being able to follow instructions was another advantage participants identified. Participants felt they developed their listening and comprehension skills, and felt more confident in their ability to talk and interact (Figure 10.3). Participants were aware of the importance and benefits of learning to speak English, particularly for gaining work, interacting with 'local' people, and building connections. While listening skills were more easily improved, one participant acknowledged that he only spoke a little because he was afraid his English was not good.

The intentional mixing of local, English-speaking volunteers from CVA with AMEP students in this Program acted as a springboard for learning and practising new words related to environmental workplaces and practices. Participants learnt about new concepts, such as native habitats and ecosystems. Participants' greater sense of confidence to engage in conversation also resonates with findings of studies of engagement in the outdoors. For example, young people and people experiencing mental health issues have reported feeling greater self-confidence as a result of outdoor programmes – giving them courage to talk to others (Barfield et al. 2021; Bowen et al. 2016).

In therapeutic landscapes theory, the capacity of a space to provide opportunities to converse with others is a key enabler of subjective health and wellbeing. In recent studies of gardeners during the COVID-19 pandemic, for example, when lockdown conditions required social distancing, people reported that gardening started a conversation with neighbours and family members (Egerer et al. 2022). The combination of outside space, shared activity, and care for the planet has been shown to help us talk with strangers (Ivory et al. 2015). In our study, the structured bringing together of people from different cultures with the intention to talk augments this effect.

Skills acquisition which may lead to employment

All participants were invited to write something on the Postcards about the activities they undertook as part of the Program. These responses captured how they prepared land, planted out plants, learnt about native trees, used a tool to pick up plastic, or sieved sand, and that they should separate plastics before putting them in the rubbish. Some examples include 'shared planting and food with classmates'; 'learnt some safety instructions and how to follow instructions'; 'learnt how to make good connection with nature and ground'; 'supported people working with plant to change the climate'; and 'collected and counted rubbish in teams'.

Participants wrote that they learnt how to work in teams, improve their communication skills, become a volunteer, and expand their social interactions. These gains were reiterated in the interviews.

> It was good. I was excited to meet new people. If you meet new people you, it is good with me, 'cause every day when you meet new people you expand your network and you might ask how to get a job and how you do things.
>
> (Student #11)

Student interviewees acknowledged that the practical tasks during each conservation activity helped them gain new skills and experience (such as how to re-pot plants and plant out small shrubs) and also expanded their knowledge about a wide range of matters, from native vegetation and food production to environmental protection. For instance, they learnt the importance of protecting trees and nature, the dangers of small plastic pieces, and a little on animal husbandry and agriculture techniques.

> I enjoyed [being] by the beach and to discover the topic for me was important, and to be conscientious on that subject and it is necessary to, when I go to supermarket now I don't use plastic bag, and I think we have to do advice, and use other materials for the bags and water containers.
>
> (Student#8)

Student interviewees saw participation in the Program as a potential pathway to work, and a way to overcome language and experience barriers to employment that particularly impact people of migrant, refugee, and asylum-seeking backgrounds in Australia (Joint Standing Committee on Migration 2013). They gained some generic skills which they may be able to apply in a work setting, such as communication skills, working in teams, and workplace safety. Participants felt they gained more knowledge about the working community and volunteering, and imagined a role they might take up in the future.

While much of the broader research on the benefits of participation in conservation programmes tends to focus on the impacts and benefits of volunteering, there is growing interest in the links between paid work and employment and environmental activity. The types of employable-worthy and transferable skills we found in our study have been identified in research on nature-based volunteering amongst disadvantaged and marginalised communities (O'Brien et al. 2011). Therapeutic landscapes studies have yet to focus on paid work as an enabler of the subjective health and wellbeing outcomes of therapeutic landscapes and programmes. Nevertheless, in the broader domain of social determinants of health studies, employment has long been accepted as an 'upstream' key driver of human health and wellbeing –in terms of both unemployment and the conditions of employment (Marmot 2015).

Improved social connections

Almost half the Postcard responses indicated that the activity gave participants the chance to meet a new person. Through the Program, participants undertook shared activities and, where possible, shared food. By interacting in these ways, participants felt they bonded and built connections with each other and acted as one 'team'. Student interviewees described the value of new connections and talked about expanding their networks. They also valued the exposure to longer-term Tasmanian residents or 'local' people. The teaching staff also indicated that the initial contacts sometimes broadened out to wider connections:

> And they may join in with the group, they might sign up and get the newsletter or the Facebook page. And so that's a difference.
>
> (Stakeholder#1)

These interactions were felt to be crucial to a 'successful' life in Tasmania, when compared with a siloed multiculturalism, where migrants primarily speak a language other than English within their cultural group.

This is consistent with literature on the benefits of environmental volunteering in general (mostly focused on non-migrants); for example, a study of Landcare volunteers in Queensland, Australia, reported that participants developed new friendships and built social networks (Gooch 2005). Our finding also aligns with the results of studies of conservation volunteering by vulnerable and disadvantaged communities who are often socially excluded (O'Brien et al.

2011). Migrants are at particular risk of loneliness and social isolation. Some nature-based programmes have demonstrated similar social connection benefits for migrant communities (e.g., Seeland et al. 2009). In the therapeutic landscapes field, there is now an established scholarship providing evidence of improved social connections from engaging with green spaces (Bell et al. 2018; Glover 2004; Kingsley et al. 2009; Marsh et al. 2018; Mygind et al. 2019). Less attention has been paid to migrant experiences, although the Harris et al. (2014) study with recently settled African humanitarian migrants who were members of a community food garden in Logan, Queensland, found improved connections within and beyond the garden spaces.

Increased understandings of the barriers faced by migrants, refugees, and asylum seekers

There are many issues facing migrants, refugees, and people seeking asylum that make settling in Australia challenging – few of which are experienced by the non-migrant community. These particular barriers were summed up by one interview participant.

> Generally, people have a lot of problems, they have no family here … they have education and qualification in their country but here they don't have it, sometimes language problems, and children's problems … cultural things. They have to carry that culture to everywhere we are going, it is hard to do, to put into this culture.
>
> (Student#1)

Before COVID-19 restrictions were implemented, the CCC Program included longer-term Tasmanian volunteers as a way to encourage interaction and connection between people from varying cultural and linguistic backgrounds. While sometimes the local volunteers were actually international students, one volunteer did feel that the cultural interactions with participants helped him broaden his understanding of the issues people from varying cultural and linguistic backgrounds faced in migrating to the state. This could be described as a benefit for the volunteer and the broader community, which indirectly could improve connections between migrants and local Tasmanians.

Gardening – particularly in the private residential gardens of the fully detached housing found in most Australian cities – has provided something of a horticultural exemplar of the impacts of increased immigration from Europe (Morgan et al. 2005). In a wonderful paper, Morgan et al. (2005) document how gardens were transformed to become sites of cultural exchange, where Australian neighbours came to learn about the lives and cultures of new migrants, and the challenges they faced. We saw this type of knowledge exchange being enacted to a small degree in this study, as AMEP students worked alongside CVA volunteers and TasTAFE teaching staff. It is not hard to envisage this occurring to a greater extent – in and through conservation practices – given the opportunity.

Environmental benefits associated with the propagation and subsequent planting of native species

Participants were aware that their conservation actions were benefitting the community and the broader environment. They were proud that they were volunteering to help protect the environment. Participants' exposure to conservation activities and ideas brought about a change in their thinking on climate protection (that is, being conscientious) and their behaviour (for instance, thinking about their own use of plastic bags). Some participants said they had not been exposed to knowledge about how to look after the environment before they came to Australia.

> My family also have a farm and so I have some experience planting plants.... And I think if we plant more trees, the way we plant, in future our children, they can get a more, more good, good enlightenment, because of us, because we plant the trees and they get a lot of things because of us.
>
> (Student#1)

Participants felt that the experience of being outside and in nature was valuable in itself. They also felt a sense of wellbeing: they developed 'a good connection' with nature and this made them feel confident, happy, relaxed. They understood that it was good for their mental health, and they built familiarity and ease with nature, apparently becoming comfortable in the bush.

> I think that volunteering is one of the most effective actions that contribute to preserving the ecosystem. That is my input.
>
> (Student #3)

Participants felt that the voluntary nature of the Program really helped them develop confidence, build a sense of achievement, create meaning, feel valued, and gain a sense of wellbeing. It also made them feel included and created a sense of belonging and of contributing. One participant commented that she felt they were doing very important work (in collecting rubbish from the beach), while other people at the beach were just swimming (Student #7).

Finally, there was a sense amongst participants that the students were making a contribution to the place, to the people in the community, and to the organisation and environment. They believed it was good to do the activities to counter the negative effects of climate change and appreciated the opportunity to do so.

This relationship between practical action for environmental benefits leading to both individual and community benefits has previously been identified in a study of 'catchment' volunteers in Queensland, Australia (Gooch 2005). Interestingly, the Harris et al. (2014) study of migrant experiences of social connection in community gardens found that shared interests extended beyond the garden to 'land care, sustainable food systems and sustainability' (p. 9210). Khorana and Sirdah (2021) found that first-generation migrants were less likely than second-generation migrants to employ conscious environmental care practices, and that education and schooling were important influences on furthering these behaviours. The CCC Program may contribute to the development of these practices in first-generation migrants. The participants' interest in sea plastics and beach waste mirrors the findings of Khorana and Sirdah's (2021) report: both first- and second-generation migrants were found to be holistically conscious about waste reduction and other minimisation measures, such as composting. These environmental behaviours are, in part, continuations of cultural practices, and of social practices brought about by years of austerity and poverty before migrating to Australia.

There is a need for the conservation and environmental movements to deliberately encourage cross-cultural engagement. Klocker and Head (2013) have argued for greater recognition of these practices as a type of cultural capital – extremely valuable cultural capital that migrants can offer their new home country as we live in a time of many environmental challenges. Dorceta Taylor (Toomey 2018) has encouraged the environmental movement, and 'green groups' in general, to stop being afraid of cross-culturalism. The development of environmental place attachment – in our case to the island of Tasmania and her natural places – demonstrates that great impacts are possible.

Miscellaneous benefits – happiness, optimism, and gratitude

The additional, or unanticipated, impacts of the CCC Program represented in both Postcard and interview data clearly focused on three things: happiness, optimism, and gratitude.

> But I really enjoy planting trees. It kind of made you happy, inside your heart, if you are planting the tree inside your heart you feel very, very pure. I don't how to explain. The smiles, the happiness, it comes from the planting trees, it's very natural.
>
> (Student #1)

Postcard responses to the question 'How do you feel today?' were overwhelmingly positive. Participants reported feeling 'happy', 'good', 'fantastic', 'excited', and 'satisfied'. They had 'fun', felt 'good' to be outside, felt 'happy' they could help the world. They enjoyed each other's company and were 'thankful' for everything (see Figure 10.4).

The responses were particularly positive about the changeable, temperate-maritime Tasmanian weather. It was generally described as 'sunny', 'cloudy', 'cold', 'warm', 'windy', and 'cool'. Of all the comments on the Postcards, only three seemed to openly express discontent with the weather, stating it was 'good but for the sun' (lack of), 'is raining, weather is not good', and 'too hot'. The activities were not without challenges or risks for the participants,

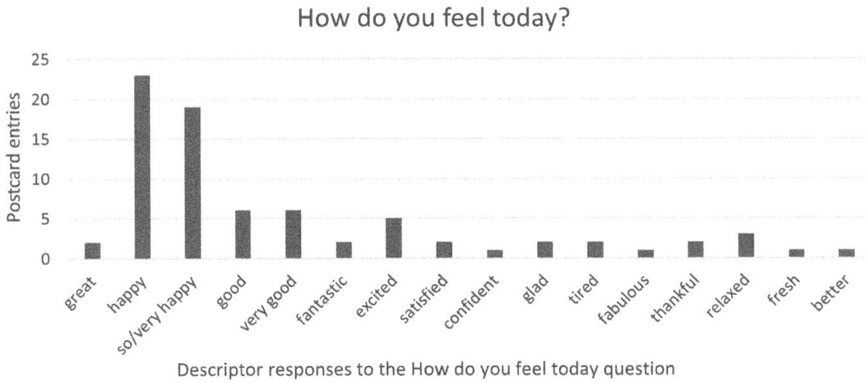

Figure 10.4 Responses to the 'How do you feel today?' prompt on the postcards.

some of which related to being outside in varying weather conditions, including one extremely hot day (31 degrees Celsius) and other cold days. Despite the challenges, most participants appreciated the weather, describing it as 'good for planting', 'stunning', 'very beautiful', 'fantastic', and 'perfect for us' (see Figure 10.5).

The level of difficulty of the work was another potential challenge for some, with several participants preferring to sit out some activities that seemed too physical for them, at least initially.

Participants felt that the experience of being outside and in nature was valuable in and of itself and said they felt happy after seeing that their achievements were improving the environment and helping animals with their habitat.

> I think looking for nature and walking in the nature, to looking at beautiful tree and beautiful flowers, yeah. I walk, when I have a problem or something sad, I live in Bridgewater. They have a river, and close to the river sometimes I go to walk, have many trees and I like to walk when I am looking for trees. Yeah.
>
> (Student #4)

Overwhelmingly, participants were grateful for the opportunities they felt they had been given through participating in the CCC Program. An advantage of the conservation activities was that it took participants to places they may not otherwise have had an opportunity to see, such as into the Australian bush, to a farm, or to the beach. They saw different life cycles, from seedlings to established ecosystems. They felt gratitude and expressed positive and strong emotions. Participants felt their work was meaningful and it made them feel 'happy', 'excited', 'satisfied', and 'deeply touched'. They were glad to join the team and enjoyed each other's company. They were able to identify and acknowledge the goodness in their lives.

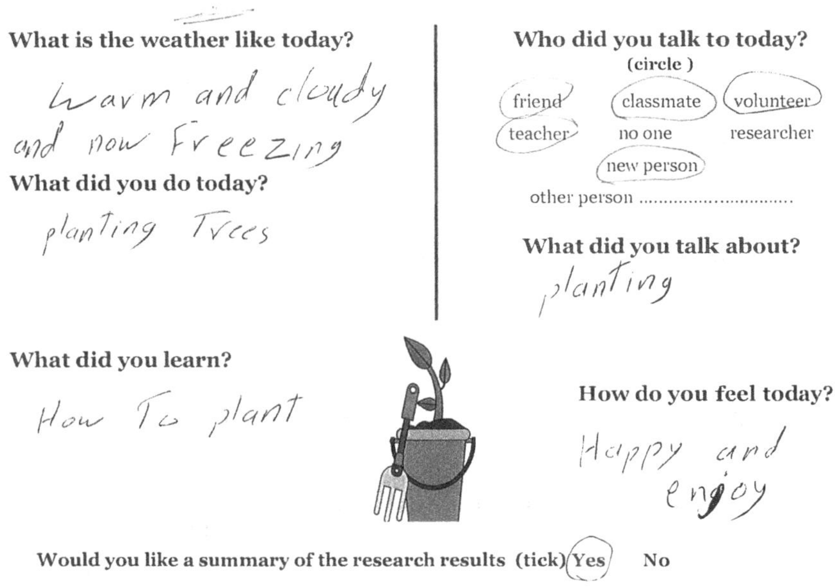

What is the weather like today?

Warm and cloudy and now Freezing

What did you do today?

planting Trees

Who did you talk to today? **(circle)**

friend classmate volunteer teacher no one researcher new person other person ....

What did you talk about?

planting

What did you learn?

How To plant

How do you feel today?

Happy and enjoy

Would you like a summary of the research results (tick) Yes No

Figure 10.5 Example of a completed postcard showing 'freezing' weather but feeling 'happy and enjoy'.

Happiness, optimism, and gratitude are important for wellbeing. These attributes are frequently found in studies of engagement in green spaces in general, across the therapeutic landscapes and conservation literatures (Birch 2005; Capaldi et al. 2014; Manner et al. 2021; Marsh et al. 2021; O'Brien et al. 2011; Pritchard et al. 2020). Feeling happy, hopeful, and thankful are important elements of environmental place-making. Our findings confirm and reiterate the centrality of emotional contributors to therapeutic landscapes of all sorts, enabled by shared cultivation of new local landscapes for people who are otherwise 'out of place'.

Limitations

The first group of participants to complete the Postcards practised writing responses in the classroom as a learning activity; the last group wrote their responses in the classroom sometime after the event. As a result, participant responses were, to some degree, influenced by teachers or peers, so are less likely to reflect direct participant experience of the Program. The COVID-19 pandemic greatly influenced the Program structure, with social distancing requirements limiting opportunities for social interactions, such as the number of CVA volunteers on site for the later events, and eliminating the shared lunches. Neither of these limitations is likely to have much bearing on the overall findings and conclusions of this study. However, they highlight the need to be flexible to accommodate unforeseen circumstances (weather, pandemics) when undertaking research in cultivated therapeutic landscapes.

Conclusions

In this study, we approached conservation sites as cultivated therapeutic landscapes, and sought to discover whether and how participation in conservation cultivation programmes and activities for people who find themselves 'out of place' due to migration can have positive benefits. The CCC Program was highly successful in that it met its aims and had a greater breadth of benefits than anticipated. This research demonstrates that programmes like the CCC – involving structured nursery and community-based conservation activities, including work in community gardens, and land care and bush care projects – can deliver positive upstream health and wellbeing benefits for migrants, including work skills, language acquisition, and social interaction.

Environmental place-making in our research was enabled by a purposeful programme designed to bring recent migrants into contact with nature and to contribute by way of 'cultivation' to its sustainability. Integrating English-speaking 'locals' into the programme further augmented its therapeutic capacity, as student participants developed confidence to speak in English with others, to meet new people, and to prepare for employment in the new host country.

This research confirms the findings of other studies on environmental volunteering, in the important contribution that programmes of this nature can generate for students from migrant, refugee, and asylum-seeking backgrounds, and for the environment broadly. The shared, purposeful conservation activities enabled a meeting of new realities, a telling of narratives of self – past, present, and future.

This research shows the need for, and value of, ongoing research involving migrant experiences of place-making via cultivation as a therapeutic activity. Future longitudinal research would be of benefit to determine whether people experience sustained benefits on employment prospects, education, wellbeing, and sense of connection to the host country.

Acknowledgements

The authors would like to acknowledge the generous support of Conservation Volunteers Australia for funding this research, and for the participants who generously gave their time and insights to the study team.

References

Australian Bureau of Statistics 2022, *Cultural diversity: Census*. https://www.abs.gov.au/statistics/people/people-and-communities/cultural-diversity-census/latest-release

Barfield, PA, Ridder, K, Hughes, J and Rice-McNeil, K 2021, 'Get outside! Promoting adolescent health through outdoor after-school activity', *International Journal of Environmental Research and Public Health*, vol. 18, no. 14, article 7223. https://doi.org/10.3390/ijerph18147223

Bell, SL, Foley, R, Houghton, F, Maddrell, A and Williams, AM 2018, 'From therapeutic landscapes to healthy spaces, places and practices: A scoping review', *Social Science & Medicine*, vol. 196, pp. 123–130. https://doi.org/10.1016/j.socscimed.2017.11.035

Birch, M 2005, 'Cultivating wildness: Three conservation volunteers' experiences of participation in the Green Gym Scheme', *British Journal of Occupational Therapy*, vol. 68, no. 6, pp. 244–252. https://doi.org/10.1177/030802260506800602

Bowen, DJ, Neill, JT and Crisp, SJR 2016, 'Wilderness adventure therapy effects on the mental health of youth participants', *Evaluation and Program Planning*, vol. 58, pp. 49–59. https://doi.org/10.1016/j.evalprogplan.2016.05.005

Braun, V and Clarke, V 2019, 'Reflecting on reflexive thematic analysis', *Qualitative Research in Sport, Exercise and Health*, vol. 11, no. 4, pp. 589–597. https://doi.org/10.1080/2159676X.2019.1628806

Byrne, D and Goodall, H 2013, 'Placemaking and transnationalism: Recent migrants and a national park in Sydney, Australia', *Parks: The International Journal of Protected Areas and Conservation*, vol. 19, no. 1, pp. 63–73.

Capaldi, CA, Dopko, RL and Zelenski, JM 2014, 'The relationship between nature connectedness and happiness: A meta-analysis', *Frontiers in Psychology*, vol. 5, article 976. https://doi.org/10.3389/fpsyg.2014.00976

Conservation Volunteers Australia 2022, 'Nature back in balance', *Conservation Volunteers Australia*. https://conservationvolunteers.com.au/

Cronon, W 1996, 'The trouble with wilderness: Or, getting back to the wrong nature', *Environmental History*, vol. 1, no. 1, pp. 7–28. https://doi.org/10.2307/3985059

Egerer, M, Lin, B, Kingsley, J, Marsh, P, Diekmann, L and Ossola, A 2022, 'Gardening can relieve human stress and boost nature connection during the COVID-19 pandemic', *Urban Forestry & Urban Greening*, vol. 68, article 127483. https://doi.org/10.1016/j.ufug.2022.127483

Egoz, S and De Nardi, A 2017, 'Defining landscape justice: The role of landscape in supporting wellbeing of migrants, a literature review', *Landscape Research*, vol. 42, no. S1, pp. S74–S89. https://doi.org/10.1080/01426397.2017.1363880

Gesler, W 2017, 'Commentary on the origins and early development of the therapeutic landscapes concept', *Medicine Anthropology Theory*, vol. 4, no. 1, pp. 1–9. https://doi.org/10.17157/mat.4.1.358

Glover, TD 2004, 'Social capital in the lived experiences of community gardeners', *Leisure Sciences*, vol. 26, no. 2, pp. 143–162. https://doi.org/10.1080/01490400490432064

Gooch, M 2005, 'Voices of the volunteers: An exploration of the experiences of catchment volunteers in coastal Queensland, Australia', *Local Environment*, vol. 10, no. 1, pp. 5–19. https://doi.org/10.1080/1354983042000309289

Harris, N, Minniss, FR and Somerset, S 2014, 'Refugees connecting with a new country through community food gardening', *International Journal of Environmental Research and Public Health*, vol. 11, no. 9, pp. 9202–9216. https://doi.org/10.3390/ijerph110909202

Ivory, VC, Russell, M, Witten, K, Hooper, CM, Pearce, J and Blakely, T 2015, 'What shape is your neighbourhood? Investigating the micro geographies of physical activity', *Social Science & Medicine*, vol. 133, pp. 313–321. https://doi.org/10.1016/j.socscimed.2014.11.041

Joint Standing Committee on Migration 2013, *Inquiry into migration and multiculturalism in Australia*, Parliament of the Commonwealth of Australia, Canberra.

Khorana, S and Sirdah, C 2021, *Young Australian migrants and environmental values: How and why do certain migrants practice care? Findings Report*, Western Sydney University, Sydney.

Kingsley, J, Townsend, M and Henderson-Wilson, C 2009, 'Cultivating health and wellbeing: Members' perceptions of the health benefits of a Port Melbourne community garden', *Leisure Studies*, vol. 28, no. 2, pp. 207–219. https://doi.org/10.1080/02614360902769894

Klocker, N and Head, L 2013, 'Diversifying ethnicity in Australia's population and environment debates', *Australian Geographer*, vol. 44, no. 1, pp. 41–62. https://doi.org/10.1080/00049182.2013.765347

Korpela, KM, Ylén, M, Tyrväinen, L and Silvennoinen, H 2008, 'Determinants of restorative experiences in everyday favorite places', *Health & Place*, vol. 14, no. 4, pp. 636–652. https://doi.org/10.1016/j.healthplace.2007.10.008

Manner, J, Doi, L and Laird, Y 2021, '"That's given me a bit more hope" – Adolescent girls' experiences of Forest School', *Children's Geographies*, vol. 19, no. 4, pp. 432–445. https://doi.org/10.1080/14733285.2020.1811955

Marmot, M 2015, *The health gap: The challenge of an unequal world*, Bloomsbury, London.

Marsh, P, Brennan, S and Vandenberg, M 2018, '"It's not therapy, it's gardening": Community gardens as sites of comprehensive primary healthcare', *Australian Journal of Primary Health*, vol. 24, no. 4, pp. 337–342. https://doi.org/10.1071/PY17149

Marsh, P, Diekmann, LO, Egerer, M, Lin, B, Ossola, A and Kingsley, J 2021, 'Where birds felt louder: The garden as a refuge during COVID-19', *Wellbeing, Space and Society*, vol. 2, article 100055. https://doi.org/10.1016/j.wss.2021.100055

Morgan, G, Rocha, C and Poynting, S 2005, 'Grafting cultures: Longing and belonging in immigrants' gardens and backyards in Fairfield', *Journal of Intercultural Studies*, vol. 26, no. 1–2, pp. 93–105. https://doi.org/10.1080/07256860500074094

Mossabir, R, Milligan, C and Froggatt, K 2021, 'Therapeutic landscape experiences of everyday geographies within the wider community: A scoping review', *Social Science & Medicine*, vol. 279, article 113980. https://doi.org/10.1016/j.socscimed.2021.113980

Mygind, L, Kjeldsted, E, Hartmeyer, R, Mygind, E, Bølling, M and Bentsen, P 2019, 'Mental, physical and social health benefits of immersive nature-experience for children and adolescents: A systematic review and quality assessment of the evidence', *Health & Place*, vol. 58, article 102136. https://doi.org/10.1016/j.healthplace.2019.05.014

O'Brien, L, Burls, A, Townsend, M and Ebden, M 2011, 'Volunteering in nature as a way of enabling people to reintegrate into society', *Perspectives in Public Health*, vol. 131, no. 2, pp. 71–81. https://doi.org/10.1177/1757913910384048

Pillemer, K, Fuller-Rowell, TE, Reid, MC and Wells, NM 2010, 'Environmental volunteering and health outcomes over a 20-year period', *The Gerontologist*, vol. 50, no. 5, pp. 594–602. https://doi.org/10.1093/geront/gnq007

Pritchard, A, Richardson, M, Sheffield, D and McEwan, K 2020, 'The relationship between nature connectedness and eudaimonic well-being: A meta-analysis', *Journal of Happiness Studies*, vol. 21, pp. 1145–1167. https://doi.org/10.1007/s10902-019-00118-6

Seeland, K, Dübendorfer, S and Hansmann, R 2009, 'Making friends in Zurich's urban forests and parks: The role of public green space for social inclusion of youths from different cultures', *Forest Policy and Economics*, vol. 11, no. 1, pp. 10–17. https://doi.org/10.1016/j.forpol.2008.07.005

Sen, A and Nagendra, H 2020, 'Local community engagement, environmental place-making and stewardship by migrants: A case study of lake conservation in Bengaluru, India', *Landscape and Urban Planning*, vol. 204, article 103933. https://doi.org/10.1016/j.landurbplan.2020.103933

Shinew, KJ, Glover, TD and Parry, DC 2004, 'Leisure spaces as potential sites for interracial interaction: Community gardens in urban areas', *Journal of Leisure Research*, vol. 36, no. 3, pp. 336–355. https://doi.org/10.1080/00222216.2004.11950027

Taylor, DE 2014, *The state of diversity in environmental organizations*, Green 2.0, Michigan.
Toomey, D 2018, 'How green groups became so white and what to do about it' (*Yale Environment*360, 21 June 2018). https://e360.yale.edu/features/how-green-groups-became-so-white-and-what-to-do-about-it
van Holstein, E and Head, L 2018, 'Shifting settler-colonial discourses of environmentalism: Representations of indigeneity and migration in Australian conservation', *Geoforum*, vol. 94, pp. 41–52. https://doi.org/10.1016/j.geoforum.2018.06.005
Williams, A 2007, *Therapeutic landscapes*. Routledge/Taylor & Francis, London.

11 Eradicating malnutrition through small-scale, diverse, and local food production

Bruce French and Anthea Maynard

Introduction

Our main focus has been the 80 countries in the world where, each year, more than one in every 50 children die before their fifth birthday (UNIGME 2021). These are not just statistics, but personal family tragedies for those involved. Of these child deaths, about 45% cent, or one every 10 seconds, is due to nutrition-related factors (World Health Organization 2020). To the best of my knowledge (BF), locally well-adapted and highly nutritious edible plants are already growing in every location in the world. Most of the women seeking to feed their families well have little nutritional knowledge, apart from the supplying energy-rich, stomach-filling, starchy staples. Around the world, 84% of the farms are two hectares or less in size, and most food in the world is grown by women (Lowder, Skoet and Raney 2016), This is the group that needs information and support.

Those who are concerned about good stewardship of the planet and the possibility of feeding our increasing global population well have a range of ideas for consideration. Globally, most not-for-profit international organisations are moving away from promoting the intense production of one variety of one crop towards something that will provide better nutrition and greater sustainability. These groups include Plant Resources Of Tropical Africa (RPOTA), Plant Resources Of South East Asia (PROSEA), Education Concerns for Haiti Organisation (ECHO), Food and Agricultural Organization of the United Nations (FAO), Bioversity International, Asian Vegetable Research and Development Center (AVRDC), Botanic Gardens Conservation International, Horticultural Innovation and Learning for Improved Nutrition and Livelihood in East Africa (HORTINLEA), International Peace Information Service (iPIS), Frontiers in Nutrition, MS Swaminathan Research Foundation (n.d.), and several others. I have been in touch with most of them, and usually our organisation (Food Plants International 2001) sends each of them an electronic copy of our database. These organisations claim that it is time to rethink our approach to agriculture and global food production. A Bioversity International report, entitled 'Diversifying food and diets', argued that it was time to bring all the world's food plants to the table (Fanzo et al. 2013). In

DOI: 10.4324/9781003355731-15

recent years, there have been several International Conferences on Neglected and Underutilised Species (NUS) to address food shortages, fight hidden hunger, and alleviate poverty. Most of the organisations listed above have published reports focusing on local plants, nutrition, diversity, and sustainability as essential ingredients for sound food production. Hunger and malnutrition are undoubtably compounded by historical and contemporary monocultural agriculture practices (Siddique, Li, and Gruber 2021). Once people are informed that there are many local plants that are rich in nutrients, there is an opportunity for people to grow, eat, and utilise them for the benefit of their families and communities.

In 1966, during my studies for my Bachelor of Agricultural Science, I went to volunteer on a mission farm in the highlands of Papua New Guinea (PNG). PNG is one of the world's most diverse countries, in terms of language and culture, as well as in plants (Camara-Leret et al. 2020). I fell in love with the place. A few years later, I returned to PNG to work as an agricultural officer in the jungles on the island of Bougainville. In hindsight, this was the best possible introduction to the agricultural systems and needs of the rural poor in a tropical region. Many of the people I encountered had never heard of machines, electricity, telephones, or agricultural chemicals. They farmed effectively using local plants and natural methods. Later, when I was invited to teach Food Crop Production at what is now the Papua New Guinea University of Natural Resources and Environment (PNGUNRE n.d.) in East New Britain, the students strongly suggested that I focus on local edible plants and methods that suited their country. My first book was a compendium on the edible plants of Papua New Guinea (French 1986), which was the first step in writing a series of books to promote local food plants in Papua New Guinea. Since then, I have done consultancies with the United Nations FAO, the World Bank, Australia's Commonwealth Scientific and Industry Research Organisation (CSIRO), Bioversity International, and others. My main focus as an Agricultural Extension Officer has been to collate information on edible plants of the world in plain English and with highly illustrated posters so that rural people can be encouraged to value their local nutritious food plants to feed themselves and their families well. For several countries, including the Solomon Islands, PNG, and Zimbabwe, I have written books on local edible plants and appropriate production methods (French 2006, 2010, 2013).

As far as I know, the Food Plants International database is the only one that is moving towards recording all the known edible plants of the world (Food Plants International). At the time of writing (September 2022), our database has over 34,000 edible plant species and we continually record their occurrence in every country of the world (French and Maynard 2019). A quick check of our database reveals its range: it includes 300 edible species that are adapted to deserts, 157 for mangroves, 23 for atolls, 109 suited to the Arctic and Antarctic, over 18,000 edible species for the tropics, 2,200 edible legumes in the bean (Fabaceae) family, which are usually high in protein and improve soil nitrogen, and many other choices. Naturally, additional information and refinement is needed for precise selection for the many biomes of the globe, recognising that often plants become established and develop increasing varietal richness when they are in a suitable location. By contrast, monoculture agricultural practices demand a maximum yield of one selected variety, usually for economic reasons. Consequently, any other plant is regarded as an intruder and therefore a 'weed'. (I would argue that the best definition of a 'weed' is a plant for which we haven't yet found good uses.)

The history of the 'Green Revolution' for rice production in Asia reveals the tragedy of this approach. Over 600 edible plants, suited for swamps, were removed, creating nutritional deficiencies in the population, including vitamin A, iron, and others. Thankfully, Indian agricultural scientist Dr MS Swaminathan recognised this, writing about it in his book, *Sustainable agriculture: Towards an evergreen revolution* (Swaminathan 1996, 2010). The MS Swaminathan Research Foundation now promotes this more enlightened approach. The Food and Agricultural Organization of the United Nations (FAO) has hosted several studies on the potential of Genetically Modified Organism (GMO) methods, noting their advantages and limitations; they declared that GMOs, such as 'Golden rice', are not the way of the future (FAO 2001).

The scientific names of plants are continually changing with increasing knowledge, but there is some stability to the identity of the plants currently being eaten by people in some locations in the world. Where and how they are eaten is very random and flexible, however. For example, about 110 of the foraged edible plants in Britain that are supplied to elite restaurants in London currently grow in my home state of Tasmania, Australia (Łuczaj, Wilde and Townsend 2021). Sadly, we call them 'weeds' and spray them with toxic chemicals. Many are highly nutritious and sometimes underutilised. For example, blackberry nightshade (*Solanum nigrum*) is eaten almost worldwide and in our database has 241 science references to its use as a food! Stinging nettles (*Urtica dioca*) have an ingredient useful to men's prostate conditions. Fat hen (*Chenopodium album*) seeds have good qualities like Quinoa (*Chenopodium quinoa*). Dog rose (*Rosa canina*) fruit are very rich in vitamin C. In a tropical location, monocultured rice is prioritised, while ignoring nutritious plants nearby. One taro leaf growing on the edge of the rice paddy could provide enough vitamin A for three children (Rashmi et al. 2018)! To make the pro-vitamin A in plants accessible to humans, the plants need to be cooked in oil. There are at least 650 plants from which oil can easily be squeezed.

Fifty-three years ago, I was a temporary science teacher in a school. I still remember a book that claimed to be able to help discern whether students were more suitable for science or the arts (Getzels and Jackson 1962). Its authors suggested that those students who had a reductionist approach would decrease options available, enabling them to solve critical problems; therefore, they had a future in science. Those students who had a divergent approach would be fascinated by a range of options which allowed them more flexibility and creativity; they would be better suited to the arts. In my view, the reductionist mindset is what created the global climate crisis and unsustainable narrowing of the plant genetic resources for food and agriculture. Instead of exploring and utilising the 34,000 edible plant species (French and Maynard 2019), with hundreds or thousands of cultivars or varieties, we have narrowed cultivation down to about 30 species, after which we seek to prop these up with technology suited to every different biome. We call this 'modern' agriculture!

A reductionist mindset in agriculture can overlook cultivars and varieties which create resilient food systems that are well adapted to their local environment. Three cases that bear this out in the context of Papua New Guinea are yams, sugar cane, and sweet potato. In PNG, there are hundreds of cultivars of both Greater Yam (*Dioscorea alata*) and Lesser Yam (*Dioscorea esculenta*) and to simply choose one because it has a high yield means that meals lack a range of tastes. Additionally, pest and disease problems become greater. Papua New Guinea is the world's home of sugarcane, with a fascinating range of varieties of chewing sugarcane (Tom, Braithwaite and Kuniata 2017). Some of these have been taken around the world for the global sugar industry (Bourke 2009). Establishing a commercial sugar industry in Papua New Guinea, based on one selected variety bolstered by 'modern' methods, means that it has been a struggle to overcome endemic pests and diseases (Kuniata, Chandler and Korowi 2001). The shift away from promoting a diversity of food plants is also evident in reducing the 3,000 or more sweet potato varieties in the country to just a small number (Chang et al. 2017; Lyons 2006; Roullier et al. 2013). Historically, diverse varieties were well suited to the ecological, cultural, and social preferences of people from different regions, although the emphasis on commercialisation and Western preferences has lowered the status of these interesting cultivars (Fujinuma et al. 2018). The commercial promotion of one soft yellow variety of sweet potato, as it is rich in pro-vitamin A (Lyons 2006), overlooks resilient nutrition strategies. A simpler approach would be to inform the village people that the young leaves of all varieties of sweet potato can be cooked in oil to make a readily available supply of vitamin A. In some regions of PNG, these young leaves are popular. Thankfully, people locally are currently discovering that edible oil can easily be extracted from 60 edible plants in this country. These oils are now being sold in local food markets.

In Australia, a high-tech and reductionist approach is currently being used in a multi-million-dollar research project which aims to produce 'one' banana cultivar, selected from the 279 kinds currently growing in Uganda, that is rich in vitamin A (Martin 2017). Uganda already has 680 plant species with edible leaves, most of which would be rich in vitamin A when cooked in oil. At least 77 species of plants in Uganda have edible oil (Food Plants International 2022). The bonus of promoting a more divergent diet using these edible leaves would be that the iron deficiency anaemia which affects about 33% of the mostly female farmers would be addressed (Nankinga and Aguta 2019).

For a short time, the winged bean (*Psophocarpus tetragonolobus*) received the world's attention as one of the best beans for tropical locations. It was even called the 'supermarket' on a stalk because of the attractive, tasty and nutritious leaves, flowers, pods, seeds, and roots it produces (Tanzi et al. 2019).

Unfortunately, scientific research and interest quickly fell away when a dwarf variety could not be found (this was needed to allow mechanised production). Such a variety did not flower or produce pods outside the tropics, where most Western scientists and research locations exist. A more genuine response would have been to supply some of the very wide range of seeds of varieties (and appropriate inoculation procedures) to those living in the tropics, which contain close to 50% of the world's population (Hance 2014). The winged bean is exceptionally good for nitrogen fixation and can be made an attractive hedge around homes. Other beans, such as Yardlong bean (*Vigna unguiculata* var. *sesquipedalis*), can be allowed to climb up corn (maize) stems, maintaining good production of both food plants.

A study published many years ago explained how much more energy positive the traditional mixed cropping farming systems was in Papua New Guinea (Morrison et al. 1994). An average garden maintained at least 40 different edible species, usually occurring with a wide range of cultivated varieties to maintain richness in the diet. This also ensures food security. Because this system had plants with a range of heights and growth patterns, it not only helped ensure a spread of different foods for regular meals throughout the year, but also maintained the soil's nutrient value, while ensuring the soil retained its moisture. Nutrients that leached deeply into the soil were recycled by deep-rooted plants such as Pigeon pea (*Cajanus cajan*); in many locations, this reduced increased salt accumulation near the surface or salinity caused by moisture evaporating towards the soil surface (NSW Department of Planning and Environment 2022).

Trees are essential for good soil management and care of the environment. Over 10,000 tree species have edible parts (Food Plants International 2022; Romm 1993). Some of these have unique properties. For example, coastal almond (*Terminalia catappa*) grows on almost every tropical beach in the world. It is a lovely shady tree under which people sit to fix fishing nets and for relaxation. Meanwhile, children climb the tree to collect the small edible kernels or nuts of the fruit. These are incredibly rich in zinc, which is needed by over 100 enzymes in the human body (Ladele et al. 2016). Stunting is the most common sign of low zinc levels, but many other body functions can be affected. Children collecting the nuts often do so incidentally; if people understood the importance of zinc, they would encourage and value this important food source. Good levels of zinc in diets, along with vitamin A, can significantly improve resistance to diseases, such as malaria and tuberculosis (TB). Trees are currently receiving a major focus due to climate change. Rather than simply giving over large areas to forestry, there is an opportunity to grow trees to perform multiple functions, such as addressing climate change through carbon storage, providing firewood for village household cooking, and providing food. Over 700 trees produce edible nuts, 6,700 species yield edible fruit, and 1,600 species have edible leaves (French and Maynard 2019). Because of their deep roots, trees restore nutrients to the surface that have washed deep into the soil. As well, where mist and low rain clouds drift by, trees act as a 'nuclei' focus for raindrops to form and fall, increasing rainfall. Many trees have good timber for house construction. But sometimes people remove trees for other reasons. The removal of trees in the Amazon so that people could grow soybean to marble beef in Britain has been well documented as an unproductive and destructive process (Greenpeace 2016). Marbling beef simply enhances its taste without providing any more food in a world facing a food crisis. Further, when the Amazon rainforest is cleared, the soil cannot be productively used for cropping after a second year of cultivation. Tropical soils deteriorate very rapidly when exposed.

About 33% of the women of the world are anaemic or iron deficient. Dark green leaves are rich in iron and can be found in just under 10,000 species with edible leaves (French and Maynard 2021). About 1,000 of these grow in arid areas where farming is difficult and malnutrition is often common (French and Maynard 2021). Folates (or vitamin B9) as chemicals are linked, even by name, to 'foliage'. Because our bodies do not store folates, it is vital that women of reproductive age eat dark green leaves every day. This is essential to provide folates to prevent birth defects and other problems. These are serious issues in many low-income countries. Trying to fortify staple foods is one option being widely introduced. A much more practical and low-cost approach is to inform women of the benefits of eating dark green leaves daily. There are over 1,000 spice plants that can be added to enhance the flavour. As well, using a mixture of leaves can greatly enhance the attractiveness of the meal. In *Traditional food plants of Kenya*, Patrick Maundu provides very practical advice on using a mixture of edible leaves (Maundu et al. 1999). Mixing a drier leaf, such as cassava leaves (once well cooked) and a more slimy leaf, such as sunset hibiscus (*Abelmoschus manihot*), can provide a better texture. According to the World Vegetable Centre, the latter plant, popular in Asia and the Pacific, is by far the richest plant analysed for folates (pers. comm. with Dyno Keating, head of the World Vegetable Centre in Taiwan.)

Globally, the world's most frequent cause of death in the last 20 years is malnutrition in the form of obesity (McCarthy and Sánchez 2020). This is caused by those living in affluent countries eating a narrow diet of over-refined foods. Often obese people are called 'salad avoiders' and, from my observations, this seems to be true. So, eating a diverse range of edible plants is important in all locations.

About 200 years BC, Herodotus suggested that we make food plants our medicine and our medicine food plants (Rhodes et al. 2022). In the early 1990s, JR Lupien (head of the nutrition section of the United Nations Food and Agriculture Organisation in Rome) suggested to me that, for the rural poor in the tropics, we need to keep nutrition information simple so that people can understand it. His suggestion was to encourage people to eat a wide range of food plants in order to ensure that, if a nutrient was missing in one plant, it would be found in another. As many of the rural poor only have one large cooking pot, mixing a wide range of plants as a stir-fried dish or a stewing pot can provide a balanced diet. As well as providing nutrients, many plants also have other health benefits. These are called 'functional foods' because they have multiple functions. Many herbs and spices have these additional health benefits and should be eaten regularly, but not in large amounts. The chemistry of the additional health benefits is almost overwhelming to anyone but a highly trained chemist. Thankfully, two of the top international horticultural organisations, the International Society for Horticultural Science (ISHS) and the World Vegetable Centre, have recently highlighted that food plants often match colours to nutrients (Ojiewo et al. 2013), and recommended that people eat all the colours of the spectrum. As an example, plants or fruit that are blue are usually rich in antioxidants.

But a focus simply on nutrition will not address the major hunger and malnutrition problems in the world. All of this must be undergirded by ecologically sound and environmentally sustainable agriculture. This is where horticulture and agriculture and food production become fascinating activities. Instead of trying to lock seeds away in seed banks, I propose that we should learn to live with diversity in gardens and communities and local environments. The best way to conserve valuable resources and their diversity is to help people to become informed and aware of the value of plants. For example, when I was involved with studying and collecting banana varieties in Papua New Guinea, the older residents were a rich wealth of information and experience. The specialist with me was looking for banana seeds. Many of the traditional diploid banana varieties have seeds in the fruit lower on the bunch. But bananas are a fascinating polypoid hybridised series, meaning that two species have crossed and some have doubled or increased their number of chromosomes. Papua New Guinea alone has 650 cultivated varieties (Australia and Pacific Science Foundation 2002). Many of the world's scientists claimed that one of the original parents (the BB species) had disappeared from history, but local farmers in PNG claimed to have both parent species still in their gardens (Sharrock 1995). We accepted the farmers' kind invitation to visit their gardens and found that their claim was true. They even had a name for both species, as well as the many other kinds, in their local language.

A common approach amongst agronomists is to select the highest-yielding variety and grow it. This is often the most bulky or largest variety and is comparatively low in food value. Replicated and randomised variety trials are usually used to select the highest yielding kind. Often, however, asking the village farmers who have spent their lives in food gardens would be a simpler technique. I remember an embarrassing personal incident when I was getting to know banana in one of the banana-dominant cultures in Papua New Guinea. I asked the farmers why they were still growing a very small cultivar. Their answer was that it was their baby food: it was easy to crush while in the skin and then squeeze into the baby's mouth. Then I asked why there weren't growing lots of a very, very large cultivar. Their answer was that it was so bulky and low in energy that you couldn't do a day's work if you ate that one, so they fed that one to the pigs! Instead of telling them to throw away their baby food and eat pig food, I simply asked them to teach me!

The reductionist scientific approach to plant selection also makes the gardening and farming system ecologically unstable. Reducing the diversity of the garden makes it vulnerable to pests and disease, and makes food production and diet susceptible to irregularities in rainfall, etc. I have seen virus diseases in taro (*Colocasia esculenta*) and bacterial wilt in potatoes (*Solanum tuberosum*) wipe out whole gardens when this reductionist monocultural approach was adopted based on one high-yielding variety. The spread of fungal diseases such as taro blight (*Phytophthora colocasiae*) becomes much less serious when a range of different plants is grown together in one garden. As well, insects that normally remain at manageable levels in mixed gardens can soon multiply to plague levels when monoculture is practised. When I see an invasion of armyworm (*Mythimna seperata*) in a local food garden, I normally suggest looking for someone monoculturing a cash crop nearby. This can often be of a coffee or cacao crop.

Soils are meant to be one of the most biologically rich environments on Earth. One gram of good healthy soil contains a multitude of micro-organisms (Gans, Wolinsky and Dunbar 2005). One of the negative impacts of excessive pesticide use is that soil is turned into lifeless dirt. The World Health Organization (2018) reports that, in local gardens, people are spraying insecticides daily in their food gardens. Judith Schwartz, a journalist who explores nature-based solutions to global environmental challenges, states, 'There are as many living organisms in a teaspoon of healthy soils as there are people on the planet' (NPR 2013). Some plant diseases, such as *Elsinoe* scab on sweet potato, are a sign of depleted soil; good soil management reduces the significance of this disease. In more contemporary, large-scale agriculture, there is a move towards Conservation Agriculture, where laser-guided automatic tractors allow planting between the rows of the previous year's crop without the need for extensive soil cultivation (Fischer and Hobbs 2019). This helps maintain soil structure, as well as the richness of the micro-organisms in the soil.

Legumes have an important role for soils and for food production. Legumes form a symbiotic relationship with soil bacteria called rhizobia, which attach to the plant roots and fix atmospheric nitrogen in a form that plants can use. But it is important to have the correct rhizobia to suit a particular legume: introducing a familiar bean or legume to an area where it is not traditional to the biome often means that the correct rhizobia may not be present. For example, soybean suits the subtropics but, without considering biomes, is commonly promoted in tropical regions. Similarly, peanuts are more commonly promoted than Bambara groundnuts, although the latter suit more arid regions. This is why it is important to 'go local' and find information about beans and legumes already growing in and adapted to an area. The bonus is that not only is soil nitrogen improved, but the protein level of the food is also increased. In Africa, for example, intercropping with leafy herbs or beans, either amongst corn (maize) or sorghum not only improves soil fertility but reduces the spread of Striga as an underground infection of these cereals. With these factors in mind, there are about 1,400 tropical legumes with edible parts that are worthy of greater attention (Food Plants International 2022).

In my observations, soil acidity/alkalinity or pH is one aspect of soils that is often poorly understood by subsistence farmers. Often, a portion of land is simply left vacant because it is either too acidic or too alkaline. In the Pacific, alkaline soils can occur on atolls and adding lots of seaweed can be at least a short-term solution. In arid areas, where water is scarce, food gardeners are inclined to re-use water containing detergents after washing clothes. In both cases, I have seen plants showing aluminium toxicity symptoms. Ladies gardening near the seashore in the Solomon Islands were reluctant to show me their gardens as they felt it was a sign that they were bad gardeners. It is possible to purify the washing water by draining it through specially built pits filled with organic matter (Queensland Government 2022). There are about 300 edible plant species that grow well in acidic soils and over 200 that can grow in alkaline soils (Food Plants International 2022). This is an additional aspect of matching plants to their environment.

In the past, smallholders in the tropics often practised a 'bush fallow' farming system, where land is cleared, crops are grown, and then everything is burnt. Farmers then move to a fresh spot, allowing the old garden to return to bush. Population increase and land pressure mean this system is no longer viable. As well, burning provides a rich supply of potash (needed by root crops) but nitrogen and sulphur are lost into the atmosphere in the form of smoke. Consequently, plants develop pale-yellow, young leaves (from sulphur deficiency) and dark brown older leaves (from nitrogen deficiency). Reducing burning and using old plant material for mulch (or composting) can give dramatic increases in garden productivity (Planet Natural Research Centre n.d.). Groups in Zimbabwe have discovered this approach means a food supply for their family of 20 bags of corn (maize) or sorghum in place of the four bags they have been used to. Similar dramatic yield increases have been demonstrated in Central America by using velvet bean and other beans as a soil protective cover between crops (Bunch and López 1995). These protective legumes are then slashed and used as mulch at planting time for the food garden.

Some crops are preferred in diets and gardening systems across cultures. In many areas of Africa, for example, corn (maize) is a preferred cereal (McCann 2005). But, as a crop with a high rainfall requirement, it can fail if the rains don't come. Africa has thousands of sorghum varieties, and this plant performs better in lower rainfall areas. Additionally, finger millet or bulrush millet can grow with much lower rainfall. Consequently, a wise recommendation is to plant some of each of these cereals every year. If the rains come and the corn (maize) grows well, then the others can either be used in mixed cereal dishes or for animal feed. If the rains fail, the family will not starve to death as the other cereals (such as millets) will survive and can be the staple diet. Cereals such as teff are popular in Ethiopia, and millets in some areas of India. When wheat is grown in arid areas, such as in Australia, the gluten content of the grain increases (Mkhabela et al. 2022). Consequently, coeliac disease, the autoimmune condition triggered by gluten, has become a major illness in some Western populations.

Crops are often grown and used in one area without a global awareness of the adoption of better varieties in other places. For example, in Ghana, hi-tech irradiation is being undertaken to improve yields of lesser yam (*Dioscorea esculenta*) as only small and low yielding varieties are available locally (Banson and Danso 2013). For this species, Papua New Guinea has a wide range of the best varieties in the world. These could be carefully disseminated to other seasonally wet/dry locations in the tropical world. There are 170 species of edible yams (*Dioscorea spp.*) that are important foods in seasonally wet/dry locations in the tropics (Food Plants International 2022). Often, these foods have great ceremonial importance at harvest time. Yams have not received the same attention from scientists and horticulturalists as cereals, even though 66 million tons of yams are grown and eaten each year (Ritchie and Roser 2020).

Although technology is amazing, it is not always the answer. At an international agriculture conference I attended in Florida, a speaker was discussing all the practical issues of trying to establish irrigation systems in an arid area in Haiti (ECHO Community 2010). At the same time, I was reading a book that simply said, 'Arid plants suit arid lands'. Sometimes the information is known to scientists but is not made accessible to village farmers. For example, an excellent article on fruit trees that grow in arid areas in Miombo woodlands in Africa also mentions that they produce their fruit in the season when food is in short supply and hunger is common (Campbell 1996). Unfortunately, this article was in such technical scientific language that even the teachers and agriculturalists in that region could probably not understand it. At Food Plants International (n.d.) we have made a fully illustrated lay summary of this article available on our website and it has been downloaded thousands of times. My task is still agricultural 'extension', as an intermediary between research and farmers, and I remain deeply grateful to good technical scientific research.

It is probably from simple survival experience that many older village farmers have accumulated a rich knowledge of edible plants and ecologically sound methods. But, as many of these farmers cannot read or write, this wisdom could be lost. *Ethnobotany Research and Applications* and other similar journals make it their focus to collect, record and make available this knowledge. For things that can be observed, a good farmer has a wealth of knowledge. But sometimes subtleties, such as nutrition deficiency symptoms, diseases, and small sucking insects, remain a mystery. (By contrast, I have found that, for larger insects, village people can explain the whole insect lifecycle, and probably also explain which ones are edible!) A lack of accessible information about the dangers of using pesticides in subsistence gardens is also a major concern, as small children forage for edible insects and fruit all day in these gardens.

Reducing the range of edible plants to about 30 species has necessitated the extensive transfer of plants and food between locations and countries. This has been very energy demanding. This narrow range of edible plants includes wheat, rice, corn (maize) and soybean, which are also supplied to famine-affected tropical countries. This has a long-term effect on local food production, especially in Africa. People in these regions have been used to tropical- and arid-region foods (such as teff, fonio, and millets) and now are suddenly being introduced to foods from temperate or monsoonal regions during periods of famine. When the famine ends, people in these regions now want to have these new foreign foods and are very slow to return to what was previously the normal farming and production of local foods. As well, temperate plants such as wheat, corn (maize) and soybean have developed a status that is unjustified, either by their nutritional value or their suitability for the local climate.

Another way that the status of diverse, nutritious local food plants is undermined is the influence of Western and wealthy consumers in local markets. Women in local markets are very observant of the food plants wealthy people buy. Many Western tourists and visitors go through tropical markets buying temperate plants with which they are familiar, such as cabbage and lettuce. Local women therefore presume these are the best foods. However, the low food value of ball cabbage can actually increase malnutrition in children: it has bulk (so children don't feel hungry) but is almost devoid of nutrients. Often, nutritionists recommend that people avoid these low food value plants while, at the same time, foreign agriculturalists are teaching farmers how to grow them!

Tropical swamps remain a difficult environment for food production and good nutrition. Sago palms grow well in these lowland regions but can develop too many suckers and, being a spiny palm, can be difficult to kill with herbicides. A report by a palm specialist on sago production for Papua New Guinea highlights this problem (Connell and Hamnet 1978). After reading this careful scientific article, the next time I was visiting a sago swamp in the Sepik Valley in Papua New Guinea I asked the local farmers. A local sago swamp gardener shared how to thin out palms: you simply cut a small hole in the unwanted sucker with your bush knife and six weeks later the sucker was dead due to the activity of very tasty and nutritious sago grubs. Unfortunately, such information is not written down for the benefit of the next generation who are at school. A starchy, sago-based diet results in very low birth weights of babies; finding suitable nutritious leaves, fruit, nuts, and seeds that will grow in sago swamps remains a problem.

Conclusion

The future of the planet depends on a more ecologically sound system of food production. The tragic global malnutrition problem can only be addressed by the appropriate use of a richly diverse and locally adapted range of food plants. Thankfully, there are over 34,000 edible plant species that are worthy of far greater attention. In our Food Plant International database (French and Maynard 2021), references for all the information are carefully documented for each plant, but such detailed documentation is inappropriate for books and posters provided for schools and clinics. It can be checked on our website or when database copies are supplied. See www.foodplantsinternational.com

References

Australia and Pacific Science Foundation 2002, *Tracing antiquity of banana cultivation in Papua New Guinea*. http://www.apscience.org.au/pbf_02_3/

Banson, K and Danso, K 2013, 'Improving the size and market value of an underutilised yam (*Dioscorea esculenta*) in Ghana: Implications for crop breeding and production choices', *Journal of Life Sciences*, vol. 7, no. 7, pp. 732–741.

Bourke, M 2009, 'Sweet potato in Oceania' in GL Loebenstein and G Thottappilly (eds), *The Sweetpotato*, Springer, Dordrecht, pp. 489–502. https://doi.org/10.1007/978-1-4020-9475-0_22

Bunch, R and López, G 1995, *Soil recuperation in Central America: Sustaining innovation after intervention*, International Institute for Environment and Development, London.

Camara-Leret, R, Frodin, DG, Adema, F et al. 2020, 'New Guinea has the world's richest island flora', *Nature*, vol. 584, no. 7822, pp. 579–583.

Campbell, BM (ed) 1996, *The Miombo in transition: Woodlands and welfare in Africa*, CIFOR, Bogor. https://doi.org/10.17528/cifor/000465

Chang, HSC, Villano, R, Irgving, D et al. 2017, 'Understanding consumer preferences for sweet potato in Papua New Guinea', *Australasian Agribusiness Perspectives*, vol. 20, pp. 59–77.

Connell, J and Hamnet, MP 1978, 'Famine or feast: Sago production in Bougainville', *Journal of the Polynesian Society*, vol. 87, no. 3, pp. 231–241.

Echo Community 2010, *ECHO International Agriculture Conference Booklet 2010*. http://edn.link/92eap9

Fanzo, J, Hunter, D, Borelli, T and Mattei, F (eds) 2013, *Diversifying food and diets: Using agricultural biodiversity to improve nutrition and health*, Routledge, Abingdon.

FAO (Food and Agriculture Organization of the United Nations) 2001, 'GMOs and the environment' in FAO, *Genetically modified organisms, consumers, food safety and the environment. FAO Ethics Series, vol. 2*, FAO, Rome. https://www.fao.org/3/X9602E/x9602e07.htm

Fischer, T and Hobbs, P 2019, 'Tillage: Global update and prospects' in J Pratley and J Kirkegaard (eds), *Australian agriculture in 2020: From conservation to automation*, Agronomy Australia and Charles Sturt University, Wagga Wagga, pp. 3–20.

Food Plants International 2022, 'Our Database'. https://foodplantsinternational.com/plants/

Food Plants International n.d., 'Miombo woodland fruit'. https://foodplantsinternational.com/free-resources-for-download/zambia/

French, BR 1986, *Food plants of Papua New Guinea: A compendium*, Australia Pacific Science Foundation.

French, BR 2006, *Food Crops of Papua New Guinea: An introduction to the crops and their importance and distribution*, Food Plants International Inc, Burnie. https://foodplantsolutions.org/wp-content/uploads/2018/06/5-Food-Crops-introduction.pdf

French, BR 2010, *Food plants of Solomon Islands: A compendium. Part 1*, Food Plants International Inc, Burnie. https://foodplantsinternational.com/download/food-plants-of-solomon-islands-a-compendium-part-1/

French, BR 2013, *Basic food gardening in Zimbabwe*, Food Plants International Inc, Burnie. https://foodplantsinternational.com/download/basic-food-gardening-zimbabwe-2015/

French, BR and Maynard, AR 2019, *Food Plants International Database*, Food Plants International. https://foodplantsinternational.com/

Fujinuma, R, Kirchhof, G, Ramakrishna, A, et al. 2018, 'Intensified sweet potato production in Papua New Guinea drives plant nutrient decline over the last decade', *Agriculture, Ecosystems & Environment*, vol. 254, pp. 10–19, https://doi.org/10.1016/j.agee.2017.11.012

Gans, J, Wolinsky, M and Dunbar, J 2005, 'Computational improvements reveal great bacterial diversity and high metal toxicity in soil', *Science*, vol. 309, pp. 1387–1390. https://doi.org/10.1126/science.1112665

Getzels, JW and Jackson, PW 1962, *Creativity and intelligence: Explorations with gifted students*, Wiley, New York.

Greenpeace 2016, '10 years ago the Amazon was being bulldozed for soy – Then everything changed' (*Greenpeace: Stories & Victories*, 22 May 2016). https://www.greenpeace.org/usa/victories/amazon-rainforest-deforestation-soy-moratorium-success/

Hance, J 2014, 'Booming populations, rising economies, threatened biodiversity: The tropics will never be the same' (*Mongabay*, 7 July 2014). https://news.mongabay.com/2014/07/booming-populations-rising-economies-threatened-biodiversity-the-tropics-will-never-be-the-same/

Kuniata, LS, Chandler, KJ and Korowi, KT (2001), 'Management of sugarcane pests at Ramu, Papua New Guinea', *Proceedings of the XXIV Congress of the International Society of Sugar Cane Technologists*, vol. 2, pp. 382–388. https://www.cabi.org/isc/abstract/20023165595

Ladele, B, Kpoviessi, S, Ahissou, H, et al. 2016, 'Chemical composition and nutritional properties of *Terminalia catappa* L. oil and kernels from Benin', *Comptes Rendus Chimie*, vol. 19, no. 7, pp. 876–883. https://doi.org/10.1016/j.crci.2016.02.017

Lowder, SK, Skoet, J and Raney, T 2016, 'The number, size, and distribution of farms, smallholder farms, and family farms worldwide', *World Development*, vol. 87, pp. 16–29. https://doi.org/10.1016/j.worlddev.2015.10.041

Łuczaj, Ł, Wilde, M and Townsend, L 2021, 'The ethnobiology of contemporary British foragers: Foods they teach, their sources of inspiration and impact', *Sustainability*, vol. 13, no. 6, article 3478. https://doi.org/10.3390/su13063478

Lyons, G 2006, *Screening and field trials of high-carotenoid sweet potatoes in Solomon Islands and Papua New Guinea to improve human vitamin A status. Final Report*, ACIAR, Canberra.

Martin, N 2017, 'QUT develops golden banana' (*The Medium*, 13 July 2017). https://medium.com/thelabs/qut-develops-golden-bananas-high-in-pro-vitamin-a-7835ced4b0b4

Maundu, PM et al. 1999, *Traditional food plants of Kenya*, Kenya Resource Centre for Indigenous Knowledge, National Museums of Kenya, Nairobi.

McCann, JC 2005, *Maize and grace: Africa's encounter with a New World crop, 1500-2000*, Harvard University Press, London and Cambridge, MA.

McCarthy, J and Sánchez, E 2020, 'Malnutrition is the leading cause of death globally' (*Global Citizen*, 14 May 2020). https://www.globalcitizen.org/en/content/global-nutrition-report-2020/#:~:text=The%20Global%20Nutrition%20Report%202020,of%20death%20and%20illness%20worldwide

Mkhabela, M, Bullock, P, Sapirstein, H et al. 2022, 'Exploring the influence of weather on gluten strength of hard red spring wheat (*Triticum aestivum* L.) on the Canadian Prairies', *Journal of Cereal Science*, vol. 104, article 103410. https://doi.org/10.1016/j.jcs.2021.103410

Morrison, RJ, Geraghty, PA and Crowl, L (eds) 1994, *Science of Pacific Island peoples: Education, language, patterns & policy*, vol. 4, Institute of Pacific Studies, Fiji.

MS Swaminathan Research Foundation n.d., https://www.mssrf.org

Nankinga, O and Aguta, D 2019, 'Determinants of anemia among women in Uganda: Further analysis of the Uganda demographic and health surveys', *BMC Public Health*, vol. 19, article 1757. https://doi.org/10.1186/s12889-019-8114-1

NPR 2013, '"Cows save the planet": Soil's secrets for saving the Earth' (*NPR Author Interviews*, 17 July 2013, transcript). https://www.npr.org/2013/06/17/191670717/cows-to-the-rescue-soils-secrets-for-saving-the-earth

NSW Department of Planning and Environment 2022, 'Types of salinity and their prevention'. https://www.environment.nsw.gov.au/topics/land-and-soil/soil-degradation/salinity/type-of-salinity-and-their-prevention

Ojiewo, C, Tenkouano, A, Hughes, JDA and Keatinge, JDH 2013, 'Diversifying diets: Using indigenous vegetables to improve profitability, nutrition and health in Africa' in J Fanzo, D Hunter, T Borelli and F Mattei (eds), *Diversifying food and diets: using agricultural biodiversity to improve nutrition and health*, Routledge, Abingdon, pp. 291–302.

Planet Natural Research Centre n.d., 'Adding to soil'. https://www.planetnatural.com/composting-101/how-to-use/adding-compost/

PNGUNRE (Papua New Guinea University of Natural Resources and Environment) n.d. https://unre.ac.pg/

Queensland Government 2022, 'Wetlandinfo: Recycle pits'. https://wetlandinfo.des.qld.gov.au/wetlands/management/treatment-systems/for-agriculture/treatment-sys-nav-page/recycle-pits/planning-design.html

Rashmi, DR, Raghu, N, Gopenath, TS et al. 2018, 'Taro (*Colocasia esculenta*): An overview', *Journal of Medicinal Plants Studies*, vol. 6, no. 4, pp. 156–161.

Rhodes, P, Richardson, RG, Guthrie, DJ et al. 2022, 'History of medicine', *Encyclopedia Britannica*. https://www.britannica.com/science/history-of-medicine

Ritchie, H and Roser, M 2020, 'Agricultural production: Roots and tubers', *Our World in Data* (online resource). https://ourworldindata.org/agricultural-production

Romm, JJ 1993, *Defining national security: The nonmilitary aspects*, Council on Foreign Relations Press, New York.

Roullier, C, Kambouo, R, Paofa, J et al. 2013, 'On the origin of sweet potato (*Ipomoea batatas* (L.) Lam.) genetic diversity in New Guinea, a secondary centre of diversity', *Heredity*, vol. 110, no. 6, pp. 594–604. https://doi.org/10.1038/hdy.2013.14

Sharrock, S 1995, 'Collecting the *Musa* gene pool in Papua New Guinea' in L Guarino, V Ramanatha Rao and R Reid (eds), *Collecting plan genetic diversity. Technical guidelines*, IUCN, Gland, pp. 647–658. http://cropgenebank.sgrp.cgiar.org/images/file/procedures/collecting1995/Chapter33.pdf

Siddique, KH, Li, X and Gruber, K 2021, 'Rediscovering Asia's forgotten crops to fight chronic and hidden hunger', *Nature Plants*, vol. 7, no. 2, pp. 116–122. https://doi.org/10.1038/s41477-021-00850-z

Swaminathan, MS 1996, *Sustainable agriculture: Towards an evergreen revolution*, Konark Publishers, Delhi.

Swaminathan, MS 2010, 'For an evergreen revolution' in MS Swaminathan (2017), *50 years of green revolution: An anthology of research papers*, World Scientific, Singapore, pp. 397–405.

Tanzi, AS, Eagleton, GE, Ho, WK et al. 2019, 'Winged bean (*Psophocarpus tetragonolobus* (L.) DC.) for food and nutritional security: Synthesis of past research and future direction', *Planta*, vol. 250, no. 3, pp. 911–931. https://doi.org/10.1007/s00425-019-03141-2

Tom, L, Braithwaite, K and Kuniata, LS 2017, 'Incursion of sugarcane smut in commercial cane crops at Ramu, Papua New Guinea', *Proceedings of the Australian Society of Sugar Cane Technologists*, vol. 39, pp. 377–384.

UNIGME (United Nations Inter-Agency Group for Child Mortality Estimation) 2021, 'Levels & Trends in Child Mortality: Report 2021, Estimates developed by the United Nations Inter-Agency Group for Child Mortality Estimation', United Nations Children's Fund, New York. https://childmortality.org/wp-content/uploads/2021/12/UNICEF-2021-Child-Mortality-Report.pdf

Witkamp, RF and van Norren, K 2018 'Let thy food be thy medicine … when possible', *European Journal of Pharmacology*, vol. 836, pp. 102–114. https://doi.org/10.1016/j.ejphar.2018.06.026

World Health Organization 2018, 'Pesticide residue in food'. https://www.who.int/news-room/fact-sheets/detail/pesticide-residues-in-food

World Health Organization 2020, 'Children: Improving survival and well-being' (*World Health Organization Newsroom*, 8 September 2020). https://www.who.int/news-room/fact-sheets/detail/children-reducing-mortality

12 Nurturing soil-adarities

Growing multispecies justice in therapeutic landscapes

Bethaney Turner

Introduction

In this chapter, I consolidate and extend nascent calls (Biglin 2020; Gorman 2017a, 2017b, 2019) to bring therapeutic landscape practices and conceptual literature into dialogue with posthuman and multispecies approaches. Consequently, while therapeutic landscapes have traditionally been defined by the health and wellbeing benefits they induce for humans, this chapter explores the ethical imperative to consider how these landscapes generate benefits or harms for all human and nonhuman entities (otherwise referred to as more-than-human entanglements) that co-produce these spaces. It does this through a focus on food-producing gardens. Gardens have been shown to have significant therapeutic potential for human participants, with work on food-producing sites demonstrating that these places can be safe spaces that nurture a sense of self-efficacy for people from diverse backgrounds (Green and Phillips 2014; Sanchez and Liamputtong 2017; White 2011). Food-producing gardens have also been identified as providing refuge from everyday stresses which offer opportunities to attune to 'nature'. This attunement to nature has been found to generate positive benefits for humans who find reassurance in the resilience of nature, particularly in times of crisis; for example, during the COVID-19 pandemic (Atkinson 2020; Marsh et al. 2021). However, while identification has grown in recent years of therapeutic spaces being relationally produced through complex and diverse more-than-human assemblages – revisiting and revitalising Gesler's (1992, p. 735) initial vision of therapeutic spaces as being produced through the interaction of 'environmental, individual and societal factors' – limited attention has been paid to the health and wellbeing of the multispecies co-constituents (the nonhuman microbes, critters, and vegetation) of these places.

Within the therapeutic landscape literature, nonhumans (notably 'nature') are typically represented as resources capable of providing therapeutic benefits to humans. This chapter argues that such an approach risks perpetuating material and discursive practices that exploit and/or damage the more-than-human world. Such human-centric logic is also shown to limit the realisation of broader therapeutic benefits for human and nonhuman participants that could arise through recognition of our shared imbrication in mutual relations of care

DOI: 10.4324/9781003355731-16

(Puig de la Bellacasa 2010, 2017). Pitt's (2014, 2018) important work on community gardens, however, shows that inequity permeates how care is given and who or what benefits. Pitt (2018) argues in her exploration of the characteristics of human/nonhuman relating in gardens that instrumental modes of care which induce benefits for humans are dominant in these sites. To this, I add the observation that the capacity for more-than-human care in gardens is also unequally distributed and experienced. Access to land and uncontaminated soil to produce food in urban sites has been shown to be less available to those from lower socio-economic and minority racial and cultural backgrounds (Aelion et al. 2013; McClintock 2015). Through a focus on how attentive and responsive relationships with multispecies garden companions can be cultivated, this chapter explores how particular modes of attending to soil supported through a multispecies justice framework could play a critical role in maximising the potential therapeutic benefits of gardens by supporting human *and* multispecies flourishing.

Rooting ethical ecological relations in the soil

The rich interdependencies of soil–human health and wellbeing make the role of soil a rich focus for work exploring therapeutic landscapes. Soil is foundational to human and planetary health and wellbeing. It supplies 95% of the world's food, contains more than 25% of the planet's biodiversity, and is critical to the effective functioning of ecosystem services, such as water filtration, nutrient and energy cycling, and carbon sequestration (UN-FAO et al. 2020). Healthy soil has been identified as underpinning at least seven of the 17 United Nations (UN) Sustainable Development Goals related to poverty (SDG-1), hunger (SDG-2), water quality (SDG-6), and climate mitigation (SDG-11, 12, 13 and 15) (Lal et al. 2021). Soil is also a growing site of bioprospecting, where scientists seek out new microbial strains helpful to human health. Urban greening initiatives designed to mitigate the effects of climate change and improve citizens' health and wellbeing are also heavily reliant on soil. Evidence suggests that a biodiverse soil microbiome not only supports the growth of vegetation but can itself induce physical and mental health benefits for humans (see Liddicoat et al. 2020). Consequently, recognition by the UN that 33% of the world's soil is degraded poses a significant threat not only to the lives and livelihoods of agricultural producers, but also to broader human and planetary health and wellbeing (Lal et al. 2020, 2021; UN-FAO et al. 2020). While urban soils in Australia have attracted limited attention, contamination in home and community gardens in inner city areas has been found to be widespread (Rouillon et al. 2017).

Recently, soil has also become a potent symbol of hope in contemporary struggles to secure climate-resilient, healthy futures through a growing global focus on soil repair, restoration, and regeneration. While much of this activity, discourse, and related policy is focused on broadscale rural agricultural lands, there is great potential for increased recognition of soil–human interdependencies in urban gardens and public spaces to improve soil health and enhance the therapeutic benefits of these sites and the ecosystems they produce. A growing body of literature in science and technology studies, philosophy, and the environmental humanities identifies soil–human relations as a productive arena for developing more ethical human–environment relations in the Anthropocene (Krzywoszynska and Marchesi 2020; Puig de la Bellacasa 2017; Salazar et al. 2020). Puig de la Bellacasa's (2010, 2017) work on soil, particularly focused on permaculture, proposes a schema for ethical ecological thinking and everyday action that is rooted in relations of care. Building on the feminist ethics of care approach, this work decentres humans, identifying that both humans and non-humans are able to care and co-produce mutual flourishing. Such an approach challenges persistent human–environmental divisions and instead identifies that 'humans are not the only ones caring for the Earth and its beings – we are in relations of mutual care' (Puig de la Bellacasa 2010, p. 164).

Therapeutic landscape literature focused on cultivated landscapes has long recognised that these places are produced through interactions between social, physical, and symbolic factors. Kearns and Milligan (2020, p. 3) note that Gesler's seminal research has consistently implied 'a need to be alert to both human and non-human interrelation'. However, the work on therapeutic landscapes and the practices it informs has been primarily focused on realising therapeutic benefits for humans with little engagement with the literature emerging out of the posthuman turn and the growing body of multispecies justice work. Limited attention has been paid to the ethical obligations that the humans who design, tend, manage, and benefit from these spaces might have to the multispecies entities that co-produce the therapeutic benefits.

Accessing therapeutic benefits in gardens: The importance of place experience

Gardens have a long history of being identified as therapeutic landscapes capable of supporting health, wellbeing, and recovery for people, particularly those who are ill, vulnerable or disadvantaged. They have been used in a range of settings, from hospitals and respite centres to private, school and community gardens. Food-producing gardens, particularly in communal settings, have been found to be therapeutic for two key reasons: first, because they enable contact with 'nature' (Jiang 2014; Marsh et al. 2021; Sampson and Gifford 2010); second, because they can support the emergence of place-based belonging that can be strengthened through building the social trust that underpins a sense of community (Kingsley et al. 2020; Mouin-Doos 2014). The ability of these sites to build the sense of self-efficacy of gardeners has also been identified as critical to the realisation of significant therapeutic benefits (Green and Phillips 2014; White 2011). Pitt (2014), drawing on Csikszentmihalyi's (2002) notion of flow, highlights the importance of place to explain how community garden spaces can become therapeutic. Using the concept of 'emplaced flow', Pitt (2014, p. 85) demonstrates that gardening which involves immersive bodily actions allows gardeners to get lost in activities where 'time passes quickly and one ceases to feel separate from task or world'. However, the lack of equitable access to sites such as urban community gardens can restrict both who is able to experience these therapeutic benefits and the multispecies make-up of these sites.

Barriers to access and participation for marginalised groups in community gardens tend to revolve around three key elements: physical garden features; forms of garden management; and the tendency for gardens and garden programmes to privilege particular values, aesthetics, and practices that fail to acknowledge multicultural and First Nations' alternatives (Turner et al. 2021). The potential health and therapeutic benefits of gardens can be limited by: the inability to access gardens on foot or via public transport; the lack of inclusive design for differently-abled participants; insecure land tenure; and the presence of contaminated soil. Overt and incidental exclusion of minority groups from gardens has been shown to emerge from how these spaces are organised and managed (Guthman 2008; Nettle 2016). This can include a lack of support for culturally and linguistically diverse participants in the form of translations or in shaping how the gardens are governed (Nettle 2016). Expectations related to how gardens are cultivated to meet preferred or dictated garden and urban landscape aesthetics can also limit culturally and socio-economically diverse participation (Strunk and Richardson 2019) and may perpetuate colonial logic. Such preferences can restrict the multispecies make-up of the sites by limiting what is able to be grown, as well as how plants and plots are cared for. Consequently, the way public food gardens are designed and managed impacts not only who has access to the health and wellbeing benefits that can be cultivated in these sites, but also which multispecies inhabitants are present, how these are cared for, and the possible human/more-than-human caring assemblages that can emerge.

While appreciation of permaculture and the need to increase biodiversity in urban spaces is growing in the minority world (wealthier nations with only a small percentage of the global population), for many accustomed to producing crops in neat, monocultural rows (including many municipal governments and organisations with oversight of public gardens), these cultivation practices are not always welcome and can raise concerns related to compliance with health and safety mandates (Strunk and Richardson 2019; Turner et al. 2021). Managing diverse cultivation styles, aesthetic preferences and risk can be a difficult balancing act. Assumptions about how a garden should look and how it is made safe and accessible are informed by pre-existing expectations that may not adequately reflect the culturally diverse and specific ecological and multispecies needs of particular sites. As Aptekar (2015, p. 212) writes, gardens can 'reflect normative notions of community and civility, as well as local memory and identity, and are inextricably connected to social inequalities and power hierarchies beyond the garden gates'. The persistence of these normative notions can restrict both how gardens can become therapeutic landscapes and who (and what) is able to benefit from these. Myers (2017, p. 298) observes that:

> Garden infrastructures matter, and not only to the plants. They not only enforce biopolitical regimes that dictate who can and cannot live inside or outside the enclosure (think weeding, pesticides, etc.). Garden infrastructures also shape how plants and people get entrained to one another's lives.

That is to say, the material features of gardens make certain human/more-than-human relations and affective experiences possible, while restricting others. If, as Gesler (1996, p. 96) outlines, it is the interplay between human and nonhuman elements that produces therapeutic landscapes through the generation 'of an atmosphere which is conducive to healing', then explorations of how spaces become therapeutic must pay careful attention to the diverse more-than-humans present.

Therapeutic gardening and green space literature overwhelmingly privileges the assumption that contact with nature is likely to generate positive health and wellbeing outcomes for humans, despite evidence of diverse experiences (see, e.g., Milligan and Bingley 2007; Plane and Klodawsky 2013). Chessbrough et al. (2019, p. 43) note that these assumed benefits tend to draw on understandings of 'nature as an un-differentiated whole'. Given the variations in socio-cultural perspectives and individual experiences in gardens, coupled with encounters with the limits of human control in cultivated spaces, gardening can lead to confronting, disturbing, and unexpected encounters. These encounters are often a response to ecological conditions. This can include frustration at pests eating newly planted crops and ripe produce; fear related to the presence of risky nature (think snakes and spiders); and experiences with extreme weather events where crops can be destroyed. Crop failures, particularly for new gardeners or for food producers with a background growing in different ecological and climatic conditions, can cause distress and challenge a positive sense of self-efficacy (Abramovic et al. 2019). These are also embodied, multisensorial experiences (see Biglin 2020) capable of generating powerful affective responses that – as identified in Gorman's (2017b, p. 317) work on smell – 'can create and facilitate a therapeutic engagement with place, whilst simultaneously intruding and disrupting therapeutic processes'.

The role of more-than-humans and the complexity of the relational and multisensorial processes that produce place experiences permeates understandings of therapeutic landscapes as assemblages or 'enabling places' (Duff 2011). As Bell et al. (2018, p. 128) note in their scoping review of therapeutic landscape literature:

> The interest in notions of assemblage and 'enabling' places stems in part from their relational and situated approaches to wellbeing, acknowledging the therapeutic nature of space as emergent in the context of dynamic sociocultural-material-affective-sensuous configurations involving both human and non-human actors.

An assemblage approach demonstrates that humans are not the only actors in therapeutic places. While this recognition can yield therapeutic benefits for humans (as identified in work noting the reassurance experienced by some people when bearing witness to nature's resilience in gardens; e.g., Atkinson 2020; Marsh et al. 2021), experiences with nonhuman agency can also undermine realisation of improved health and wellbeing for human participants in these sites. Subsequent human responses to these 'sociocultural-material-affective-sensuous challenges or threats' (Bell et al. 2018, p. 128) can lead to actions that have a detrimental impact on the health and wellbeing of some more-than-humans. This includes enactment of regimes of care that support the flourishing of some species at the expense of others. This can have wide-ranging impacts, including site biodiversity (What is planted, cleared, baited and/or trapped?); the toxins present (Which pesticides and herbicides are used? What testing and remediation of past contamination has been carried out?); and how resources such as water are used. Consequently, while certain gardening activities and experiences may produce therapeutic benefits for some human participants, the modes of care used in these sites determine the make-up of the multispecies communities present and whether these more-than-humans are beneficiaries of this care or are harmed by it (Pitt 2018). This then impacts on the human and more-than-human assemblages of care that can emerge.

Therapeutic landscapes and multispecies wellbeing

A growing body of academic literature across the last two decades has directed attention to the entangled relational ecologies through which humans and more-than-humans coexist in the world across a range of disciplines (notably geography, cultural studies, social science, and philosophy). This work covers a wide range of engagements, such as human–animal relations (see, e.g., Buller 2013; Candea 2010; Haraway 2007); human–plant interactions in 'vegetal geography' and critical plant studies (see Head, Atchison and Phillips 2015; Marder 2013), and work exploring human–microbiome encounters through 'microbiopolitics' (Paxson 2008), 'symbiopolitics' (Helmreich 2009) and 'probiotic environmentalities' (Lorimer 2020). By focusing on the affective, visceral, and entangled dimensions of human and more-than-human relations and entities, this work exposes the flawed basis of claims to human exceptionalism which have fuelled our contemporary experiences of ecological crisis.

It is vital to highlight here that not all those constituting the Anthropos of the Anthropocene have engaged equally in practices that threaten human and planetary health and wellbeing; neither has a nature-culture binary been foundational to all cultures. First Nations peoples' ways of knowing, being, and doing have always been grounded in relational and reciprocal connections between humans and nature (Somerville et al. 2021). In Australia, where the research informing this chapter was conducted, this manifests through the vitality of Country which can be understood as that which 'encompasses th[e] vibrant and sentient understanding of space/place which becomes bounded through its interconnectivity. Country and everything it encompasses is an active participant in the world, shaping and creating' (Bawaka Country et al. 2015, p. 270). While there is growth in decolonial research practices, and there have been some explicit efforts to explore the role of therapeutic landscapes in relation to Indigenous human–nature relations (e.g., Mcintosh et al. 2021), more Indigenous-led work in this area and exploration of how this connects to the development of posthuman and multispecies justice research across disciplines is needed.

Therapeutic landscape literature recognises nonhumans as co-producers of 'healing atmospheres' (Gesler 1996) and 'enabling places' (Duff 2011), but has not adequately attended to the health and wellbeing of nature, Country, or the multispecies entities that enliven these sites. As Gorman (2019, p. 321) writes, this oversight risks perpetuating 'anthropocentric politics' and may prevent 'more equitable framings' of human and more-than-human relations. Gorman's (2017a, 2017b, 2019) own work on care farms (where activities are designed to induce a range of therapeutic benefits for human participants) attempts to redress this gap. His research highlights the ways in which therapeutic benefits are produced through embodied, sensorial experiences which may induce varied therapeutic impacts on human participants, depending on their individual experiences and socio-cultural context. Gorman's research also identifies the failure of therapeutic programmes and literature to acknowledge and attend to the needs and wellbeing of the more-than-humans that are critical to the production of therapeutic places. To remedy this, Gorman (2017b, p. 317) advocates a multispecies approach which identifies that 'there is a need to decentre humans in discussions of therapeutic spaces and instead begin to consider a post-human and multispecies approach that rejects the prioritization of human-centric norms, assumptions, behaviours and practices'. This chapter builds on Gorman's efforts to bring multispecies and posthuman-informed work into dialogue with therapeutic landscape literature and extends the focus beyond animals to the multispecies, geological, and chemical entity that is soil. In so doing, this work highlights the ethical imperative of attending to nonhuman needs and desires in therapeutic landscape work and argues for the necessity of taking a multispecies justice approach. That is, this chapter contends that therapeutic landscapes must not solely be designed, managed, and assessed in relation to the benefits they might generate for human participants, but also the potential benefits for the more-than-humans which are critical to the co-constitution of these places and their therapeutic atmospheres.

Multispecies justice in therapeutic garden landscapes

Multispecies justice (MSJ) is an emerging academic field that seeks to embed conceptual and practical ethical approaches and decision-making related to human and more-than-human interactions in policy and everyday practice (Haraway 2018; Tschakert et al. 2021). MSJ is also concerned with how multispecies entities are configured in and through institutions, their related discursive regimes, and the actions these authorise. Consistent with the emerging literature in the broader more-than-human, multispecies and posthuman spaces, MSJ adopts a relational ontology but recognises that notions of justice and the shaping of institutions that interpret and practise justice remain human domains. Consequently, MSJ 'seeks to understand the types of relationships humans ought to cultivate with more-than-human beings so as to produce just outcomes' (Celermajer et al. 2021, p. 120). MSJ is pragmatic in its recognition that not everyone will subscribe to a relational ontology. It therefore recognises the vital importance of engaging varied publics in debate and action in divergent ways to provide the conditions for mutual flourishing (Celermajer et al. 2021). Rather than being solely reliant on individual ethical transformation, MSJ is also concerned with achieving just ends by inscribing the tenets of multispecies justice into fair processes and practices.

In therapeutic garden landscapes, a multispecies justice approach necessitates a shift away from a principal focus on meeting dominant, instrumentalist human needs towards recognition of the need to support mutual more-than-human flourishing. This represents a challenge to the human/nature binary that permeates much work in the therapeutic landscapes literature. An MSJ approach to decision-making and cultivation of care practices in gardens necessitates the cultivation of what Haraway (2007, p. 71) refers to as response-ability, 'a relationship crafted in intra-action through which entities, subjects and objects, come into being'. It also necessitates an openness and sensitivity to Latour's (2004, p. 205) call 'to learn to be affected, meaning "effectuated", moved, put into motion by other entities, humans or nonhumans'. Working within an MSJ framework also requires a decolonising, intersectional and inclusive approach that recognises that individual healing and therapy must not perpetuate inequity, violence, and damage to others. Consequently, food-producing gardens, and the soil–human relations that enable them to flourish, provide an important focus for exploring how particular affordances, garden design, and management could be cultivated to realise multispecies justice.

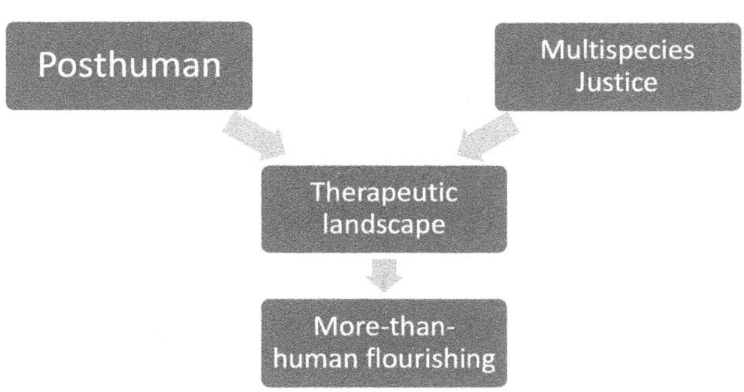

Figure 12.1 Theoretical and conceptual frameworks.

In the case of public food gardens, an MSJ approach does not mean that all entities will thrive or be welcome. Gardens are sites where some desired multispecies entities are encouraged to live, while others regarded as pests and weeds are killed, removed, or controlled. However, MSJ provides a framework for reflecting on and guiding where, how, and why the boundaries and borders are drawn between what is cared for, included, and loved and what is not. Overarchingly, these decisions must move beyond a human-centric focus towards taking seriously multispecies flourishing. Soil, as the foundation of most green spaces and food-producing gardens, directly shapes the multispecies health, wellbeing, and therapeutic impacts of gardens. How soil is cared for in these spaces – and the possible ways it can then contribute to caring for others – plays a role in activating potential therapeutic benefits.

In the next section, I draw on fieldwork in home and community gardens to explore some of the ways soil–human relationships manifest. I then briefly touch on how an MSJ approach to the design, management and care guidelines for therapeutic food-producing gardens could induce more ethical forms of human and multispecies care. In so doing, I explore how fostering particular modes of soil–human relations could be critical to maximising the potential therapeutic benefits of gardens by better supporting human and planetary health and wellbeing to foster mutual flourishing.

Soil–human relations: Caring-with multispecies communities in gardens

Throughout my decade-long ethnographic work in home and community gardens, I have observed, been told about, and experienced numerous manifestations of soil–human relations that gardeners have (sometimes unwillingly) entered into. These are rarely straightforward interactions grounded in a desire to connect with the fundamental life-giving force and life-filled entity that is soil. Instead, they are relations often fraught with tension. This is evident, for example, in a backyard garden where soil infected with the pathogen *fusarium oxysporum* thwarts efforts to move towards organic self-sufficiency. These tensions are also palpable in a community garden site where refugees and new migrants encounter soil with poor water-holding capacity that does not respond well to their traditional cultivation techniques. While the form of friction varies in different soil–human encounters, these relations are always embodied, multisensorial, and responsive. Gardeners engage with the smell, feel, and, sometimes, the taste of soil (Ferguson et al. 2016; Kelley 2020). The soil's vitality – its capacity to act and do things – is consistently acknowledged, irrespective of the ways gardens are tended. While the outcomes of the soil's agency may not always be welcome, it is noticed (Tsing 2015) and it does call gardeners into action. The gardeners I research with consistently practise and refine 'arts of attentiveness' (van Dooren et al. 2016, p. 6), which involves becoming attuned to the multispecies entities present by 'paying attention to others and meaningfully responding'. This attentiveness and responsiveness may play a key role in the capacity of gardens to become therapeutic places.

The intimacy of soil–human relations and its affective force pervades my research. This is exemplified by one organic community gardener's claim to 'know every grain of soil … personally' in the plot he has tended for 18 years. Another, who had put considerable bodily effort into improving his plot's soil by carting all of his food waste on site to compost was surprised by the closeness of the resulting relationship:

> It's great to get down and dig and feel the consistency of it [the soil] now and the worms in it. It's something that really strikes you. I never thought that I would be quite turned on by putting my hand in the soil and looking at it. I guess the older you get different things appeal.

These are intense relations fuelled by 'passionate immersion' (Tsing, 2010; van Dooren et al. 2016). One gardener declared that she 'couldn't leave' her soil, while another noted he 'would have to die' before leaving his. A newly arrived refugee expressed a deep need to garden and 'touch' soil, observing that, 'People in Australia go to gyms or do sports or go to the park or that and we do gardening [laughter]' and, 'Even in the apartment where I'm living, I do not have enough space but I put pots there.… We love to touch the soil. [Laughter]. It is our character. We need to do this'.

Other gardeners express a more distant relationship with soil, emphasising their inability to 'know' it and its mysterious workings. This unknowability does not reflect a less attentive or intense relationship. These gardeners also invest heavily in noticing and responding to the physical, chemical, and multispecies components that make up soil. One new organic community gardener assigned a plot with poor soil observed that, while she was doing her best to add organic materials, 'You can't be too precious about some things'. She was of the belief that 'the land and the soil, sort of, they always come to the fore, to the party'. As we gardened and chatted alongside each other in her newly created beds, she marvelled at what she saw as the almost magical capacity of soil to sustain newly planted seedlings. The recognition of the agency of soil and its multispecies inhabitants, and the consequent acknowledgement of the limits of human control, was keenly felt by these gardeners. As one home gardener and keen composter explained, she was attuned to the fact that there are always 'things moving in the garden'. She worked hard to compost all her food waste so she could 'feed the garden' and support 'the whole micro diversity thing'.

Being attentive in the garden involves feeding and caring for often-unseen others. By understanding the garden as an assemblage of humans and nonhumans, gardeners become aware of the intimacy and inextricability of these more-than-human entanglements. Some emerging relationships may not be positive or desired by the gardeners, prompting them to act in ways that can cause harm to some nonhumans; for example, they might remove some plants, bait some pests, or smother some soil with weed mat. But by noticing and responding to the 'things moving in the garden', gardeners demonstrate awareness of their shared reliance on – and vulnerabilities to – the more-than-human relations that give gardens life. In my work, I consistently find gardeners adjusting their tastes, preferences, and gardening habits to work with what can and will grow in their plots. The benefits – or joys – for these gardeners are not solely generated by filling pre-determined instrumental human goals in these spaces. Instead, they derive from practising the 'arts of attentiveness' (van Dooren et al. 2016) which, through its need for active, ongoing, and place-based noticing and responsiveness to the dynamic and sometimes unexpected conditions and inhabitants of the garden, can generate 'emplaced flow' (Pitt 2014). While 'emplaced flow' can yield therapeutic benefits for humans, its place-based focus can also sharpen the skills and capacities of these gardeners to adjust and adapt to enter into mutual relations of care with somewhat uncharismatic and difficult-to-love others.

Recognition of multispecies agency, coupled with acknowledgement and acceptance of the limits of human control in gardens, could be seen to arise from a position of relative privilege. Most of those involved in my research have rarely been reliant on their garden produce to meet the majority of their nutritional needs or provide an income. Thus, they have been more able to accept lost or failed crops, as well as more willing to provide care for plants and soil that do not meet their immediate needs. But even those motivated to garden to reduce financial hardship have demonstrated a keen awareness of the limits of human control in the garden and the need to work with its multispecies inhabitants. Building soil health is always the principal focus of these activities. While the intended outcomes of such human efforts include increased production or growth of foods with higher nutritional value to benefit growers, this goal is rarely mentioned as the main driver of these activities. One gardener with a vegetable plot in a disused clay tennis court described his 'biodynamic composting' efforts as being motivated by a desire to provide better care for plants. His aim was to 'garden in a way that naturally strengthens the plant immune system'. This focus on the health of the soil to enhance its capacity to care for plants typifies how gardeners in my fieldwork spoke of their actions. Many also connected their modes of garden care to broader ecosystem impacts, with gardeners observing and commenting on relationships with surrounding insect, bird, reptile, and invertebrate life (even when this was not desired). This was exemplified by one community gardener (motivated to garden for financial and health reasons) who linked her zeal for organic practices to life beyond her family's plot.

> The kids have found in there so many more bugs and insects ... which they've never seen [before]. We now have frogs in the garden, we have maybe two blue tongue lizards ... My husband and I were in two different spots and we both yelled out 'I caught a frog' and we both did it at the same time. It was the funniest thing ... We're only in the third year [of the community garden] but it's increasing in its natural habitat being there and magpies catching baby mice and things like this, you would never [normally] see.

While improved soil care in food producing gardens can induce physical health and therapeutic benefits for humans, this care can often induce benefits for multifarious nonhumans and is not *only* given for instrumentalist reasons. In fact, gardeners commonly observe that the time and effort they invest in nurturing soil and their plants are not repaid in terms of yield. Some have told me it would be much easier and cheaper to buy-in all their own food. While Pitt's (2018) typology of care in gardens makes a significant contribution to knowledge in the area by demonstrating that humans wield the dominant power in the garden and that not all care is equal (see also Turner and Tam (2022) for a typology of care in food rescue practices), my research does not concur with her finding that instrumentalist human needs and desires are the primary driver of the modes of care offered. Instead, I regularly encounter human gardeners who feel obligated to care for gardens in ways that best meet the needs of their multispecies inhabitants, rather than just those of the gardeners. While humans do decide what is cared for in gardens and how this is done (Pitt 2018), gardeners can find themselves caring for unwanted others simply because they grow well in the available conditions. To illustrate how some of these non-instrumentalist caring practices play out, I turn briefly to examples from two gardeners growing in vastly different climates, soil types and conditions.

Adjusting to multispecies needs in the garden

During a walk and water interview on a summer's evening in the midst of a drought, a community gardener in Canberra, Australia's inland capital city, showed me her much-hated, but highly productive, patch of Jerusalem artichokes (or 'fartichokes', as she called them). She has never eaten them and refuses to try them. The plants were magnificent and vast in scale. The gardener quickly pointed out that she had not planted them. As we spoke, she assiduously watered them while continuing to complain about how well they grow. She guessed that they had travelled to their current location as a thin sliver in compost-rich soil she moved to her plot from another section of the garden. The gardener described the capacity for the tubers to spread through small nodes knocked off in the soil as 'being worse than cancer'. Choosing to water these plants was remarkable because, at the time of the research, Canberra was locked in Stage 3 water restrictions. This meant that plants could only be watered by hand (irrigation systems were banned) every second day within a two-hour period in the early morning or late afternoon to evening. Resigned to the fact that she was unlikely to win any battle to eradicate the plants, this gardener chose to care for them and the soil in which they grew, seeing some value in their use as a 'wind break' to shelter other crops. As our conversation continues, it emerged that she also diligently harvested the tubers and worked hard to distribute them to people who would eat them (not an easy task, she tells me) to avoid them going to waste. The most hated crop in the plot was calling this gardener into action and receiving considerable care and attention as a result.

Further north in Sydney, Australia's most populous city, a home gardener and member of Sharewaste (a community composting initiative) spoke with me about her extensive efforts to improve her 'terrible' soil that was sandy, acidic, and high in zinc. While she constantly added compost, she was never sure what the results would be or which plants would survive. This gardener felt she had little control over how her garden looked and how the space could be used. Instead, she had to work around whatever would grow. This necessitated ongoing experimentation: 'I've kind of stuck things in thinking, "Will it grow? Will it grow?", and it grows. And then I can't move it because it's grown really well'. Both plant success and failure were met with a mixture of frustration and resignation. While this gardener continued to hope for a highly ordered and productive space, she found herself constantly having to adjust to the multispecies vitalities and particularities of the site. Her longed-for order stood in stark contrast to the uncontrollable realities of the everyday materials she encountered. These tensions also permeated her composting regime. She had strict rules about what went into both the compost heap and her worm farm. While the worm farm was primarily the recipient of coffee grounds, the gardener worked hard to make sure this was supplemented with other goodies so they had a 'balanced diet' (but she did note that they would not eat everything she offered). As a community compost host, she had developed a series of rules and instructions for food waste donors that included the need to process scraps into small pieces, a ban on corn cobs ('Put them in your green waste bin'), and the exclusion of fruit and vegetable stickers. These rules were developed to support rapid decomposition to feed the garden and to prevent plastic contaminating the bodies of the compost's multispecies labourers and of the soil it was then added to. This gardener was vigilant about sorting every bucket of food waste that was dropped off, often chopping it into smaller pieces before adding to the compost. She spoke about the labour involved in trying to keep up with the voracious appetite of the compost heap, noting that she was keen to encourage food waste donations to keep up with the heap's demands.

While these gardeners may derive therapeutic benefits from the practices of garden care they engage in, these efforts are not solely motivated by fulfilling their own needs. Indeed, this care sometimes enables undesirable entities to flourish, causing significant frustration for the gardeners and requiring them to invest time and labour (emotional and physical) in adjusting their plans to learn to live together with the nonhuman companions that co-constitute their gardens. In the examples I have offered, the intimacy of these more-than-human encounters shapes the relations of care that emerge. However, in her work in community gardens in the United Kingdom, Pitt (2018) draws attention to the limits of viewing proximity as a means of encouraging gardeners to develop ethical relations. Instead, she emphasises the need to focus on the quality of the relations, writing that, 'Community gardeners relate to nonhumans in complex ways, and do not unequivocally care as a result of close encounters' (p. 263). Intimate, regular human/nonhuman interactions certainly cannot guarantee that ethical relations will emerge, just as is the case in human-to-human interactions. However, proximity can play a role in generating the affordances needed to recognise and respond to entangled more-than-human realities and to contribute to the realisation of multispecies justice.

Here, it is important to point out that closeness, attentiveness, and responsiveness do not necessarily lead to forms of connection or attachment. Instead, these relations may be marked by desires for detachment, where the nonhumans remain unknown, unknowable, and difficult to relate to. However, ethical more-than-human relations and multispecies justice in gardens cannot be reliant on humans forming attachments to loved nonhumans or vice versa. While it is inevitably easier for humans to care for entities considered charismatic, desirable, and unthreatening – think frogs over snakes, ladybeetles (ladybirds) over slugs, lettuce over stinging nettle – our shared existence relies on the presence of the ugly, undesirable, and often unseen nonhumans which form integral parts of our ecosystems. Writing about the 'awkward' relationships London gardeners forge with slugs, Ginn (2014, p. 539) notes that 'the inevitability of detachment, not the fact of our being related, is grounds for ethics'. Abrahamsson and Bertoni's (2014) work on worm farms does not identify detachment or connection as the ethical basis for human/nonhuman relations. Instead, they observe that, 'Vermicomposting is about doing togetherness' (Abrahamsson and Bertoni 2014, p. 126) and that 'you may not know, but rather become attuned to your worms' (p. 134). Attunement to unknowable nonhumans and recognition of our inextricable entanglements (regardless of experiences of attachment or detachment) may play a key role in fostering ethical more-than-human relations (Abrahamsson and Bertoni 2014; Candea 2010; Ginn 2014). The development of the skills and capacities to navigate this uncomfortable and uncharted terrain may be nurtured more frequently and broadly in gardens than has hitherto been recognised in the academic literature.

Gardeners' efforts to control 'natural' conditions to derive instrumental benefits for themselves may be common, but the practices of care they provide are shaped by sensitivities to the limits of this control and the need to adapt to the multispecies entities enlivening their plots. Attuning to what is *moving* in the garden, a *letting go* of ultimate control, and a focus on *working with* nonhumans dominates the practices of the gardeners I research with. This openness to the affective force of nonhuman encounters and the nurturing of an ethic of response-ability in the garden is not reliant on individual ethical transformation alone. It can be encouraged and enhanced through a suite of social, material, and affective enabling resources (Duff 2012) and conditions that draw on a multispecies justice framework. This is particularly true in community gardens or other public therapeutic landscapes where there is scope to guide how these sites are designed, managed, and cultivated.

Designing therapeutic landscapes and gardening approaches for multispecies justice

Therapeutic benefits – or the creation of 'enabling places' (Duff 2011) – are not reliant on the presence of harmonious or unproblematic human/nonhuman relations. Community gardens, and gardens more generally, are sites where not all experiences are joyful or agreeable. Yet, such landscapes have been overwhelmingly viewed as capable of inducing therapeutic benefits for human participants through their capacity to enable people to escape from everyday pressures, concerns, and routines (Pitt 2014). While Pitt (2014) questions whether therapeutic benefits will be maintained if gardening becomes routinised, an ongoing focus on cultivating the skills to engage in the 'arts of attentiveness' (van Dooren et al. 2016) may surmount this issue. The process of noticing, attuning, and responding to changes in conditions and multispecies inhabitants in gardens involves constant interactions with novelty that often thwart a gardener's best laid plans. The resulting more-than-human relationships that form might have benefits beyond what has already been recognised as humans finding reassurance in the 'resilience' of nature (Atkinson 2020; Marsh et al. 2021) or the ability to grow a sense of 'emplaced flow' (Pitt 2014) when engaged in garden care. Instead, attunement to shared human/nonhuman vulnerabilities through gardening could encourage deeper understanding and sensing of our inescapable entanglement with 'nature' and imbrication in mutual relations of care (Puig de la Bellacasa 2017).

Recognition of more-than-human entanglements not only involves identifying and cultivating comfortable, cosy relations with loved multispecies entities. It also involves learning to live together with uncharismatic others in ways that recognise the complex assemblages of humans and nonhumans that make everyday life possible. Such recognition and experiences will not occur for all gardeners in all spaces, nor will it lead to all multispecies inhabitants being welcome and a universal displacement of the privileging of human needs and benefits in gardening (Pitt 2018). However, by bringing the therapeutic landscape literature and practices into conversation with posthuman and multispecies justice work, we open up the possibilities of better identifying and building-in the social, material, and affective enabling resources (Duff 2012) and conditions that could stretch and strengthen these experiences in places designed to induce therapeutic benefits. The lively multispecies entity that is soil provides a focus for building these capacities and nurturing relationships of care where human power and privilege can be unlearnt and attunement to 'togetherness through difference' can be nurtured to generate benefits for human and multispecies inhabitants (Turner 2019).

In the remainder of this chapter, I briefly gesture towards some practical steps that could be taken to embed a multispecies justice approach to soil grounded in the 'arts of attentiveness' in therapeutic urban spaces. To do this, I transpose insights from the work of Celermajer and O'Brien (in Celermajer et al. 2020) on soil in agricultural settings to urban community, school, hospital, and other public garden spaces. Equipping gardeners with the knowledge and skills to identify, understand, work with and workshop with others the multispecies, chemical, and physical aspects of soil is a critical starting point. Through my earlier discussion in this chapter, we saw that the gardeners in my research have developed an awareness and appreciation of the intensity of their relationship with soil and the realities of mutual soil–human care. While most of these gardeners developed these soil–human relations through individual trial and error, experimentation and bit-by-bit tinkering (Turner 2019), the modes of relating that underpin such caring practices can be actively fostered in therapeutic landscapes through education, design, resourcing, and support. Learning how to identify what makes soil healthy, how to assess soil needs and what cultivation styles and crops will be best suited to the soil profiles of particular sites is an important starting point. Demonstration and guided experiential learning of the links among soil, plant, and ecosystem health should be offered to those accessing these sites. This can support growth in understanding soil as 'dynamic ecologies in the becoming of which human beings are implicated, with whom they are shaped, and on which they depend' (Krzywoszynska and Marchesi 2020, p. 194). This process of engagement and education can be fun, creative, and experimental to maximise participation and potential therapeutic benefits. Examples include holding soil-based creative art workshops (from soil painting to clay modelling) or participating in citizen science experiments such as the Soil Your Undies challenge, where the decomposition rate of a pair of underpants buried in the ground is used to estimate biological activity and, thus, soil health.

Garden design, access to resources, materials and the actual gardening practices that are encouraged can all impact on how soil is cared for. At the outset, the design of sites and garden beds that enable low- or no-till cultivation methods to limit disturbance to soil structure, beneficial fungal and microbial networks, and which minimise carbon release is critical. The inclusion of accessible composting infrastructure, appropriate materials and support for ongoing compost management should be built into the design of garden sites to allow continuous feeding of the soil. This might require the use of multiple styles of composting (from enclosed rotating bins to open heaps), plans for how materials that aid decomposition will be supplied (garden waste alone will be slow to breakdown), and the provision of compost training to enable participants to learn about and understand the process. Building in opportunities for gardeners to access the required resources and to develop the skills needed to care for compost is a low-cost and effective way to improve soil care, reduce waste and support closed-loop production. Composting is also key to sustaining the gardening practices that best nurture mutual more-than-human caring relations. Organic and permaculture practices that avoid the use of toxic chemicals, encourage biodiversity, and contribute to effective ecosystem functioning are grounded in practices of caring well for soil. While these approaches to gardening should be supported, they must be inclusive of diverse, multicultural and First Nations' gardening styles that are aligned with these principles. This requires ongoing, place-specific, and culturally appropriate engagement with local communities and the traditional custodians of the land.

Getting arty with soil, prioritising composting, and advocating permaculture principles might improve soil care but these practices can be aesthetically confronting for some. Given that sensory appeal is a significant concern within the garden-focused therapeutic landscape and wellbeing literature, we can identify a potential misfit between what is assumed to induce therapeutic benefits for humans and what constitutes good multispecies soil care in these sites. As discussed earlier in the chapter, gardens are sites where socio-cultural backgrounds and the privileging of some lives over others – and some ways of life over others – can come to the fore. Dominant aesthetic sensibilities and risk-averse approaches to public garden design persist in urban locales. This tends to manifest in a focus on maintaining lines of site and encouraging neat, orderly garden practices that eliminate weeds and involve the careful training and control of plant and other multispecies entities. In Singapore, for example, Montefrio et al. (2020) outline two overarching aesthetic sensibilities that dominate approaches to community gardens: the need to be 'orderly' and to have 'beauty'. 'Orderly' gardens must maintain their 'cleanliness', 'tidiness' and 'neatness', often requiring 'a uniform spatial pattern of growing plants, such as planting and trimming in straight lines'; the presence of colour is the main determinant of 'beauty' (Montefrio et al. 2020, p. 1466). Gardens where some areas of the soil are 'resting' (rather than being under cultivation) have been found to fail to meet these aesthetic objectives (Montefrio et al. 2020).

A preference for order and beauty can be found in the ways most public and therapeutic garden spaces are designed and managed around the globe. These views privilege particular renderings of what a garden and city should look like to induce therapeutic benefits for humans. In Melbourne, gardeners with plots in a community garden at the Collingwood Children's Farm, which was demolished in early 2022, have suggested it was destroyed due to perceptions that it was aesthetically unpleasing. The garden was marked by a patchwork of plots with extensive use of recycled materials and diverse cultivation styles, with soil built up, by some, over decades. The garden lacked an overarching design, and neat, orderly rows were few and far between. Officially, the site was removed due to health and safety concerns, including the use of uncapped star pickets and the potential dangers presented by snakes lurking underneath recycled materials. While the Children's Farm has committed to rebuilding the site, a different model of management and access to plots has been proposed (a move from individually cultivated plots to communal ones) to promote greater access. A sleek, uniform, accessible design is being workshopped. Public gardens do need to be accessible for all, but how the new design will address issues of multispecies justice – starting with soil – remains to be seen. Previous gardeners on the site engaged in significant activism and protests to try and stop the removal of their gardens. As the bulldozers loomed, one lamented the loss of the soil she had invested in building, stating that, 'You actually literally invest in the soil, you regenerate the soil, it's a relationship to place' (Dexter 2022).

Conclusion

There is a need for therapeutic landscape literature and the design of therapeutic landscape sites to engage more deeply with posthuman and multispecies approaches. Doing so would improve the capacity for therapeutic landscape work to engage more ethically with the field's recognition that health and wellbeing benefits emerge through interactions between humans and nonhumans. Stretching this to attend to multispecies justice in the design, management, and activation of therapeutic landscapes would convey the message that instrumentalist human benefits should not be the sole focus of these sites. Adequately addressing multispecies health and wellbeing (including of uncharismatic, unknowable, and difficult-to-love entities) and understanding the inescapable entanglement of our more-than-human relations should be foundational to the design, management, and everyday practices supported in gardens intended to induce therapeutic benefits. As Gorman (2017b, p. 329) observes, 'There is a need to examine the roles of more-than-human elements and actants in creating the relations which lead to space "becoming therapeutic"'. Doing so requires reflection on socio-cultural and historically informed aesthetic and sensory assumptions about what makes a space therapeutic. A focus on the grounded, emplaced actions of gardeners and support for approaches to gardening that nurture multispecies justice is critical. Soil, as the foundation of gardens, provides a fertile focus for efforts that set out to maximise mutual more-than-human flourishing.

References

Abrahamsson, S and Bertoni, F 2014, 'Compost politics: Experimenting with togetherness in vermicomposting', *Environmental Humanities*, vol. 4, no. 1, pp. 125–148. https://doi.org/10.1215/22011919-3614962

Abramovic, J, Turner, B and Hope, C 2019, 'Entangled recovery: refugee encounters in community gardens', *Local Environment*, vol. 24, no. 8, pp. 696–711. https://doi.org/10.1080/13549839.2019.1637832

Aelion, CM, Davis, HT, Lawson, AB, Cai, B and McDermott, S 2013, 'Associations between soil lead concentrations and populations by race/ethnicity and income-to-poverty ratio in urban and rural areas', *Environmental Geochemistry and Health*, vol. 35, no. 1, pp. 1–12. https://doi.org/10.1007/s10653-012-9472-0

Aptekar, S 2015, 'Visions of public space: Reproducing and resisting social hierarchies in a community garden', *Sociological Forum*, vol. 30, no. 1, pp. 209–227. https://doi.org/10.1111/socf.12152

Atkinson, J 2020, 'The impulse to garden in hard times has deep roots' (*The Conversation*, 1 May 2020). https://theconversation.com/the-impulse-to-garden-in-hard-times-has-deep-roots-137223

Bawaka, C, Wright, S, Suchet-Pearson, S, Lloyd, K, Burarrwanga, L, Ganambarr, R, Ganambarr-Stubbs, M, Ganambarr, B and Maymuru, D 2015, 'Working with and learning from Country: Decentring human author-ity', *Cultural Geographies*, vol. 22, no. 2, pp. 269–283. https://doi.org/10.1177/1474474014539248

Bell, SL, Foley, R, Houghton, F, Maddrell, A and Williams, AM 2018, 'From therapeutic landscapes to healthy spaces, places and practices: A scoping review', *Social Science & Medicine*, vol. 196, pp. 123–130. https://doi.org/10.1016/j.socscimed.2017.11.035

Biglin, J 2020, 'Embodied and sensory experiences of therapeutic space: Refugee place-making within an urban allotment', *Health & Place*, vol. 62, article 102309. https://doi.org/10.1016/j.healthplace.2020.102309

Buller, H 2013, 'Animal geographies I', *Progress in Human Geography*, vol. 38, no. 2, pp. 308–318. https://doi.org/10.1177/0309132513479295

Candea, M 2010, '"I fell in love with Carlos the meerkat": Engagement and detachment in human-animal relations', *American Ethnologist*, vol. 37, no. 2, pp. 241–258. https://doi.org/10.1111/j.1548-1425.2010.01253.x

Celermajer, D, Chatterjee, S, Cochrane, A, Fishel, S, Neimanis, A, O'Brien, A, Reid, S, Srinivasan, K, Schlosberg, D and Waldow, A 2020, 'Justice through a multispecies lens', *Contemporary Political Theory*, vol. 19, no. 3, pp. 475–512. https://doi.org/10.1057/s41296-020-00386-5

Celermajer, D, Schlosberg, D, Rickards, L, Stewart-Harawira, M, Thaler, M, Tschakert, P, Verlie, B and Winter, C 2021, 'Multispecies justice: Theories, challenges, and a research agenda for environmental politics', *Environmental Politics*, vol. 30, no. 1–2, pp. 119–140. https://doi.org/10.1080/09644016.2020.1827608

Cheesbrough, AE, Garvin, T and Nykiforuk, CI 2019, 'Everyday wild: Urban natural areas, health, and well-being', *Health & Place*, vol. 56, pp. 43–52. https://doi.org/10.1016/j.healthplace.2019.01.005

Csikszentmihalyi, M 2002, *Flow: The classic work on how to achieve happiness*, Rider, London.

Dexter, R 2022, 'Stand in front of those plots: Community gardeners rally before bulldozers roll in' (*The Age*, 12 February 2022).

Duff, C 2011, 'Networks, resources and agencies: On the character and production of enabling places', *Health & Place*, vol. 17, no. 1, pp. 149–156. https://doi.org/10.1016/j.healthplace.2010.09.012

Duff, C 2012, 'Exploring the role of "enabling places" in promoting recovery from mental illness: A qualitative test of a relational model', *Health & Place*, vol. 18, no. 6, pp. 1388–1395. https://doi.org/10.1016/j.healthplace.2012.07.003

Ferguson, H and Northern Rivers Landed Histories Research Group 2016, 'More than something to hold the plants up: Soil as a non-human ally in the struggle for food justice', *Local Environment*, vol. 21, no. 8, pp. 956–968. https://doi.org/10.1080/13549839.2015.1050659

Gesler, W 1996, 'Lourdes: Healing in a place of pilgrimage', *Health & Place*, vol. 2, no. 2, pp. 95–105. https://doi.org/10.1016/1353-8292(96)00004-4

Gesler, WM 1992, 'Therapeutic landscapes: Medical issues in light of the new cultural geography', *Social Science & Medicine*, vol. 34, no. 7, pp. 735–746. https://doi.org/10.1016/0277-9536(92)90360-3

Ginn, F 2014, 'Sticky lives: Slugs, detachment and more-than-human ethics in the garden', *Transactions of the Institute of British Geographers*, vol. 39, no. 4, pp. 532–544. https://doi.org/10.1111/tran.12043

Gorman, R 2017a, 'Smelling therapeutic landscapes: Embodied encounters within spaces of care farming', *Health & Place*, vol. 47, pp. 22–28. https://doi.org/10.1016/j.healthplace.2017.06.005

Gorman, R 2017b, 'Therapeutic landscapes and non-human animals: The roles and contested positions of animals within care farming assemblages', *Social & Cultural Geography*, vol. 18, no. 3, pp. 315–335. https://doi.org/10.1080/14649365.2016.1180424

Gorman, R 2019, 'What's in it for the animals? Symbiotically considering "therapeutic" human-animal relations within spaces and practices of care farming', *Medical Humanities*, vol. 45, no. 3, pp. 313–325. https://doi.org/10.1136/medhum-2018-011627

Green, GP and RG Phillips (eds) 2014, *Local food and community development*, Routledge, New York.

Guthman, J 2008, 'Bringing good food to others: Investigating the subjects of alternative food practice', *Cultural Geographies*, vol. 15, no. 4, pp. 431–447.

Haraway, D 2007, *When species meet*, University of Minnesota Press, Minneapolis and London.

Haraway, D 2018, 'Staying with the trouble for multispecies environmental justice', *Dialogues in Human Geography*, vol. 8, no. 1, pp. 102–105. https://doi.org/10.1177/2043820617739208

Head, L, Atchison, J and Phillips, C 2015, 'The distinctive capacities of plants: Re-thinking difference via invasive species', *Transactions of the Institute of British Geographers*, vol. 40, no. 3, pp. 399–413. https://doi.org/10.1111/tran.12077

Helmreich, S 2009, *Alien ocean*, University of California Press, Los Angeles.

Jiang, S 2014, 'Therapeutic landscapes and healing gardens: A review of Chinese literature in relation to the studies in Western countries', *Frontiers of Architectural Research*, vol. 3, no. 2, pp. 141–153. https://doi.org/10.1016/j.foar.2013.12.002

Kearns, R and Milligan, C 2020, 'Placing therapeutic landscape as theoretical development in Health & Place', *Health & Place*, vol. 61, article 102224. https://doi.org/10.1016/j.healthplace.2019.102224

Kelley, L 2020, 'Geophagiac: Art, food, dirt' in JF Salazar, C Granjou, M Kearnes, A Krzywoszynska and M Tironi (eds), *Thinking with soils: Material politics and social theory*, Bloomsbury, London, pp. 191–209.

Kingsley, J, Foenander, E and Bailey, A 2020, '"It's about community": Exploring social capital in community gardens across Melbourne, Australia', *Urban Forestry & Urban Greening*, vol. 49, article 126640. https://doi.org/10.1016/j.ufug.2020.126640

Krzywoszynska, A and Marchesi, G 2020, 'Toward a relational materiality of soils: Introduction', *Environmental Humanities*, vol. 12, no. 1, pp. 190–204. https://doi.org/10.1215/22011919-8142297

Lal, R 2020, 'Managing soils for resolving the conflict between agriculture and nature: The hard talk', *European Journal of Soil Science*, vol. 71, no. 1, pp. 1–9. https://doi.org/10.1111/ejss.12857

Lal, R, 2021, 'United Nations food systems summit: What is the role of soil health in putting the Sustainable Development Goals on track?', *Journal of Soil and Water Conservation*, vol. 76, no. 6, pp. 105A–107A. https://doi.org/10.2489/jswc.2021.1013A

Latour, B 2004, 'How to talk about the body? The normative dimension of science studies', *Body & Society*, vol. 10, no. 2–3, pp. 205–229. https://doi.org/10.1177/1357034X04042943

Liddicoat, C, Sydnor, H, Cando-Dumancela, C, Dresken, R, Liu, J, Gellie, NJ, Mills, JG, Young, JM, Weyrich, LS, Hutchinson, MR, Weinstein, P and Breed, MF 2020, 'Naturally-diverse airborne environmental microbial exposures modulate the gut microbiome and may provide anxiolytic benefits in mice', *Science of the Total Environment*, vol. 701, article 134684. https://doi.org/10.1016/j.scitotenv.2019.134684

Lorimer, J 2020, *The probiotic planet: Using life to manage life*, vol. 59, University of Minnesota Press, Minneapolis.

Marder, M 2013, *Plant-thinking: A philosophy of vegetal life*, Columbia University Press, New York.

Marsh, P, Diekmann, LO, Egerer, M, Lin, B, Ossola, A and Kingsley, J 2021, 'Where birds felt louder: The garden as a refuge during COVID-19', *Wellbeing, Space and Society*, vol. 2, article 100055. https://doi.org/10.1016/j.wss.2021.100055

McClintock, N 2015, 'A critical physical geography of urban soil contamination', *Geoforum*, vol. 65, pp. 69–85. https://doi.org/10.1016/j.geoforum.2015.07.010

McIntosh, J, Marques, B and Mwipiko, R 2021, 'Therapeutic landscapes and Indigenous culture: Māori health models in Aotearoa/New Zealand' in JC Spee, AJ McMurray and M McMillan (eds), *Clan and tribal perspectives on social, economic and environmental sustainability: Indigenous stories from around the globe*, Emerald Publishing Limited, Bingley, pp. 143–158.

Milligan, C and Bingley, A 2007, 'Restorative places or scary spaces? The impact of woodland on the mental well-being of young adults', *Health & Place*, vol. 13, no. 4, pp. 799–811. https://doi.org/10.1016/j.healthplace.2007.01.005

Montefrio, MJF, Lee, XR and Lim, E 2020, 'Aesthetic politics and community gardens in Singapore', *Urban Geography*, vol. 42, no. 10, pp. 1459–1479. https://doi.org/10.1080/02723638.2020.1788304

Moulin-Doos, C 2014, 'Intercultural gardens: The use of space by migrants and the practice of respect', *Journal of Urban Affairs*, vol. 36, no. 2, pp. 197–206. https://doi.org/10.1111/juaf.12027

Myers, N 2017, 'From the anthropocene to the planthroposcene: Designing gardens for plant/ people involution', *History and Anthropology*, vol. 28, no. 3, pp. 297–301. https://doi.org/10.1080/02757206.2017.1289934

Nettle, C 2016, *Community gardening as social action*, Routledge, London.

Paxson, H 2008, 'Post-Pasteurian cultures: The microbiopolitics of raw-milk cheese in the United States', *Cultural Anthropology*, vol. 23, no. 1, pp. 15–47. https://doi.org/10.1111/j.1548-1360.2008.00002.x

Pitt, H 2014, 'Therapeutic experiences of community gardens: Putting flow in its place', *Health & Place*, vol. 27, pp. 84–91. https://doi.org/10.1016/j.healthplace.2014.02.006

Pitt, H 2018, 'Questioning care cultivated through connecting with more-than-human communities', *Social & Cultural Geography*, vol. 19, no. 2, pp. 253–274. https://doi.org/10.1080/14649365.2016.1275753

Plane, J and Klodawsky, F 2013, 'Neighbourhood amenities and health: Examining the significance of a local park', *Social Science & Medicine*, vol. 99, pp. 1–8. https://doi.org/10.1016/j.socscimed.2013.10.008

Puig de la Bellacasa, M 2010, 'Ethical doings in naturecultures, *Ethics, Place and Environment*, vol. 13, no. 2, pp. 151–169.

Puig de la Bellacasa, M 2017, *Matters of care: Speculative ethics in more than human worlds*, University of Minnesota Press, Minneapolis.

Rouillon, M, Harvey, PJ, Kristensen, LJ, George, SG and Taylor, MP 2017, 'VegeSafe: A community science program measuring soil-metal contamination, evaluating risk and providing advice for safe gardening', *Environmental Pollution*, vol. 222, pp. 557–566. https://doi.org/10.1016/j.envpol.2016.11.024

Salazar, JF, Granjou, C, Kearnes, M, Krzywoszynska, A and Tironi, M (eds) 2020, *Thinking with soils: Material politics and social theory*, Bloomsbury, London.

Sampson, R and Gifford, SM 2010, 'Place-making, settlement and well-being: The therapeutic landscapes of recently arrived youth with refugee backgrounds', *Health & Place*, vol. 16, no. 1, pp. 116–131. https://doi.org/10.1016/j.healthplace.2009.09.004

Sanchez, EL and Liamputtong, P 2017, 'Community gardening and health-related benefits for a rural Victorian town', *Leisure Studies*, vol. 36, no. 2, pp. 269–281. https://doi.org/10.1080/02614367.2016.1250805

Somerville, W, Turner, B and Markulin, K 2021, 'Decolonizing research: Collaborating with Indigenist, posthuman, and new materialist perspectives' in P Liamputtong (ed), *Handbook of social inclusion: Research and practices in health and social sciences*, Springer, Cham, pp. 1–18.

Strunk, C and Richardson, M 2019, 'Cultivating belonging: Refugees, urban gardens, and placemaking in the Midwest, USA', *Social & Cultural Geography*, vol. 20, no. 6, pp. 826–848. https://doi.org/10.1080/14649365.2017.1386323

Tschakert, P, Schlosberg, D, Celermajer, D, Rickards, L, Winter, C, Thaler, M, Stewart-Harawira, M and Verlie, B 2021, 'Multispecies justice: Climate-just futures with, for and beyond humans', *Wiley Interdisciplinary Reviews: Climate Change*, vol. 12, no. 2, e699. https://doi.org/10.1002/wcc.699

Tsing, A 2010, 'Arts of inclusion, or how to love a mushroom', *Manoa*, vol. 22, no. 2, pp. 191–203.

Tsing, AL 2015. *The mushroom at the end of the world.* Princeton University Press, Princeton.

Turner, B 2019, *Taste, waste and the new materiality of food*, Routledge, London.

Turner, B, Abramovic, J and Hope, C 2021, 'More-than-human contributions to place-based social inclusion in community gardens' in P Liamputtong (ed), *Handbook of social inclusion: Research and practices in health and social sciences*, Springer, Cham, pp. 1–19.

Turner, B and Tam, D 2022, 'Moving from risky to response-able care', *Antipode*, vol. 54, no. 3, pp. 914–933. https://doi.org/10.1111/anti.12804

UN-FAO, ITPS, GSBI, SCBD, and EC 2020, *State of knowledge of soil biodiversity: Status, challenges and potentialities. Report 2020*, FAO, Rome. https://doi.org/10.4060/cb1928en

Van Dooren, T, Kirksey, E and Münster, U 2016, 'Multispecies studies: Cultivating arts of attentiveness', *Environmental Humanities*, vol. 8, no. 1, pp. 1–23. https://doi.org/10.1215/22011919-3527695

White, MM 2011, 'Sisters of the soil: Urban gardening as resistance in Detroit', *Race/Ethnicity: Multidisciplinary Global Contexts*, vol. 5, no. 1, pp. 13–28.

13 Tending the wilds inside

Cultivating healing at the unruly edges of the garden

Alice McSherry and Robin Kearns

Introduction

In the settler-colonial context of Aotearoa New Zealand, within which we both find ourselves planted, gardens tend to be sites that are spatially bounded and micro-worlds of imposed order. As such, they are expressions of human control over a more-than-human world. Clear delineations of which species belong (and where) and which do not generally prevail. In fact, much of the act of gardening is not just the cultivation of the intended species, but also the keeping out of incursions of those deemed unruly, coded as 'weeds'. For gardeners, pride in one's garden often derives from a pair of dual successes: first, the growing; and second, the banishing (in which botanical separatism is achieved through acts such as pulling, digging, and poisoning). There are, however, many other forms of therapeutic gardening that are practised and enabled when more relational ontological perspectives are tended and given space to flourish.

In this chapter, we explore the therapeutic possibilities of embracing botanical unruliness in the garden. While we acknowledge the potential for wellbeing yields that might lead the conventionally ordered garden to serve as a therapeutic space, these yields generally arise from the physical exertion of 'weeding' and the maintenance of gardened order. Alternatively, they could flow from emotional collateral in the form of the pride that arises from achieving order and witnessing curated beauty within a garden plot. However, we seek to take a different tack, and head to the edge of orthodox gardening practices. We seek to disrupt the cultivated edges of elegance commonly sought in everything from lawns to public gardens (Wilson 2019). Instead, we explore the fruits of embracing a feral disposition by, quite literally, thinking-with-weeds. Through this positioning, we seek to engage with the therapeutic possibility of relating across a broader sweep of species to acknowledge that all plants have an inherent place-in-the-world.

Recent years have seen an increasing interest in the role and agency of the more-than-human world in the cultivation of belonging in place (see, e.g., Bennett 2010; Puig de la Bellacasa 2017). The call to think beyond the human

DOI: 10.4324/9781003355731-17

invites creative, and sometimes contentious, forays into thinking-with (rather than just about) the millions of other-than-human species comprising the 'garden' – whether conceived of at the scale of a raised garden bed, a sprawling field, or perhaps even the berms (verges) of public spaces in an urban environment. In this spirit of inclusivity, we choose to frame 'the garden' as a close-to-home ecosystem of which we, as humans, are a part, albeit as frequently in-part curators. When thought of in a network across space, however, the 'garden' can also be perceived as a microcosm within a larger 'planetary garden', where the whole Earth is seen as a dynamic ecosystem of constant relationality and reciprocity between agentic species (Clément 2011). We do this not to dismiss or disrupt the conventional notion of the garden, but rather to extend its reach as a space of concern and care, much as a 'park' is not only an urban public green space, but also – potentially – a wilderness area.

We contend that, from this vantage point, the garden can be transformed into a site of ecological co-becoming. This emergence occurs when material relations between human and more-than-human species are embodied (that is, through the planting, harvesting, and sharing of crops) and when immaterial relations of healing and belonging can be enacted and given space to flourish. From this perspective, the garden comprises a co-constitutive landscape of tending and mending and shifts from being a site of extraction to one of an interspecies relationship. To do this, we adopt a vegetal phenomenology. In this chapter, we combine eco-philosophical reflections relating to relationality, conviviality, and biophilia in the garden with elements of autobiographic reflection from the two authors. The first (Alice) is a practising community herbalist, amateur permaculturalist[1] and cultural geographer; the second (Robin) is a health geographer and home gardener. The result will doubtlessly defy the orderliness and predictability of the typical garden but, in its variegated and unpredictable character, somewhat mirrors the diverse character of the garden-world we seek to consider.

Positioning in literature: A stake in the garden

The therapeutic landscape literature that forms a bedrock for this edited book offers a scholarly starting point for our chapter and for thinking about gardens with specific reference to human–plant relations. At the root of the therapeutic landscape concept lies two recognitions: first, a convergence of material, social, and symbolic elements which create focused place-based contexts that are generative of wellbeing; and, second, the reputational evolution of these places, such that they become known and increasingly attractive for visitation and engagement (Bell et al. 2018; Jiang 2014; Kearns and Milligan 2019). Gesler's foundational case studies (e.g., 1992, 2003) involved single-site, strongly reputational places that attracted high volumes of visitors, some in an explicit search for healing. Subsequent adoptions and adaptations of the therapeutic landscape construct have sought to apply it across a wide spectrum of places, including domestic settings (e.g., Donovan and Williams 2007).

The interior spaces of the home are potentially, if not actually, therapeutic in the sense that they can offer sanctuary from the stresses of the outside world through their familiarity, safety, and personalisation. Gardens, arguably, constitute an extension of the home. They can be regarded as a buffer zone between the dwelling and public spaces. What distinguishes gardens is that they are not as controllable as inanimate, interior spaces but are, nonetheless, sites within the boundedness of private property that are usually both personalised and curated. In other words, gardens are commonly an extension of the domestic space of the home, with plants grown for both aesthetic (e.g., flowers) and productive (e.g., vegetables) reasons.

The wellbeing collateral derived from domestic cultivation is well documented (e.g., Bingley, Milligan and Gattrell 2004). In part, it is the active work of tending a garden that is regarded as therapeutic. There is a long history to this recognition. In the monastic tradition, for instance, the labour of gardening was woven into the daily spiritual routine. Enclosed gardens were also incorporated into the options for convalescence in hospital grounds (Gerlach-Spriggs et al. 1998). Arguably, it is the active process of 'working' in a garden, rather than the more passive action of 'looking' at it, that has the most yields in terms of wellbeing (Stigsdotter and Grahn 2002). Yet, simply being in the midst of a garden is potentially meditative and hence therapeutic. In contemporary times, this emphasis on the co-benefits of gardening has become a particular focus in concerns around ageing. With the time availability offered by retirement from employment and the imperatives of social connection, as well as the maintenance of physical fitness, gardens and gardening are seen to be rich in health-promoting potential (Detweiler et al. 2012).

The idea of the 'healing garden' has been widely recognised and implemented in institutional environments. Moving away from the practice of cultivation, the belief in promoting wellbeing here refers to green outdoor spaces associated with healthcare facilities, where there is an opportunity to find relief from the stress and preoccupation of illness or injury for patients, staff, and families (Eckerling 1996). This connection between gardens and potential stress reduction is pervasive, with environmental psychologists noting that, while environmental stressors (e.g., crowding, noise) can elicit substantial disturbance in people, visual access to nature shows effects on stress recovery (Ulrich et al. 1991; Ulrich and Parsons 1992). Beyond hospitals, those afflicted by particular conditions have found solace in gardens. Examples range from 'scented gardens for the blind' to 'wander gardens', spaces designed for those afflicted by dementia. In the latter, safety is maintained through all plants being edible, the garden being enclosed, and paths all leading back to entrances (Detweiler et al. 2012). Invariably, however, the pervasive focus within scholarship at the garden-therapy nexus – regardless of whether plant life is on display or actively cultivated – has two deeply embedded and implicit assumptions that both involve control: first, that gardens are firmly bounded spaces; second, that only selected species are welcome. Our chapter concerns the therapeutic possibilities of engaging with the edges of these assumptions. What are the wellbeing benefits of moving our focus to growth at the edge of the orderliness and species-ism of 'good horticultural practice'? How does our vantage point shift when we listen to our life journeys?

Box 13.1 Sentinel moments of biophiliac encounter

Alice Ever since I was a young child, I have found great pleasure in thumbing through the tomes of the old *materia medica* (herbal encyclopaedias) that sat on the bookshelves in my family home. Some were European in origin, while others explored plants from my Chinese heritage. The stories of each plant were diverse, yet there was always one thing that perplexed my young mind, that so many of the healing plants that our ancestors relied upon – and indeed, contemporary herbalists still use frequently – are the very same ones that I see as the unwelcomed plants that the countless herbicides lining the shelves of garden stores in contemporary times seek to control, and in some cases, eradicate completely. This dissonance presented me with a great deal of confusion as a child. As I embarked on the herbalist path, I found myself often becoming a 'weed-activist', getting into heated arguments about how often one should mow the lawn and sharing views to encourage others to expand their understandings of questions such as: How is it that we have lost our appreciation and kinship with these healing plants? And how might we reckon and remember our once sacred relationships with the pharmacopoeia of plants that have been codified as 'weeds' under colonial modernity?

Robin One of my earliest recollections as a pre-schooler was a family picnic beside a pond in the part of Scotland my mother was from. My English grandfather was visiting with us and, from the comfort of the rug covered in cushions and lunchtime delicacies, he asked in a rhetorical tone, 'What do you think that weed in the pond is?'. Without hesitation, my four-year-old self replied 'amphibious persecaria'. Given the pre-Google times, the reference was *The Observer's Book of Wildflowers* and those assembled established I was right. To this day, I don't know how or why I had retained the name, but that utterance set in train a love of wild and beautiful dwellers at the threshold of orderliness.

As suggested in Box 13.1, the weaving of story and relationship between the human and more-than-human world has, for both of us, cultivated a deep sense of belonging to plant-filled places. Without these influences, this sense of belonging might not otherwise have developed. Such stories contribute to a lifelong trajectory of empathy for the vegetal world and, through the therapeutic connection with both plants and others, the engagement in a dance of healing – deepening our relationship to the plant world through the simple act of seeing plants as vital, alive, and replete with their own connections. Perhaps it is not too much of a stretch to suggest that plants themselves call us together; indeed, there is significant scholarship emerging from disciplines such as critical plant studies and vegetal geographies (see Lawrence 2022) that gestures in exactly this direction. From this perspective, each plant – regardless of their molecular healing properties or socially constructed 'identity' within an ecosystem – is transformed into a sentient being within their own right – capable of relationship with the human 'self', rendering codes such as 'native' or 'introduced' simply as concepts created via our anthropocentric spatial reckonings of the geographical imagination. What, then, might be some downstream benefits to rethinking the culturally bounded (and predominantly Eurocentric) terms of 'wild', 'cultivated', and 'weed' in the context of whole, lively ecosystems?

Healing in the planetary garden

In her research on community gardens, Virginia Webb (2020) explores the complex nature of how gardens are transformed into places of community cohesiveness and belonging. Gardens are increasingly sites not only of cultivation, but also – perhaps more importantly – deployed as tools for enacting a kind of 'biosocial becoming'. In this term, coined by anthropologists Tim Ingold and Gisli Palsson (2013), we can discern an immanence of intersection between the human and non-human worlds, where the gardener and the garden are interdependent and work together to 'become' one another. In a similar vein, landscape architect Gilles Clément (2011) invites gardeners, in his essay 'In praise of vagabonds', to consider the diverse forms and functions of a plant before rushing to categorise it as a 'weed'. He goes on to argue that, by adopting a more expansive view of understanding plants-in-context, new relations emerge between the gardener and the garden, and allow for a more nuanced and symbiotic relationships to manifest and be tended. Webb (2020, p. 38) builds on this, writing, 'Gardeners act as intermediaries for encounters among species and cultures that are not necessarily, *a priori*, destined to meet'. Taking seriously Clément and Webb's direction thus invites gardeners to see themselves as participating in a wider ecosystem of lively relationality. Each act of tending, whether it is growing food crops, adding compost, or 'weeding', can be seen as an expression of active engagement with a sentient place, even with the plant beings who may be unexpected or undesirable additions to the previously envisioned landscape.

Clearly, gardens are good for bees and other invertebrates, for instance, but to introduce the human into the ecology and close the separations of power and control, a garden can have agency on its own terms in assembling human-botanical community. In forming community around a garden, there is arguably an enhanced possibility of seeing community within a garden. As individuals, we are prone to purposeful achievement (getting the garden dug, pruned, watered, and so on). But as a community forms around a garden, we are more likely to talk, pause, remark on what we encounter, stop the 'work' of gardening and discuss the sinews of connection linking species between themselves and with us as tenders of tendrils and all that dwells in and around what is growing – regardless of its perceived usefulness or uselessness.

Box 13.2 An unlikely interspecies encounter

It was the day of the lifting of the *kumara* [traditional sweet potato]. There was a good turnout of about ten at the *maara* [communal garden] adjacent to the *marae* [Māori community gathering place]. We were rolling the dense tangled mass of the surface stems and foliage of the *kumara* plants into liftable piles to take to the composting area. But then there was a discovery. A child had seen a frog. All work stopped. There the frog sat, motionless in a clump of weeds between rows of *kumara*, only a shade brighter green than the weedy foliage. Conversations ensued. Where was the nearest pond or waterway? How did it get into this large, netted enclosure? Where should it be taken? What else might lurk within the weeds? The frog was suddenly the feral centre of a communal expression of curiosity. There was a sense of being uplifted by its amphibious allure, its statuesque stillness, both in its in-place yet out-of-place-ness. A month later and the frog is still remembered. No one had seen one there before, and no one has seen one since. The unexpected garden resident had stopped intentional labour in the garden, brought us together in hushed awe, and still travels with us in story. The world felt like a healthier place for the encounter. The frog offered a therapeutic encounter among the weeds that had offered it a temporary home.

Box 13.3 Wisdom at the garden's edge

An extended 'moment' of challenge for **Robin** was, at the age of 18, being hired by my home city's Parks Department during the university summer break to tend a rose garden. Lawn edges needed to be immaculately straight, any sign of random germination on the exposed soil bed eradicated, flowers plucked as soon as soon as they were past their best, and spray applied to the first sign of any insects except useful bees and attractive butterflies. The rose garden was a resolutely ordered and monocultural space; a site of display whose very existence spoke of civic pride. Just being there led me to occasionally lean on my rake and wistfully recall a similarly ordered garden space at my former high school, where I was complicit in an act of youthful sabotage. Two classmates had, respectively, brought to school a bag of cabbage seed (the son of a market gardener) and a sack of manure (the son of a horse breeder). My role was quantity surveyor. Under small bushes and specimen plants, we furtively installed our seed and horse excrement combination. As the dreary winter weeks unfolded, we watched cabbages grow in the flower gardens, much to the bewilderment of teachers and grounds-people. That desire to mix things up and to break out of the imposed order of the school day led to an early act of edginess and disruptive gardening. I still apply horse manure, but these days wait with intrigue and engage with whatever germinates among intended plantings.

Some years ago, **Alice** was privileged to attend a series of *Rongoā Māori* [traditional Māori medicine] *wānanga* [a dedicated space for learning and deepening knowledge]. I attended with curiosity to engage with *Te Ao Māori* [the Māori world, including Māori philosophy and practice] and meet some of the Indigenous plant medicines available to us in the forest. However, within the first day, I realised that my journey into *Rongoā Māori* was going to be less about plant identification and far more about the ethic of human–plant relationships. Healing – and by proxy, the act of cultivating therapy – was a relational dance between the ethics of reciprocity, respect, responsibility, and redistribution, as much as it was about the literal plants themselves. I will never forget our *kaiako* [teacher] when he started a group walk through the regenerating bush:

> Always be attentive to the edges, because that is where the medicine is. The strongest plants, the most resilient ones, and therefore, quite likely, the ones that humans need, find the edges – because they want to be seen … and quite literally, they want to help us.

I have taken this pearl of wisdom into my practice as a herbalist to this day, for there is a whole world to be discovered in the at thresholds of places.

Dandelion and kikuyu: A tale of convivial reconciliation

Two plants growing vivaciously in a typical garden and lawn on Waiheke Island, where Alice and Robin both live, are dandelion (*taraxacum officinale*) and kikuyu (*pennisetum clandestinum*). Both plants carry stories of migration and appreciation at various stages of global history. Yet, both are also commonly viewed with disdain as a 'nuisance' to gardeners in contemporary times. Indeed, one walk around a garden store in New Zealand reveals shelves lined with various herbicides that have intentionally been designed to starve the long taproots of dandelion, or the kill the far-reaching rhizomatic tendrils of kikuyu that allow it to spread so effectively across vast areas of land. Yet, in an alternative aisle, one can also find the odd seed packet, surreptitiously labelled as 'chicken greens' that contain a mix of common wildflowers and other 'weeds', including both dandelion and kikuyu seeds. Having said this, the number of herbicides certainly outweighs the packets of seeds, and it quickly becomes evident that there is an inherent paradox at play in terms of the value and attitudes that we, as humans in a settler-colonial world, hold towards these plants. Confronting the complexity of contemporary values towards non-native 'invasive' plants thus sheds light not only on our collective socio-political histories, but also perhaps on our own positionality as individual gardeners and plant-tenders as well, thus placing the human within Clément's concept of the 'planetary garden' as part of a dynamic, unbounded site of relations.

The dandelion is a humble little plant (see Figure 13.1) treasured in the apothecaries of herbalists since time immemorial, primarily for the bitter compounds that act as an effective diuretic and detoxifier for the human body (Lawrence 2020) – so much so that one of the original names for dandelion in French was *pis-en-lit*, quite literally, 'to piss the bed' (Grieve 1931). It is not known for sure when dandelion was first introduced to New Zealand soils, but

Figure 13.1 Morning meetings with Dandelion (with some Kikuyu alongside) in the first author's garden. Photographs by Alice McSherry.

it is thought to have been brought by European settlers as a nutrient-dense food crop. Thanks to its evolutionary capacity for efficient seed dispersal, dandelions quickly find a home in environments of all manners, rural, urban, or otherwise (Massey University 2022). Dandelion is known as a persistent plant, often forcing itself through the cracks of concrete, and able to sprout new vegetation from anywhere along its deeply embedded taproot. To some, this may disrupt a desired aesthetic (for example, in the cracks of pavement, or in a neatly mown lawn), but the biological and metaphorical story of dandelion points to the resilience of life, regardless of how inhospitable an environment may be.

Dandelion's medicine extends to other-than-human bodies as well. It is said that dandelions boost a cow's milk yield, as well as being an excellent nitrogen-fixer due to their symbiosis with numerous micro-organisms (such as beneficial bacteria and fungi) that attach to the roots, reducing the need for other mineral fertilisers that can have harmful downstream effects to the existing landscape (Lowenfels and Lewis 2006). The point here is that dandelion occupies a multiplicity of ecological functions and identities in the garden, despite the dominant insistence that it ruins the pristine and 'pure' aesthetic of an immaculate lawn; arguably a material manifestation of the imperial mindset that seeks to impose anthropocentric conformity upon a living Earth (Ignatieva et al. 2015). The expansion of one's perspective with dandelion therefore reveals some unexpected and novel ways of relating to a sentient landscape and shifts our relationship with the garden from passive admiration to active engagement with a dynamic, therapeutic world.

Every morning, Alice visits her patch where Dandelion grows abundantly on the outskirts of her garden (see Figure 13.1). Here, she nibbles on a few bitter leaves, which seem to miraculously appear overnight, and is able to engage in a ritualistic meeting with Dandelion that allows Alice to reflect on the day ahead, as expressed through Dandelion's intimate role in her garden and place in the world.

Box 13.4 Dandelion magic

Alice remembers, as a small child, being obsessed with dandelions, unwittingly helping the plant reproduce by blowing the cotton-like seed heads out as a thousand wishes into the world. She also remembers being fascinated by the white 'goo' that emerged from a broken stem, and often pasted it on her and her friend's skin as a 'cream' when simulating a doctor–patient role-play. Later, as an apprentice herbalist, she found out that this was in fact reminiscent of an old folk healing practice, as the milky latex in a dandelion stem is proven as an effective remedy for all manner of skin complaints, ranging from warts to sores (Lawrence 2020).

Kikuyu, on the other hand, is not known for its medicinal qualities,[2] but reveals other gifts for human–plant reflection and relation in the garden. Kikuyu was introduced to Aotearoa New Zealand in the 1920s by the Department of Agriculture as a fast-growing grass that could provide easy fodder for cows, as demand for pasture to sustain New Zealand's agriculture industry dramatically increased (Howell, 2008). Agricultural expansion is inherently tied to New Zealand's identity as a settler-colonial nation, transforming a living landscape into a tradable commodity. This history is inextricably linked to the history of colonisation in New Zealand, where vast tracts of land was confiscated from Māori *iwi* [tribes] and *hapū* [sub-tribes] to further the wealth of the British Crown (Gilling 2001). Kikuyu thus became a vegetal extension of the colonial project, and often presents an uncomfortable reckoning to sit with in contemporary times. Like dandelion, kikuyu quickly became a fast-growing, tenacious grass, capable of extending its rhizomes up to three metres underground without sunlight. Entering any garden in New Zealand, an inevitable conversation revolves around a shared disdain for kikuyu's persistence and its penchant for smothering young seedlings and overtaking a garden. While there is truth to this, when we become acquainted with kikuyu in a more intimate way, we discover that the plant is in fact a drought-resistant perennial, covering any bare earth that is left untended, in order to keep what little moisture is left in the soil and work to distribute what little nutrients are left in the ground to other micro-organisms who require them. Furthermore, kikuyu, when extracted and then fermented, transforms into a nutrient-dense fertiliser that can later be applied to the garden to redistribute essential minerals back into the soil. Kikuyu, seen from this perspective, thus teaches its human co-conspirators some precious lessons in the garden. For while it is, in many senses, a 'coloniser', it also embodies a capacity to literally compost itself and regenerate the ecosystem within which it finds itself. This can be translated metaphorically to the gardener's mind where, in witnessing kikuyu, we are able to become intimate with the unruly, often unrecognised aspects of being humans under settler-colonialism, and work to compost these to (re)imagine more equitable relations and futures.

Dandelion and kikuyu are unlikely friends to the gardener: one spreads laterally, while the other delves deeply into the ground, both are persistent and difficult to control. Despite their dominant identity as a 'weed' in contemporary times, however, both plants work in symbiotic relationship with the land upon which they find themselves. A slight shift in perspective from weeds-as-nuisance to weeds-as-beings reveals complex therapeutic relationships and often undervalued resources that are made available to gardeners, should they take the time to observe and be-with them. To be a 'weed tender' must not be conflated with the dominant assumption that one does not care about the preservation and flourishing of the more endangered, Indigenous flora in an ecosystem. The point remains that, in these unruly gardening relationships, we find unique ways of coming back to a state of 'right relationship' with land and the nature of living systems themselves. Perhaps weeds, in this way, can act as the tellers of 'cautionary tales' of what happens in a landscape when the sacred contracts of reciprocity and harmony between human and land are dishonoured (Yunkaporta 2019). Kikuyu has, after all, only spread to the extent it has due to intensive and often harmful agricultural practices, while dandelion is only vilified because it is seen to disrupt the homogenous aesthetic of a 'pristine' lawn. The practices of intensive agriculture and lawn-keeping, however, are inextricably linked to the project of settler-colonialism. To think-with weeds may therefore provide gardeners with a unique opportunity to simultaneously dissolve our deeply entrenched binary logic around what belongs and does not belong in the garden, as well as reckon with our shared colonial past. In this way, dandelion and kikuyu beckon us to seek relationship, not just with the material unruly edges of our gardens and landscapes but perhaps, by extension, with the unruly aspects of our human psyche, opening alternative pathways to 'right relationship' with place and body as a radical act of ecological reconciliation in compromised times.

From extractive gardening to interspecies flourishing

We draw this brief reflection to a close by observing that the usual act of gardening, at least in Aotearoa New Zealand, is in a similar style to pastoral farming, based on a binary way of thinking, in which 'the weed' is a social construction based on an active process of othering. 'Weeds' become the collective noun for enemy of the cultivated and cultivator, that which invades, intrudes, chokes, and colonises. The weed has, in Mary Douglas' (1966, p. 44) framing, an equivalency to dirt: it is 'matter out of place'. Paradoxically, if one thumbs through any old *materia medica*, many of the plants that appear in the pages are the very plants that contemporary Western society spends so much attempting to eradicate from our gardens, berms, and 'wild' spaces.

Probiotic, life-affirming perspectives on the garden invite us to engage with a deeper, non-material understanding of our place in the world (Lorimer 2020). This vantage point demands that we reflect on our own place-in-the world and reflect on the complex and dynamic ecosystem thinking that is centred upon a relational ethics of respect, reciprocity, responsibility, redistribution, and reverence. From this place, the garden can then become a site of resistance and reconciliation. But in order for such generative processes to come about, we require intimate co-becomings with the vegetal other, for it is the play of such processes that creates the therapeutic benefits and, ultimately, the therapeutic landscape.

Gardens, we conclude, are as much political as therapeutic spaces. We are challenged by the weedy world, either to be complicit in a vegetative version of the war on otherness or to embrace and transform the messiness of plant legacies and the subsequent attitudes that arise in the process. There is a diversity of life forms in any garden, and we are but one of them (Yeo 2021). To heal-with weeds as co-conspirators within an 'interspecies community' of health and healing is perhaps an expression of revolting against empire and its capitalist and patriarchal underpinnings. It is to weave new strands of relationship into existing conceptualisations of the garden as both a planetary and a therapeutic space. The garden thus holds potential to be transformed into a dynamic place of healing and belonging, where we are encouraged through contemplation and engagement into an acceptance of unruliness and interdependency, rather than upholding our engrained colonial impulses for neatness and perfection.

Notes

1 Permaculture is a body of thought and practice which emerged in the 1980s that uses whole-system thinking to cultivate productive food gardens (sometimes known as food forests). The co-originators of the concept are Bill Mollison and David Holmgren. Their work was first published in *Permaculture One: A Perennial Agriculture for Human Settlements* (1978) and has since turned into a worldwide movement.

2 Kikuyu is native to the African continent, and there may very well be undocumented uses within the oral lore of peoples who have been living-with kikuyu for a much longer time. However, in a New Zealand context, Kikuyu is conceived primarily as an invasive grass that has a tendency to overtake garden spaces and is difficult to remove.

References

Bell, SL, Foley, R, Houghton, F, Maddrell, A and Williams, AM 2018, 'From therapeutic landscapes to healthy spaces, places, and practices: A scoping review', *Social Science & Medicine*, vol. 196, pp. 123–130. https://doi.org/10.1016/j.socscimed.2017.11.035

Bennett, J 2010, *Vibrant matter: A political ecology of things*, Duke University Press, London.

Bingley, A, Milligan, C and Gattrell, A 2004, '"Cultivating health": Therapeutic landscapes and older people in northern England', *Social Science & Medicine*, vol. 58, no. 9, pp. 1781–1793. https://doi.org/10.1016/S0277-9536(03)00397-6

Clément, G 2011, 'In praise of vagabonds' trans, J Skinner. *Qui Parle*, vol. 19, no. 2, pp. 275–297. https://doi.org/10.5250/quiparle.19.2.0275

Detweiler, MB, Sharma, T, Detweiler, JG, Murphy, PF, Lane, S, Carman, J and Kim, KY 2012, 'What is the evidence to support the use of therapeutic gardens for the elderly?', *Psychiatry Investigation*, vol. 9, no. 2, pp. 100–110. https://doi.org/10.4306/pi.2012.9.2.100

Donovan, R and Williams, A 2007, 'Home as therapeutic landscape: Family caregivers providing palliative care at home' in A Williams (ed), *Therapeutic landscapes*, Ashgate, Aldershot, pp. 199–220.

Douglas, M 1966, *Purity and danger: An analysis of the concepts of pollution and taboo*, Routledge, London.

Eckerling, M 1996, 'Guidelines for designing healing gardens', *Journal of Horticultural Therapy*, vol. 8, pp. 21–25.

Fardon, R 2013, 'Citations out of place', *Anthropology Today*, vol. 29, no. 1, pp. 25–27. https://doi.org/10.1111/1467-8322.12007

Gerlach-Spriggs, N, Kaufman, RE and Warner, SB 1998, *Restorative gardens: The healing landscape*, Yale University Press, New Haven.

Gesler, WM 1992, 'Therapeutic landscapes: Medical issues in light of the new cultural geography', *Social Science & Medicine*, vol. 34, no. 7, pp. 735–746. https://doi.org/10.1016/0277-9536(92)90360-3

Gilling, B 2001, 'Raupatu: The punitive confiscation of Māori Land in the 1860s' in AR Buck, J McLaren and NE Wright (eds), *Land and freedom: Law property rights and the British diaspora*, Ashgate, Aldershot, pp. 117–134.

Grieve, M 1931, *A modern herbal*, Stone Basin Books, USA.

Howell, C 2008, *Consolidated list of environmental weeds in New Zealand*, Department of Conservation Research and Development Series 292, Science and Technical Publishing, Department of Conservation, Wellington.

Ignatieva, M, Ahrné, K, Wissman, J and Eriksson, T 2015, 'Lawn as a cultural and ecological phenomenon: A conceptual framework for transdisciplinary research', *Urban Forestry & Urban Greening*, vol. 14, no. 2, pp. 383–387. https://doi.org/10.1016/j.ufug.2015.04.003

Jiang, S 2014, 'Therapeutic landscapes and healing gardens: A review of Chinese literature in relation to the studies in western countries', *Frontiers of Architectural Research*, vol. 3, no. 2, pp. 141–153. https://doi.org/10.1016/j.foar.2013.12.002

Kearns, R and Milligan, C 2019, 'Placing therapeutic landscape as theoretical development in Health & Place', *Health & Place*, vol. 61, article 102224. https://doi.org/10.1016/j.healthplace.2019.102224

Lawrence, AM 2022, 'Listening to plants: Conversations between critical plant studies and vegetal geography', *Progress in Human Geography*, vol. 46, no. 2, pp. 629–651. https://doi.org/10.1177/03091325211062167

Lawrence, S 2020, *Royal Botanic Gardens Kew Witch's Garden: Plants in folklore, magic and traditional medicine*, Welbeck Publishing Group, London.

Lorimer, J 2020, *The probiotic planet: Using life to manage life*, University of Minnesota Press, Minneapolis.

Massey University 2022, 'Dandelion', *Massey University Weeds Database*. https://www.massey.ac.nz/massey/learning/colleges/college-of-sciences/clinics-and-services/weeds-database/dandelion.cfm

Puig de la Bellacasa, M 2017, *Matters of care: Speculative ethics in more than human worlds*, University of Minnesota Press, Minneapolis.

Stigsdotter, U and Grahn, P 2002, 'What makes a garden a healing garden?', *Journal of Therapeutic Horticulture*, vol. 13, no. 2, pp. 60–69.

Ulrich, RS and Parsons, R 1992, 'Influences of passive experiences with plants on individual well-being and health' in D Relf (ed), *The role of horticulture in human well being and social development: A national symposium*, Timber Press, Portland, pp. 93–105.

Webb, V 2020, '"We just have to get them growing their own food": The cultural politics of community gardens', Unpublished PhD Thesis, Massey University, New Zealand.

Wilson, S 2019, 'Planting the berm – The gift of a public garden' (*NZ Herald*, 15 November 2019). https://www.nzherald.co.nz/nz/simon-wilson-planting-the-berm-the-gift-of-a-public-garden/ENJKUARI7EWAIUS776NIBQO4AM/

Yeo, P 2021, 'Can we allow *planta non-grata* to become *planta conviva*? A reconciliation ecology approach to invasive species', *Biodiversity*, vol. 22, no. 3–4, pp. 143–145. https://doi.org/10.1080/14888386.2021.2006082

Yunkaporta, T 2019, 'Lessons from stone – Indigenous thinking and the law' (*The Conversation*, 6 September 2019). https://theconversation.com/friday-essay-lessons-from-stone-indigenous-thinking-and-the-law-122617?fbclid=IwAR3SRSxG5H_e5TGuaeMQKHaFQNJ22HpTqyI9oO_4fAUq9ekZTOolrU4coO4

14 Horti-cultural geographies

Situating the garden as an assemblage of health and wellbeing

Ronan Foley

Introduction: the garden as taskscape

Some years ago, I led Year 2 Geography students through the Jardin du Monastère de Cimiez in Nice after a brief lecture on therapeutic landscapes. As they watched a gardener carefully clipping plants in the early spring sun overlooking the city, I overheard one student say, 'What a great job. I'd love to do that for a living' – only to be countered by a second student saying, 'No way I could do that. I have terrible hay fever'. This short vignette reflects David Conradson's (2005) comment on how any imbrication in therapeutic space works differently for different people and may not always be positive. Yet, this book provides many excellent examples of imbrications in gardens as representative cultivated therapeutic landscapes that flag up the breadth of health-enabling potential contained within such (mostly) green spaces and associated therapeutic practices. They also echo Tim Ingold's idea of a 'taskspace … an ensemble of tasks, in their mutual interlocking' (Dunkley 2009, p. 89; Ingold 2000), framing gardens as classic therapeutic assemblages, as mobile and always becoming spaces of active practice, with different shapes, shades, and forms, that in turn provide core physical and mental health and wellbeing benefits. These include everything from physical exercise, attention-restoration, stress-reduction, and socialisation through to more individual outcomes, from respite and solace to self-discovery and self-esteem (Brindley et al. 2018; Sia et al. 2020). In addition, across this book, the garden's wider role as foodscape, third space, setting for environmental awareness, eco-politics, and a range of direct more-than-human engagements, speaks to a wider 'one health' focus as well (Pitt 2018). This concluding chapter will not explicitly discuss the ideas and themes identified within the excellent individual book chapters, but rather consider some theoretical, applied, and methodological contributions to ongoing research that frame gardens and gardening as horti-cultural therapeutic assemblages of health and wellbeing (Page 2020).

Digging in

Recent methodological approaches within therapeutic landscapes research utilise different embodied, emotional, and experiential ways by which health can

DOI: 10.4324/9781003355731-18

be unearthed, a digging with the pen (Heaney 1966),[1] that reflects how both gardens and research often emerge in unpromising spaces through sheer hard work and disciplined practice. Given the many different types of spaces and practices that emerge across this book, the specific ways in which active healing practices are excavated within these therapeutic settings are everyday, contested, and complex (Williams 1999, 2007). Gardens never emerge automatically, with different models and strands of cultivation always relational – to locale, location, and sense of place (Massey 2005) – but also to wider place, ecological, cultural, and political flows and connections (Eizenberg 2013). The many different approaches used in both this book and the wider interdisciplinary literature document a range of qualitative methods, from traditional interviews and participant observations to archival work and narratives from written/recorded diaries, both collective and individual, which document digging in, a specific version of immersion that requires the researcher to be equally immersive. Hard/Graft, a Dublin-based communal gardening project with precarious and excluded societal groups, taught people how to graft tree branches to build a small community orchard, reflecting the literal hard graft of growing and growth as a continuous assemblage across the lifecourse (O'Sullivan 2017). While less commonly used to date with garden research, the growth in *in-situ* tools and technologies has opened up new directions for geo-narratives (Bell et al. 2015; Foley et al. 2020); experiential accounts recorded in place by wearable technologies that can simultaneously trace mobility, encounter, and change over time alongside physiological and psychological response to places (Osborne 2019). Digging in also draws from recent healthy green and blue space research on different aspects of productive leisure/labour, wherein gardens and allotments involve productive engagements of flowers, plants, and food that blur the boundary between both labour and leisure, but that also, crucially, spark joy and fun – essential and criminally under-regarded components in the measurement of wellbeing (Phoenix and Orr 2014). Scales of analyses within research will also reflect the size and scale of gardens, from balconies and road verges to urban development plans and large cultivated destination garden settings (Kingsley et al. 2021). In estimating the value of such spaces, it is important not to conflate scale with matched effects on health and wellbeing; health outcomes being much more relational with personal, cultural, and structural components as well (Foley et al. 2022). Finally, methods should always respect problematic components within garden research to consider the spaces, weeds and all, and take care around exclusion and removal, to fully reflect the complex assemblages encountered within such work (Bhatti and Church 2000).

Assembling health and wellbeing

Across the book, there is a real sense that both older and newer understandings of therapeutic landscapes continue to have value when applied within individual reviews and case studies (Gesler, 1992). Garden spaces and the people who cultivate them exemplify recent writing on assemblages of health, a theoretical underpinning appropriately inspired by the Deleuzian idea of the rhizome (Duff 2014). Such rhizomatic thinking, a tracing of underground roots and routes, which grow in often unpredictable directions across space, has been used to explain relational geographies of nature in terms of becoming and stabilisation that reflect recurring ecological cycles and flows, but also of humans across space and time (Head and Atchison 2009). Assemblage thinking has been utilised within medical/health geographies to consider different routes to health and wellbeing (physical and mental) that have social, affective, and material dimensions (Duff and Hill 2022). Such thinking can also be used to consider the role of ecological, sustainable, and more-than-human actants to provide a 'one-green' perspective on the garden as a therapeutic space, recognising an important natural agency beyond the human (Bell, Leyshon and Phoenix 2019). It also recognises the role of unpromising nature, where the rhizome emerges unexpectedly – for example, in post-earthquake Christchurch – to create spaces reclaimed for urban guerrilla gardening ('gap-fillers') that promoted well-being and recovery, but also in the speed by which weeds, roots, and branches grew over previously manicured suburban lawns, yards, and bungalows that remained as visible material markers of loss and trauma (Dickinson 2019). Wider therapeutic assemblage thinking sees both spaces and health status as always in formation, coalescing at key moments (seasons, events, communities, structural change and wider social geographies) across time and space. Such formations and coalescences also need to be managed and sustained, reflecting this book's wider interests in illness prevention and restoration, health promotion and community health, and how they emerge in garden spaces (see Introduction). Indeed, the verbs 'assembling' and 'cultivating' have much in common, using different but parallel approaches and levels of intent. Intentionality is another key driver of the assemblage, sometimes characterised through the concept of meshwork, usable in gardening terms as an analogy for both equal and unequal apportionments of land but stressing and emphasising how the interlocking nature of that wider grid is made stronger by its mobile components (Duff and Hill 2022). Intentionality is also an important process in the making and sustenance of gardens, with Pitt (2015) identifying flow and control within communal gardens as both positive and negative in terms of active experience; both representing important dimensions of a wider sustainability of natural spaces that helps deepen and sustain health and wellbeing connections to and with nature, even if these are intermittent and momentary (Richardson et al. 2021).

Structures, cultures, and behaviours

Horticulture considers the art or practice of garden cultivation and management, processes and behaviours that are always shaped by the cultural, economic, and ecological political structures within which that cultivation occurs. These, in turn, create allowances for how a set of spatial practices shapes both individual and collective health and wellbeing in multi-scalar ways, both managed and unmanaged. Structurally, political attitudes to the settings within which gardening takes place, from relatively intensive to micro scales, can vary from indifference and neglect to over-regulation and a free-market approach to land that is differentially inclusive, with spatially variable public and private ownership of land and space. This reflects recent concerns within critical health geography work (Brown et al. 2017) on how health and wellbeing are shaped by wider structural forces of globalisation, neo-liberalism, intensive food production, and climate change – all unsustainable models under considerable current pressure. In contrast, examples in this book describe some small-scale organic gardening methods, while other examples like the *CittàSlow* movement in Germany and Italy reframe the garden (and indeed the garden kitchen) as a quality food source in terms of provenance and sustainable produce (Sept 2021). Structural governmentality also emerged throughout the COVID-19 pandemic, in terms of access to garden spaces, with a global assemblage of differing national rules and lockdown practices and regulations, which produced an almost uncanny voluntary governmentality (Garrido Cumbrera et al. 2021; Philo 2012). In another sense, structures are also emergent in the micro-geographies of soil, earth, *terrain* (in its viticultural sense), and the micro-management of gardens and their biological inhabitants. Here experimentation, trial and error, discovery, loss and gain, and working with and against nature are all essential elements of gardening practice that reflect the management of wellbeing (Marsh et al. 2021).

How gardens are managed also reflects societal cultures and practices, about how land and earth should be negotiated with for optimal outcomes for humans and non-humans alike. Indigenous Australian and Māori cultures make no distinction between the two, recognising a co-ownership and co-identity for people and environment (*Country* and *hauora*, respectively) (Olive and Wheaton 2021; Wilson 2003). The design of aesthetically pleasing gardens can also be about controlling nature, and for all that therapeutic design can also be involved, there is a distinctly commodified and aspirational side to some forms of manicured cultivation of gardens; though countered by the many examples of communal and collective garden spaces, practices, and communities discussed in the book. These more hopeful examples represent, as the editors note, a deliberate cultivation of equity, especially important given inequality often stems from a lack of cultivation and a form of careless neglect at a structural level; for example, the regular pressure to develop scarce urban garden spaces into building land (Eizenberg 2013). Gardens emerge best from a deeply embodied engagement with elemental forces (mud, rain, wind, sun) that are literally no place for the squeamish; true nature-connection involves literally getting down and dirty with the microbes and more-than-human cultures of the worm and slug (Franco and Robson 2022). That nature-connection, often the product of physical exertion and co-management with nature, can be essential for older bodies (Scott et al. 2015), although Wiles et al. (2021) suggest thinking beyond exercise and mindfulness of garden spaces (important as they are) to wider more-then-therapeutic benefits around agency, performativity, identity, and adaption across lifecourses that support healthy ageing (Milligan et al. 2004).

The garden as a space/place for all

The place of the garden has been explored in multiple forms in the chapters of this book, identifying gardens as space/place for all: children and adults, insider and outsider, established and new citizen, semi-professional to beginner in cultivation knowledge terms (see Introduction). Historically, public urban gardens like the Vauxhall Gardens in London were liminal sites of pleasure and socialisation (Conlin 2008), while other market spaces (Covent Garden in London, Les Halles in Paris) were proto foodscapes where city met country (Smith 2002). More significantly, the country house landscape was scaled down and became a central element of built environments the world over, with the private domestic garden (now estimated as covering about 4.3 million km^2 across Britain, as a single country example) available to those who could afford one. While some face outwards to the street, most are inward-facing, protected, and private spaces that are often central to lived lives, yet that very slight shift in spatial orientation generates a small incremental space loss that creates inaccessibility and makes a part, apart. In other built environments, communal and shared gardens were designed to create community, although these require shared commitments to tending and care that are not always evident (Eizenberg 2013). Recent research on the strong protective relationship of private garden ownership against a range of measures of premature mortality, including cardiovascular and respiratory disease, highlights the benefits of gardens in ways that are not accessible to all (Roscoe et al. 2022).

In thinking about the different boundaries, both sharp and blurred, between public and private gardens, that, too, can be shaped by active practices that make place open or closed, that consider access in quite direct and directed ways. In the text, the term ‘tending’ is often used in a positive sense as a practice of care, paying close attention to the flora, fauna, and human inhabitations and how those relations are respectful to all needs and deepening a sense of equality. Yet, an older Latin root and derived meaning for the word ‘horti’ is to grasp or enclose. From England, the original great gardens for all, the Commons, were one of the first to be enclosed from the seventeenth century on, something musician Chris Wood identifies as a domestic precursor to Britain’s global colonial project (Wood 2008). The enclosure of the Commons, set alongside the colonial expansions between the fifteenth and nineteenth centuries, set in place many still-current global issues, in terms of sustainability, globalisation, migration, and climate change, most of which had the effect of displacing poor populations or deepening inequalities across society. That enclosing instinct still exists, and sometimes itself needs tending; keeping fences up, keeping unwanted ecologies out, privatising previously public space and marking an ongoing contestation of garden spaces and a lingering governmentality across many countries and spaces (Eizenberg 2013; Williams, see Chapter 2).

While it is sometimes easier to map out inequality than to generate equalities in applied and practical terms, there are many exemplars in this book of good practices and initiatives, mostly and literally built from the ground-up with an open sensibility. That ongoing link between commons and community lies at the heart of shared community gardens, of collective as well as public ownership, and a deep commitment to co-production in what Eizenberg (2013) in her work on urban gardening in New York refers to as an 'actually existing commons' populated by what she refers to as 'organic citizens'. While the development and restoration of green cultivated spaces in the city is often a deeply contested one, especially in the often-desperate struggle for space, such spaces also form an important set of nodes within the therapeutic assemblage (Corcoran, Kettle, and O'Callaghan 2017). In many cases, gardens thrive but also die off, or reflect in their material produce different cultural and ecological trends. They also involve assembling and at times disassembling varying communities of gardening practice, such that the citizen too has an organic life from their immersions in that space. Equally, and a strong presence in much gardening literature, is the role of the garden as a welcoming space for migrant groups and other excluded/precarious groups in society, building in new forms of citizenship in an organic way. Here, the cultivation is, in part, introducing one culture to another, but also in allowing cultures to survive and thrive in new surroundings, which also leaves space perhaps for a necessary spatial separation as a safe cultural space for those migrant groups (Ekstam et al. 2021; Harris 2014). As Turner, Abramovic, and Hope (2021) note in relation to access to community gardens, this can be a competitive process with long waiting lists, where it's hard for migrants to get on the list. Drawing also from research with swimmers and surfers, the power of a community of practice within green or blue space is just that, a shared activity that brings people loosely together and where community, if found, is also for newcomers, where such a welcome may not be as accessible within a workplace or neighbourhood (Britton and Foley 2021). Equally, not all communities adhere and in terms of the political impacts and networks beyond the yard, there are other examples where urban agriculture brought two side of the political divide together in Belfast in the late noughties (Corcoran and Kettle 2015), while other Irish work identifies new communities – often younger women – challenging the 'dis-assembling' effects of modern urban living and the challenge of austerity by reintegrating life and food into revived allotment spaces, acting in the creation of a commoning instinct to promote health and society (Kettle 2014).

Gardening in a time of contagion

Gardens have received renewed attention in the now three-year-long COVID-19 pandemic, with people and communities at all levels rediscovering nearby green spaces, including gardens, allotments, pocket parks, and even some unpromising edge-spaces (Hawkins et al. 2013; Houghton and Houghton 2015). In the rapidly expanding research on COIVD-19, nature, and health, Andrews et al. (2021) include considerable numbers of health geographers writing on how COVID-19 and indeed other pandemic futures are shaping health and place. More broadly, a wide range of enhanced health and wellbeing benefits have been listed for garden spaces in this time, including stress-reduction and nature-connection (Egerer et al. 2021), physical and mental health for older people (Corley et al. 2021), and refuge and respite (Foley et al. 2022; Marsh et al. 2021). In an incredibly disruptive time in citizens' lives, and in the face of a global infectious disease outbreak, the importance of space outdoors was central to health and wellbeing, with many of the studies identifying complex emotional, embodied, and experiential evidence of that value. These included a very strong narrative around escape, refuge, and respite (Foley et al. 2022; Marsh et al. 2021) from the pressures of confined homes and confined activity spaces during public lockdowns. In addition, the reduction of individual activities and activity spaces made the garden a crucial space for physical health, with gardening work central to a renewed embodied space of nature encounter requiring strong shoulders, sore backs, and muddy fingers. Equally, the capacity of garden spaces to promote mental wellbeing and rest emphasised the passive as well as active components of the therapeutic assemblage, within which memory, grief and loss could at least temporarily feel held in place. Across the globe, the role of spaces like gardens as sites to get away from it all, but also as ongoing site of new experiences, was constantly reframed as children spent more time with parents and learned about nature together. Several studies identified a renewal of the senses in terms of an awareness especially of bird song during the strictest lockdowns; the birds were not singing any louder, we just heard them better without the normal sonic competition (Foley et al. 2022; Marsh et al. 2021).

Other COVID-19 studies look at the importance of elements like a view of a garden or other natural spaces, as well as the relative importance, on a micro-spatial level, of indoor or built environment of plants, flowers, and grow-bags as part of coping strategies (Finlay et al. 2021; Garrido Cumbrera et al. 2021). Drawing from the famous Ulrich study of views of nature from hospital beds, Garrido and colleagues (2022) identified, from a study carried out across Spain, England, and Ireland, direct benefits from having access to such views, in the reduction of anxiety and depression, although they also noted a much lower comparative ownership of garden spaces in Spanish homes. More broadly, research on how different geographies emerged within the differential health impacts of COVID-19 found a global trend towards greater impacts in deprived parts of cities, often affecting those with least access to green spaces and characterised by larger proportions of ethnic minorities (Jassi and Dutton 2022). Results from a German study (Lehberger, Kleih, and Sparke 2021) indicated that garden owners had substantially greater life satisfaction and mental wellbeing than non-garden owners, a process deepened by COVID-19 in terms of access to public garden spaces as well. How home and garden connect externally also emerged in COVID-19 studies that recognised flow as a mobility component that lubricates therapeutic assemblage, but which became interrupted and blocked during the pandemic (Foley et al. 2022). A lack of flow along the rhizomatic routes on and under garden spaces can also lead to a certain atrophy within spaces, an ecological choking, suggestive of wider links with contagion and communality within a neglected garden or allotment, in turn a potential breeding ground for pests or weeds and as a danger to other gardeners. But in the many places kept open and alive during and after the pandemic, the importance of multiple scales of connection, from pot, windowsill, and balcony up to large organic farm and garden sites, shows how multi-scalar mobilities are also part of a wider assemblage that holds essential restorative powers in times of public emergency and fear.

The garden as a site of maintenance and repair

Coming to the central role of cultivation and tending, the garden as a space for therapeutic practice fits nicely within a slow but deepening interest in geographies of maintenance and repair (Greenhough et al. 2022; Mattern 2018; Millington and Lawhon 2019; Schmid 2019). This is evident in the deep underpinning in many of the chapters by place care, both in terms of health/wellbeing and in terms of sustainable cultivation, with a genuine recognition of repair as a central theme in terms of climate change and planetary ecological pressures (Hoyle 2021). In assemblage terms, a garden is never fixed as a healing space, but is always an open and mobile space in its capacities to cultivate health and wellbeing and its complex emergence. Such capacities – human, social, ecological, climatic – are always at play through what Nelson and Bigger (2022) term 'infrastructural nature'. In tending to the infrastructure, just as with assemblages of health, lines of communication and connection must be kept open, and maintenance and repair act as a metaphor for primary care. That primary care has an always overlooked community component (Brown et al. 2017) but a core aim is to maximise health and wellbeing to reduce the risk and need for acute care across life-courses. Infrastructural nature also recognises how cultivation, too, must be kept moving, so as to remove what Fullilove (2016) describes as 'root shock', a social counter to an ecological choking that recognises how place care can be removed by neoliberal ideologies and a cultivated neglect that can be brutal in its impacts (Eizenberg 2013). Place care and cultivation of communal garden spaces act as counter and corrective to such threats, but also emerge as always contested, activist, even playful spaces; indeed, the role of gardens as spaces of play and discovery for children, in all their forms, provides an important early introduction to nature more broadly, which promotes lifelong mental health (Bratman et al. 2019). At the other end of the scale, gardens within institutional settings also reflect other constrained lives; people with dementia, mental illness, prisoners are also contained in safe spaces (Moran et al. 2021). Marsh (2017) identifies gardens as helpful third/liminal spaces for end-of-life care, for lives that are sadly no longer maintainable, but like all living things pass on.

Just as place care involves maintenance and repair across a continuum of lives and practices, so too do the complex networks of the therapeutic assemblage. As spaces where built and natural landscape intersect and overlap, they represent relational geographies, often without sharp boundaries and representing an ecological fuzziness; sometimes cultivation occurs in unpromising spaces, such as Richard Mabey's Edgelands or small overgrown patches of ground that suggest other palettes, grey, brown, red, that characterise different gardens around the world (Houghton and Houghton 2015). Whatever the size and type of garden, they are spaces that hold meaning and provide coherence in people's lives. Just as COVID-19 continues through long COVID and new variants (the fatigue associated with the former making gardening impossible for many long-haulers) to exploit cracks in physiologies and economies, so the value of therapeutic gardening practices makes them an essential salve to try and seal up those cracks as a very literal form of repair. Gardens are also deeply affective spaces and underpinning all of this chapter is the sense that a core engine in the maintenance of any therapeutic landscape is their emotional geography. Gardens, especially those in communal spaces, are often the site and source of much affective power in people's lives, as places of shared learning between grandparents and children, as favourite places, as ongoing sites of nature-connection. They are important spaces of memory and knowledge, too, wherein individual plants and trees act as triggers to a deep emotional mapping that holds wellbeing in place (Wiles et al. 2021). In our own garden, we have a deliciously named *Zantedeschia aethiopica*, an arum lily grown first in the 1940s by my grandmother, who died before I was born. Cuttings from the plant went from Tipperary, where my dad grew up, to Tramore, where he retired; then, after my mother's death in 2020, I took a cutting to my home in Dublin, where it peeped out to say hello in June 2021. Seeing that plant provides a very literal relational nature connection but roots that connection to a wider assemblage of care. It is no coincidence that the garden continues to have an ongoing role within health promotion as a site for preventative health care in community health terms. It also has a new and contested place within discussions on social prescribing, with differing positive/negative positionalities around, respectively, enhanced autonomy or a creeping individualisation; other discussions around what a correct dose of the garden might share that contested view (Bell et al. 2018). However it is framed, the garden spaces, practices, and cultures documented across this book provide rich illustrations of a cultivation that is intentionally care-full to provide, maintain, and, where possible, repair health and wellbeing within spaces that continue to add meaning and joy in everyday lives.

Note

1 https://www.poetryfoundation.org/poems/47555/digging

References

Andrews, G, Crookes, V, Pearce, J and Messina, J (eds) 2021, *COVID-19 and similar futures: Pandemic geographies*, Global Perspectives on Health Geographies, Springer Nature, London.

Bell, S, Foley, R, Houghton, F, Maddrell, A and Williams, A 2018, 'From therapeutic landscapes to healthy spaces, places and practices: A scoping review', *Social Science & Medicine*, vol. 196, pp. 123–130. https://doi.org/10.1016/j.socscimed.2017.11.035

Bell, SL, Leyshon, C and Phoenix, C 2019, 'Negotiating nature's weather worlds in the context of life with sight impairment', *TIBG: Transactions of the Institute of British Geographers*, vol. 44, no. 2, pp. 270–283. https://doi.org/10.1111/tran.12285

Bell, SL, Phoenix, C, Lovell, R and Wheeler, BW 2015, 'Using GPS and geo-narratives: A methodological approach for understanding and situating everyday green space encounters', *Area*, vol. 47, no. 1, pp. 88–96. https://doi.org/10.1111/area.12152

Bhatti, M and Church, A 2000, '"I never promised you a rose garden": Gender, leisure and home-making', *Leisure Studies*, vol. 19, no. 3, pp. 183–197. https://doi.org/10.1080/02614360050023071

Bratman, GN, Anderson, CB, Berman, MG et al. 2019, 'Nature and mental health: An ecosystem service perspective', *Science Advances Review*, vol. 5, no. 7, eaax0903. https://doi.org/10.1126/sciadv.aax0903

Brindley, P, Jorgensen, A and Maheshwaran, R 2018, 'Domestic gardens and self-reported health: A national population study', *International Journal of Health Geographics*, vol. 17, article 31. https://doi.org/10.1186/s12942-018-0148-6

Britton, E and Foley, R 2021, 'Sensing water: Uncovering health and well-being in the sea and surf', *Journal of Sports and Social Issues*, vol. 45, no. 1, pp. 60–68, https://doi.org/10.1177/0193723520928597

Brown, T, Andrews, G, Cummins, S, Greenhough, B, Lewis, D and Power, A 2017, *Health geographies: A critical introduction*, Wiley-Blackwell, Chichester.

Conlin, J 2008, 'Vauxhall on the boulevard: Pleasure gardens in London and Paris, 1764–1784', *Urban History*, vol. 35, no. 1, pp. 24–47. https://doi.org/10.1017/S0963926807005160

Conradson, D 2005, 'Landscape, care and the relational self: Therapeutic encounters in rural England', *Health & Place*, vol. 11, no. 4, pp. 337–348. https://doi.org/10.1016/j.healthplace.2005.02.004

Corcoran, MP and Kettle, PC 2015, 'Urban agriculture, civil interfaces and moving beyond difference: The experiences of plot holders in Dublin and Belfast', *Local Environment*, vol. 20, no. 10, pp. 1215–1230. https://doi.org/10.1080/13549839.2015.1038228

Corcoran, MP, Kettle, PC and O'Callaghan, C 2017, 'Green shoots in vacant plots? Urban agriculture and austerity in post-crash Ireland', *ACME*, vol. 16, no. 2, pp. 305–331.

Corley, J, Okely, JA, Taylor, AM, Page, D, Welstead, M, Skarabela, B, Redmond, P, Cox, SR and Russ, TC 2021, 'Home garden use during COVID-19: Associations with physical and mental wellbeing in older adults', *Journal of Environmental Psychology*, vol. 73, article 101545. https://doi.org/10.1016/j.jenvp.2020.101545

Dickinson, S 2019, 'Changing places: Geographies of post-disaster landscapes', *Geography*, vol. 104, no. 3, pp. 116–124. https://doi.org/10.1080/00167487.2019.12094074

Duff, C 2014, *Assemblages of health*, Springer, New York.

Duff, C and Hill, N 2022, 'Wellbeing as social care: On assemblages and the "commons"', *Wellbeing, Space and Society*, vol. 3, article 100078. https://doi.org/10.1016/j.wss.2022.100078

Dunkley, CM 2009, 'A therapeutic taskscape: Theorizing place-making, discipline and care at a camp for troubled youth', *Health & Place*, vol. 15, no. 1, pp. 88–96. https://doi.org/10.1016/j.healthplace.2008.02.006

Egerer, M, Lin, B, Kingsley, J, Marsh, P, Diekmann, L and Ossola, A 2021, 'Gardening can relieve human stress and boost nature connection during the COVID-19 pandemic', *Urban Forestry & Urban Greening*, vol. 67, article 127483. https://doi.org/10.1016/j.ufug.2022.127483

Eizenberg, E 2013, *From the ground up: Community gardens in New York City and the politics of transformation*, Ashgate, Farnham.

Ekstam, L, Pálsdóttir, AM and Asaba, E 2021, 'Migrants' experiences of a nature-based vocational rehabilitation programme in relation to place, occupation, health and everyday life', *Journal of Occupational Science*, vol. 28, no. 1, pp. 144–158. https://doi.org/10.1080/14427591.2021.1880964

Finlay, JM, Kler, JS, O'Shea, BQ, Eastman, MR, Vinson, YR and Kobayashi, LC 2021, 'Coping during the COVID-19 pandemic: A qualitative study of older adults across the United States', *Frontiers in Public Health*, vol. 9, article 643807. https://doi.org/10.3389/fpubh.2021.643807

Foley, R, Bell, S, Gittins, H, Grove, H, Kaley, A, McLauchlan, A, Osborne, T, Power, A, Roberts, E and Thomas, M 2020, 'Disciplined research in undisciplined settings: Critical explorations of in-situ and mobile methodologies in geographies of health and wellbeing', *Area*, vol. 52, no. 3, pp. 514–522. https://doi.org/10.1111/area.12604

Foley, R, Garrido Cumbrera, M, Braçe, O, Guzmán, V and Hewlett, D 2022, 'Home and nearby nature: Uncovering flows between domestic and neighbourhood greenspace in three countries during COVID-19', *Wellbeing, Space and Society*, vol. 3, article 100093. https://doi.org/10.1016/j.wss.2022.100093

Franco, A and Robson, D 2022, 'How mud boosts your immune system' (*BBC Future*, 11 October 2022). https://www.bbc.com/future/article/20220929-how-outdoor-play-boosts-kids-immune-systems

Fullilove, MT 2016, *Root shock: How tearing up city neighborhoods hurts America, and what we can do about it*, New Village Press, New York.

Garrido Cumbrera, M, Foley, R, Braçe, O, Correa-Fernández, J, López-Lara, E, Guzman, V, Gonzalez-Marín, A and Hewlett, D 2021, 'Perceptions of change in the natural environment produced by the first wave of the COVID-19 pandemic across three European countries. Results from the GreenCOVID study', *Urban Forestry & Urban Greening*, vol. 64, article 127260. https://doi.org/10.1016/j.ufug.2021.127260

Garrido-Cumbrera, M, Foley, R, Correa-Fernández, J, González-Marín, A, Braçe, O and Hewlett, D 2022, 'The importance of healthy views from home for well-being during the COVID-19 pandemic. Results of the GreenCOVID study', *Journal of Environmental Psychology*, vol. 83, article 101864. https://doi.org/10.1016/j.jenvp.2022.101864

Gesler, W 1992, 'Therapeutic landscapes: Medical issues in light of the new cultural geography', *Social Science & Medicine*, vol. 34, no. 7, pp. 735–746. https://doi.org/10.1016/0277-9536(92)90360-3

Greenhough, B, Davies, G and Bowlby, S 2022, 'Why "cultures of care"?', *Social & Cultural Geography*. https://doi.org/10.1080/14649365.2022.2105938

Harris, N, Minniss, F and Somerset, S 2014, 'Refugees connecting with a new country through community food gardening', *International Journal of Environmental Research and Public Health*, vol. 11, no. 9, pp. 9202–9216. https://doi.org/10.3390/ijerph110909202

Hawkins, JL, Mercer, J, Thirlaway, KT and Clayton, DA 2013, '"Doing" gardening and "being" at the allotment site: Exploring the benefits of allotment gardening for stress reduction and healthy aging', *Ecopsychology*, vol. 5, no. 2, pp. 110–125. https://doi.org/10.1089/eco.2012.0084

Head, L and Atchison, J 2009, 'Cultural ecology: Emerging human-plant geographies', *Progress in Human Geography*, vol. 33, no. 2, pp. 236–245. https://doi.org/10.1177/0309132508094075

Heaney, S 1966, *Death of a naturalist*, Faber and Faber, London.

Houghton, F and Houghton, S 2015, 'Therapeutic micro-environments in the Edgelands: A thematic analysis of Richard Mabey's The Unofficial Countryside', *Social Science & Medicine*, vol. 113, pp. 280–286. https://doi.org/10.1016/j.socscimed.2014.11.040

Hoyle, HE 2021, 'Climate-adapted, traditional or cottage-garden planting? Public perceptions, values and socio-cultural drivers in a designed garden setting', *Urban Forestry & Urban Greening*, vol. 65, article 127362. https://doi.org/10.1016/j.ufug.2021.127362

Ingold, T 2000, *The perception of the environment*, Routledge, London.

Jassi, J and Dutton, A 2022, *Access to garden spaces: England. The statistical significance of differences in garden access when all factors are controlled simultaneously*, Office for National Statistics, London. https://www.ons.gov.uk/economy/environmentalaccounts/methodologies/accesstogardenspacesengland

Kettle, P 2014, 'Motivations for investing in allotment gardening in Dublin: A sociological analysis', *Irish Journal of Sociology*, vol. 22, no. 2, pp. 30–63. https://doi.org/10.7227/IJS.22.2.3

Kingsley, J, Egerer, M, Nuttman, S et al. 2021, 'Urban agriculture as a nature-based solution to address socio-ecological challenges in Australian cities', *Urban Forestry & Urban Greening*, vol. 60, article 127059. https://doi.org/10.1016/j.ufug.2021.127059

Lehberger, M, Kleih, A-K and Sparke, K 2021, 'Self-reported well-being and the importance of green spaces – A comparison of garden owners and non-garden owners in times of COVID-19', *Landscape and Urban Planning*, vol. 212, article 104108. https://doi.org/10.1016/j.landurbplan.2021.104108

Marsh, P, Diekmann, L, Egerer, M, Lin, B, Ossola, A and Kingsley, J 2021, 'Where birds felt louder: The garden as a refuge during COVID-19', *Wellbeing, Space and Society*, vol. 2, article 100055. https://doi.org/10.1016/j.wss.2021.100055

Marsh, P, Gartrell, G, Egg, G, Nolan, A and Cross, M 2017, 'End-of-Life care in a community garden: Findings from a participatory action research project in regional Australia', *Health & Place*, vol. 45, pp. 110–116. https://doi.org/10.1016/j.healthplace.2017.03.006

Massey, D 2005, *For space*, Open University, Milton Keynes.

Mattern, S 2018, 'Maintenance and care', *Places Journal*. https://doi.org/10.22269/181120

Milligan, C, Gatrell, A and Bingley, AF 2004, '"Cultivating health": Therapeutic landscapes and older people in northern England', *Social Science & Medicine*, vol. 58, no. 9, pp. 1781–1793. https://doi.org/10.1016/S0277-9536(03)00397-6

Millington, N and Lawhon, M 2019, 'Geographies of waste: Conceptual vectors from the Global South', *Progress in Human Geography*, vol. 43, no. 6, pp. 1044–1063. https://doi.org/10.1177/0309132518799911

Moran, D, Jones, PI, Jordaan, JA and Porter, AE 2021, 'Does nature contact in prison improve well-being? Mapping land cover to identify the effect of greenspace on self-harm and violence in prisons in England and Wales', *Annals of the American Association of Geographers*, vol. 111, no. 6, pp. 1779–1795. https://doi.org/10.1080/24694452.2020.1850232

Nelson, SH and Bigger, P 2022, 'Infrastructural nature', *Progress in Human Geography*, vol. 46, no. 1, pp. 86–107. https://doi.org/10.1177/0309132521993916

O'Sullivan, S 2017, *HARD/GRAFT: Conversations on Planting a Commons*. http://www.seoidinosullivan.com/#/hardgraft/

Olive, R and Wheaton, B 2021, 'Understanding blue spaces: Sport, bodies, wellbeing, and the sea', *Journal of Sport and Social Issues*, vol. 45, no. 1, pp. 3–19. https://doi.org/10.1177/0193723520950549

Osborne, T 2019, 'Biosensing: A critical reflection on doing memory research through the body' in D Drozdsewski and C Birdsall (eds), *Doing memory research: New methods and approaches*, Palgrave Macmillan, Singapore, pp. 63–85. https://doi.org/10.1007/978-981-13-1411-7_4

Page, S 2020, 'Assemblage theory' in A Kobayashi (ed), *International encyclopaedia of human geography*, 2nd edition, Elsevier, Amsterdam, Netherlands, pp. 223–227. https://doi.org/10.1016/B978-0-08-102295-5.10506-2

Philo, C 2012, 'A "new Foucault" with lively implications – or "the crawfish advances sideways"', *TIBG: Transactions of the Institute of British Geographers*, vol. 37, no. 4, pp. 496–514. https://doi.org/10.1111/j.1475-5661.2011.00484.x

Phoenix, C and Orr, N 2014, 'Pleasure: A forgotten dimension of physical activity in older age', *Social Science & Medicine*, vol. 115, pp. 94–102. https://doi.org/10.1016/j.socscimed.2014.06.013

Pitt, H 2015, 'Therapeutic experiences of community gardens: Putting flow in its place', *Health & Place*, vol. 27, pp. 84–91. https://doi.org/10.1016/j.healthplace.2014.02.006

Pitt, H 2018, 'Muddying the waters. What urban waterways reveal about bluespaces and wellbeing', *Geoforum*, vol. 92, pp. 161–170. https://doi.org/10.1016/j.geoforum.2018.04.014

Richardson, M, Passmore, HA, Lumber, R, Thomas, R and Hunt, A 2021, 'Moments, not minutes: The nature-wellbeing relationship', *International Journal of Wellbeing*, vol. 11, no. 1, pp. 8–33. https://doi.org/10.5502/ijw.v11i1.1267

Roscoe, C, Mackay, C, Gulliver, J, Hodgson, S, Cai, Y, Vineis, P and Fecht, D 2022, 'Associations of private residential gardens versus other greenspace types with cardiovascular and respiratory disease mortality: Observational evidence from UK Biobank', *Environment International*, vol. 167, article 107427. https://doi.org/10.1016/j.envint.2022.107427

Schmid, B 2019, 'Repair's diverse transformative geographies: Lessons from a repair community in Stuttgart', *Ephemera*, vol. 19, no. 2, pp. 229–251.

Scott, T, Masser, B and Pachana, N 2015, 'Exploring the health and wellbeing benefits of gardening for older adults', *Ageing & Society*, vol. 35, no. 10, pp. 2176–2200. https://doi.org/10.1017/S0144686X14000865

Sept, A 2021, '"Slowing down" in small and medium-sized towns: CittàSlow in Germany and Italy from a social innovation perspective', *Regional Studies, Regional Science*, vol. 8, no. 1, pp. 259–268. https://doi.org/10.1080/21681376.2021.1919190

Sia, A, Tam, WWS, Fogel, A et al. 2020, 'Nature-based activities improve the well-being of older adults', *Scientific Reports*, vol. 10, article 18178. https://doi.org/10.1038/s41598-020-74828-w

Smith, C 2002, 'The wholesale and retail markets of London, 1660–1840', *Economic History Review*, vol. 55, no. 1, pp. 31–50. https://doi.org/10.1111/1468-0289.00213

Turner, B, Abramovic, J and Hope, C 2021, 'More-than-human contributions to place-based social inclusion in community gardens' in P Liamputtong (ed), *Handbook of social inclusion*, Springer, Cham. https://doi.org/10.1007/978-3-030-48277-0_96-1

Wiles, J, Miskelly, P, Stewart, O, Rolleston, A, Gott, M and Kerse, N 2021, 'Gardens as resources in advanced age in Aotearoa NZ: More than therapeutic', *Social Science & Medicine*, vol. 288, article 113232. https://doi.org/10.1016/j.socscimed.2020.113232

Williams, A 1999, 'Introduction' in A Williams (ed), *Therapeutic landscapes: the dynamic between place and wellness*, University Press of America, Lanham, MD, pp. 1–11.

Williams, A (ed) 2007, *Therapeutic landscapes*, Ashgate, Farnham.

Wilson, K 2003, 'Therapeutic landscapes and first nations peoples: An exploration of culture, health and place', *Health & Place*, vol. 9, no. 2, pp. 83–93. https://doi.org/10.1016/s1353-8292(02)00016-3

Wood, C 2008, *Trespasser* (sleeve notes), Music CD, RuF Records.

Index

Please note that references to Notes are indicated by the letter 'n' followed by the Note number. Page references to Figures are indicated in **bold**, while references to Tables are in *italics*.